Springer Texts in Statistics

Advisors:
Stephen Fienberg   Ingram Olkin

Springer Texts in Statistics

Alfred	Elements of Statistics for the Life and Social Sciences
Blom	Probability and Statistics: Theory and Applications
Chow and Teicher	Probability Theory: Independence, Interchangeability, Martingales. Second Edition
Christensen	Plane Answers to Complex Questions: The Theory of Linear Models
du Toit, Steyn and Strumpf	Graphical Exploratory Data Analysis
Kalbfleisch	Probability and Statistical Inference: Volume 1: Probability. Second Edition
Kalbfleisch	Probability and Statistical Inference: Volume 2: Statistical Inference. Second Edition
Keyfitz	Applied Mathematical Demography. Second Edition
Kiefer	Introduction to Statistical Inference
Kokoska	Statistical Tables and Formulae
Madansky	Prescriptions for Working Statisticians
Nguyen and Rogers	Fundamentals of Mathematical Statistics: Volume I: Probability for Statistics
Nguyen and Rogers	Fundamentals of Mathematical Statistics: Volume II: Statistical Inference
Peters	Counting for Something: Statistical Principles and Personalities
Santner and Duffy	The Statistical Analysis of Discrete Data

H.T. Nguyen G.S. Rogers

Fundamentals of Mathematical Statistics

Volume II

Statistical Inference

Springer-Verlag
New York Berlin Heidelberg
London Paris Tokyo Hong Kong

Hung T. Nguyen
Gerald S. Rogers
Department of Mathematical Sciences
New Mexico State University
Las Cruces, NM 88003-0001, USA

Mathematics Subject Classification: 62-01

Library of Congress Cataloging-in-Publication Data
Nguyen, Hung T., 1944–
Fundamentals of mathematical statistics / H.T. Nguyen and G.S. Rogers.
p. cm. — (Springer texts in statistics)
Bibliography: v. 1,
Includes indexes.
Contents: v. 1. Probability for statistics — v. 2. Statistical inference.
ISBN-13:978-1-4613-8916-3
1. Mathematical statistics. I. Rogers, Gerald Stanley, 1928– II. Title. III. Series.
QA276.12.N49 1989
519.5—dc20 89-6426

Printed on acid-free paper.

Softcover reprint of the hardcover 1st edition 1989

Camera-ready copy prepared by the authors.

9 8 7 6 5 4 3 2 1

ISBN 978-1-4613-8916-3 ISBN 978-1-4613-8914-9 (eBook)
DOI 10.1007/978-1-4613-8914-9

Preface

This is the second half of a text for a two semester course in mathematical statistics at the senior/graduate level for those who need a strong background in statistics as an essential tool in their career. To study this text, the reader needs a thorough familiarity with calculus including such things as Jacobians and series but somewhat less intense familiarity with matrices including quadratic forms and eigenvalues.

For convenience, these lecture notes were divided into two parts: *Volume I*, Probability for Statistics, for the first semester, and *Volume II*, Statistical Inference, for the second. The probability background needed for this Volume II is precisely the material in Volume I. It is important to note that the measure theory concepts introduced in Volume I were just enough for the statistical inference in this Volume II, general measure theory is not required. In this Volume II, we often refer to Volume I, such as lesson 7, part II, etc.

We suggest that the following distinguish this text from other introductions to mathematical statistics.

1. The most obvious thing is the layout. We have designed each lesson for the (U.S.) 50 minute class; those who study independently probably need the traditional three hours for each lesson. Since we have more than (the U.S. again) 90 lessons, some choices have to be made. In the table of contents, we have used a * to designate those lessons which are "interesting but not essential" (INE) and may be omitted from a general course; some exercises and proofs in other lessons are also "INE". We have made lessons of some material which other writers might stuff into appendices. Incorporating this freedom of choice has led to some redundancy, mostly in definitions, which may be beneficial.

For the first semester, Parts I, II, III (Volume I) contain 49 lessons of which 11 are INE; for the second semseter, (this Volume II) Parts IV, V, VI, contain 36 lessons (1 INE) and Part VII contains 11 (partially independent) choices.

2. Not quite so obvious is the pedagogy. First, most of the exercises are integrated into the discussions and cannot be omitted. Second, we started with elementary probability *and* statistics (Part I) because most students in a first course have little, if any, previous experience with statistics as a mathematical discipline. Just as importantly, in this part, the discussions begin

with (modified) real examples and include some simple proofs. We think that our line of discussion leads naturally to the consideration of more general formulations.

3. In this vein, we believe that today's students of mathematical statistics need exposure to measure theoretic aspects, the "buzz words", though not the truly fine points. Part II is such an introduction and is more difficult than Part I but, as throughout these notes, we blithely skip over really involved proofs with a reference to more specialized texts. In teaching, we have at times omitted some of the proofs that are included herein. Otherwise, we have tried to emphasize concepts and not details and we recognize that some instructors prefer more "problems". Part III contains special pieces of analysis needed for probability and statistics.

4. The topics in Volume II, Parts IV, V, VI are rather traditional but this listing does not show the care with which we have stated the theorems even when the proof is to be limited to a reference. The immense breadth of statistics forces all writers to be selective; we think we have a good choice of major points with some intersting sidelights. Moreover, our INE materials (in particular Part VII) give the reader (instructor) additional variety.

The following general comments on the contents summarize the detailed overviews at the beginning of each part.

Volume I: Probability For Statistics.

Part I: *Elementary Probability and Statistics.* We begin with the (Western) origins of probability in the seventeenth century and end with recognition of the necessity of more sophisticated techniques. In between, there are some practical problems of probability and examples of two principles of statistics: testing hypotheses and finding confidence intervals. As the title says, the notions are elementary; they are not trivial.

Part II: *Probability and Expectation.* As the reader is asked to take a broader view of functions and integration, this material is a good bit harder than that in Part I. Although we present most of the theory, we do so with a minimum of detailed proofs and concentrate on appreciation of the generalization of concepts from Part I. In particular, "almost surely" comes to be understood and not overlooked as some authors suggest.

Part III: *Limiting Distributions.* Herein, we complete the basic results of probability needed for "statistics". We have not addressed the problem specifically but Parts II and III should be adequate preparation for a course in "stochastic processes". Again, the emphasis is on appreciation of concepts and not all details of all proofs of all theorems are given. However, we do include an almost complete proof of one central limit theorem.

Volume II: Statistical Inference.

Part IV: *Sampling and Distributions.* Now we pursue the concepts of random sample and statistics as the bases of the theory of statistical inference; we include some sampling distributions in normal populations. We explain some useful properties of statistics such as sufficiency and completeness, in particular, with respect to exponential families.

Part V: *Statistical Estimation.* This is a discussion of some of the problems of point estimation and interval estimation of (mostly) real parameters. In some cases, the results are exact for a given sample size; in others, they are only approximations based on asymptotic (limiting) distributions. Criteria for comparing estimators such as bias, variance, efficiency are discussed. Particular care is given to the presentation of basic results regarding the maximum likelihood method.

Part VI: *Testing Hypotheses* In this part, we discuss three variations of likelihood (colloquially, probability) ratios used to derive tests of hypotheses about parameters. Most of the classical examples about normal populations are included.

Part VII: *Special Topics.* We call the topics in this part special because each is the merest introduction to its area of statistics such as decision theory, linear models, non–parametrics, goodness–of–fit, classified data. The reader is invited to choose freely and often.

We extend our thanks to: various groups of students, particularly those in our course in 1986–87, for their patience and comments; our department head, Carol Walker, for her patience and encouragement; to Valerie Reed for the original typing in T^3.

Hung T. Nguyen and Gerald S. Rogers
Las Cruces, March 10, 1989.

PS. We especially want to recognize the people at TCI Software Research, Inc. Without their generous assistance and T^3 word processing system, we would not have been able to present this work in our own style. HTN/GSR

CONTENTS

Volume II: Statistical Inference

* indicates a lesson which is "interesting but not essential" and may be omitted; this is tru for all lessons in Part VII.

PART IV: SAMPLING AND DISTRIBUTIONS

Overview

As noted in earlier lessons, a probability space associated with tossing a coin can be a model for other yes/no, success/failure, life/death experiments. In "probability", a basic question is of the type, "What is the probability that in $n = 100$ tosses of a coin the number of heads is between 70 and 80?" In "statistics", a basic question is of the type, "If in $n = 100$ tosses of a coin the number of heads is 72, what can be said about $P(H) = \theta$ on a single toss?"

One answer to the latter question is the "confidence interval":

for a binomial RV X, we can find $L(X)$, $U(X)$ such that

$$P(L(X) < \theta < U(X)) \geq .98 \text{, etc.}$$

Then (Lesson 19, Part I), each observed interval $(L(x), U(x))$ is a 98% confidence interval for θ. Before this can be formalized and generalized to other parameters in other distributions, some theory of statistics must be developed. This begins with the properties of the distributions of samples (Lesson 1).

Formally, "statistics" turn out to be functions of samples so that rules for transformations of random variables must be found; some of these are in lessons 2 and 3 and some classical examples for normal populations are in lessons 4 and 5. The particular case of order statistics is treated in lesson 6.

There are some difficulties in really understanding vectors of observations and in extracting the "information" from a sample. The term "information" has a technical meaning which is treated only briefly, and herein, "information" remains a general reference to what can be learned from a sample. In particular, lesson 7 examines "reduction" of a sample to a small number of "statistics": measurable functions of the observable random quantities (or qualities) $X_1, X_2, \cdots$.

The principal means of reduction to "sufficient statistics" allows inferences to be drawn from these statistics with the same validity as those based on the initial sample. Such theory and applications are discussed in lessons 7,8,9. A particular case, (general enough to include many of the distributions in common use, normal, binomial, etc.) is that of exponential families whose theory is laid out in lessons 10 and 11.

LESSON 1. SAMPLING AND STATISTICS

In this lesson, we describe some basic vocabulary and some assumptions needed in "probability and statistics"; in the latter, an "experiment" is simply the making of one or more observations (rather than being limited to the procedures usually thought of in laboratories). Without further structure we have nothing to study. Thus an experiment will be represented by its collection of possible outcomes Ω with an associated σ–field $\mathscr{A}$ and one probability measure P ; in other words, the statistical model for an experiment is a probability space $[\Omega, \mathscr{A}, P]$. In most cases, we will focus on some random vector: $X : \Omega \to \mathbb{R}^d$, $d \geq 1$. For example, as noted early in Part I, the effect of a medicine on an individual is important but for other conclusions we may want other functions like the numbers of good, bad, indifferent reactions, etc. The function must also be $(\mathscr{A}, \mathscr{B}_d)$ measurable,

$$X^{-1}(B) \in \mathscr{A} \text{ for all } B \in \mathscr{B}_d,$$

for then the probability distribution of X can be defined by

$$P_X(B) = P(X^{-1}(B)) \text{ for all } B \in \mathscr{B}_d .$$

The simplest form of statistical estimation is as follows. Suppose that the distribution P is not known exactly but only that P is a member of some family $\{P_1,P_2,\cdots\}$. Given the one observation ω or $X(\omega)$, which P_α is "most reasonable" ? Of course, we have to prescribe also what "most reasonable" means; this will be done in different ways in later lessons.

From a mathematical point of view, experiments are "repeatable". For example, we can always consider a product space $\Omega \times \Omega \times \Omega$, its product σ–field and a product measure generated by

$$P(A_i \times A_j \times A_k) = P(A_i)P(A_j)P(A_k).$$

Then the corresponding random vectors X_1,X_2,X_3 on $(\Omega \times \Omega \times \Omega)$ are independent:

$$\text{for all } B_1, B_2, B_3 \in \mathscr{B}_d ,$$

$$P(X_1 \in B_1, X_2 \in B_2, X_3 \in B_3) = P(X_1 \in B_1)P(X_2 \in B_2)P(X_3 \in B_3) ;$$

and identically distributed (IID):

$$\text{for all } B \in \mathscr{B}_d$$
$$P(X_1 \in B) = P(X_2 \in B) = P(X_3 \in B)\,.$$

This leads to

Definition: *A simple random sample of size* $n \geq 1$ *is a set of random variables (or vectors)* $X_1, \ldots, X_n$ *which are Independent and Identically Distributed (IID).*

In an engineering laboratory, it is reasonable to assume that "repeatable experiments" mimic the mathematical requirements. For example, we can use the *same* mixtures, the *same* drying conditions, and the *same* tests of concrete strengths; we believe that we do obtain a random sample of the population of all strength measurements for *this* mixture.

If we are looking for a sample of the yield of corn stalks in the field before us, some things are beyond our control: rainfall, sunlight, soil conditions, etc. It is often possible to take these into account by using slightly different distributions for $X_1, X_2, \cdots$; one is still left with modeling "independence". It is known that yields of stalks close to each other may not be independent; we really want "representatives" from all of the field. For example, if there are three areas of different soils in the field, we would want a random sample from each of the three areas; these are called *stratified samples*. If one area has, say 500 stalks, which we number quite arbitrarily, then a table of random numbers can be used to select, say, n = 20. In lesson 15, Part I, there is a small table; we read triples under 500, skipping some digits:

421 305 (7) 078 409 (9) 102 (8) 007 (5) 238 (5)(6)(9)
188 (5) 429 (7) 371 205 (5) 142 094(8) 462(8)(9) 322 195
(5)(5) 296 (7) (8) 466(6) 363 (8) (8) 454

Our sample will consist of stalks numbered 421, 305, 78, 409, etc. Of course, this can be programmed on even modest sized computers. Other properties of the field might suggest other divisions.

When dealing with people, there are even more side conditions to consider; for example, an obvious dependence is that of the political opinions of a married couple.

This discussion has opened the door of "design of experiments" and you can see that this is very much tied to subject matter. We will pass by this opening and focus on simple random samples. Thus we assume that the "experiments" will present us with the observed value $(x_1, x_2, \cdots, x_n)$ of the IID random observable $(X_1, X_2, \cdots, X_n)$. We sometimes see phrases like "Each of $X_1, X_2, \cdots, X_n$ is an observation of (from) the population."

But as noted in Part I, this refers to the *statistical population*, usually the common distribution of X, not some bunch of people or colony of bees; we want to know something about the corn from the field, not just where the stalks were. Naturally, if we have, say, $n = 2000$ observations each a vector of 50 measurements, we need to summarize "information" in this sample. This is accomplished by using *statistics*:

Definition: *A statistic is a (measurable) function of the observable outcome "ω" alone.*

For example, if $\Omega = \{H,T\}$ as in a coin toss, $X(H) = 6$, $X(T) = 9$, is a function of ω alone and is a statistic. However, $Y(H) = 6$, $Y(T) = 9 + K$ is not a statistic as one of its values depends upon the unknown K. Both X and Y are random variables with probability distributions, but when we perform the experiment and T occurs, we know the value of X but not the value of Y.

Exercise 1: Let $X_1, \ldots, X_n$ be a random sample (RS) from a normal distribution with mean μ and variance σ^2. Let

$$\bar{X} = \frac{1}{n}\sum_{i=1}^{n} X_i, \quad S^2 = \frac{1}{n}\sum_{i=1}^{n} (X_i - \bar{X})^2 .$$

Which of the following are statistics?

a) $\bar{X}$ b) S^2 c) $(\bar{X} - \mu)/S$ d) $(X_1 - \bar{X})/S)$ e) $(\bar{X} - \mu)/\sigma$

In the exercise above, each of $X_1, X_2, \cdots, X_n, \bar{X}, S$ is a statistic. "Before" the sample is observed (the experiment is performed), that is all we know. "After" the experiment is performed, we know the values

$$x_1, \cdots, x_n, \bar{x} = \frac{1}{n}\sum_{i=1}^{n} x_i, \quad s^2 = \frac{1}{n}\sum_{i=1}^{n} (x_i - \bar{x})^2 ,$$

etc. Sometimes the value of the statistic is also called a statistic; generally, the context will indicate which meaning is intended.

Now we make one more limitation on the distributions we will study in this Part. The family $\{P_1, P_2, \cdots\}$ of possible distributions will be limited to that indexed by one or more real–valued parameters.

Definition: *A real-valued parameter is a function of the probability distribution P .*

For example, if X is a continuous real RV, the parameter

$$E[X] = \int_\Omega X(w)dP(w) = \int_{-\infty}^{+\infty} x\, dF(x)$$

changes with a change of P . To be more specific, if X is a normal RV with variance 1,

$$E[X] = \int_{-\infty}^{+\infty} x\, e^{-(x-\mu)^2/2}\, dx/\sqrt{(2\pi)} = \mu$$

changes with μ . The entire family (of univariate) normal distributions is indexed by the parameter $\theta = (\theta_1, \theta_2)$ where θ_1 is the (real) mean and θ_2 is the (non–negative) variance. When $\theta_2 > 0$, the density function $\dfrac{e^{-(x-\theta_1)^2/2\theta_2}}{\sqrt{(2\pi\theta_2)}}$ is more commonly written as $\dfrac{e^{-(x-\mu)^2/2\sigma^2}}{\sigma\sqrt{(2\pi)}}$.

In such cases, the problem of "choosing the most reasonable P" is phrased as "estimating the parameter θ" .

Definition: *Let $\{P_\theta : \theta \in \Theta \subseteq R^m\}$ be a family of probability measures on $(\Omega, \mathcal{A})$. Θ is the parameter set (space); Ω is the sample space. If each P_θ is a probability distribution for a random variable X, then $\mathcal{X} = \{X(w) : w \in \Omega\} \subseteq R^d$ may be called the sample space.*

In most of our examples and theory, we also assume that the distribution of X is given by a density f, with respect to Lebesgue measure in the continuous case and wrt a counting measure in the discrete case; we treat no mixed cases. Then

$$\mathcal{F} = \{f(x;\theta): x \in X, \theta \in \Theta\}$$

is called a statistical model (or model for short). The problem is to find the "best" f by finding the "best" θ and all of this Part IV is devoted to proper formulation of some questions and answers for such statistical inferences.

LESSON 2. TRANSFORMATIONS OF REAL RANDOM VARIABLES

In many applications of probability and statistics, the real interest is not in the (outcomes of the) original random variable X but rather in some transformation h(X) . For example, if X is normally distributed with mean μ and variance σ^2, then one transformation of interest is

$$Y = h(X) = \frac{1}{\sigma}(X - \mu).$$

"Grouping" which occurs in many surveys entails a transformation like:

$$Y = 0 \text{ if } 0 \le X < \$2000$$
$$Y = 3 \text{ if } \$2000 \le X < \$4000$$
$$Y = 5 \text{ if } \$4000 \le X < \$6000$$

and so on.

The present lesson is devoted to obtaining the distribution of h(X) from that of X when X is a real RV; the next lesson will treat the case when X is a vector of real random variables.

Fix the probability space $[\Omega, \mathcal{A}, P]$. Let $X: \Omega \to R$ be a real valued RV, discrete or continuous, with density f_X; X is $(\mathcal{A}, \mathcal{B})$ measurable. Let $h: R \to R$ be a measurable mapping; h is $(\mathcal{B}, \mathcal{B})$ measurable. Consider the RV Y defined by $Y(\omega) = h(X(\omega))$.

For $A \in \mathcal{B}$, $P_Y(A) = P(Y \in A) = P(Y^{-1}(A))$

$$= P(X^{-1}(h^{-1}(A))) = P_X(h^{-1}(A)) .$$

If X is discrete,then

$$P(Y = y) = P[X \in h^{-1}(y)] = f_Y(y) = \sum_{x \in h^{-1}(y)} f_X(x)$$

since $h^{-1}(y) = \{x : h(x) = y\} \subset \mathcal{R} = X(\Omega)$ is at most countable.

Example: Consider the distribution:

X	-2	-1	0	1	2
f_X	1/10	3/10	2/10	2/10	2/10

Let $h(x) = x^2$ so that $Y = X^2$ and the range of Y is {0, 1, 4}.
Then,

$$P(Y = 0) = P(X = 0) = 2/10 ,$$
$$P(Y = 1) = P(X = -1) + P(X = 1) = 5/10 ,$$
$$P(Y = 4) = P(X = -2) + P(X = 2) = 3/10 .$$

Exercise 1: Let X be a Poisson RV with parameter λ . Find the density function of $Y = \frac{1}{1 + X}$.

If X is continuous, one might proceed to find the CDF of Y first, and then f_Y by differentiation.

Example: Let X be uniformly distributed over $(-\frac{\pi}{2}, \frac{\pi}{2})$ and $Y = h(X) = \tan(X)$; the range of Y is R. Then,

$$F_Y(y) = P(Y \le y) = P(h(X) \le y) = P(X \le \tan^{-1}(y))$$

$$= \int_{-\pi/2}^{\tan^{-1}(y)} dx/\pi = [\tan^{-1}(y) + \pi/2]/\pi$$

and

$$f_Y(y) = \frac{d}{dy} F_Y(y) = 1/\pi(1 + y^2) .$$

This is called the standard Cauchy density.

Exercise 2: Let X have density $f_X(x) = \frac{3}{8} x^2 I(0 < x < 2)$. Find the density of $Y = -X^2$.

In the last example, the transformation given by $h(x) = \tan(x)$ preserves absolute continuity so that Y is a continuous RV with a density. The following theorem contains a formula for finding f_Y directly when the transformation h is such that Y has this density when X has the density f_X .

Theorem 1. *Let $\mathscr{D} = \{x \in R : f_X(x) > 0\}$ be the support of f_X; and let $\mathscr{H} = h(\mathscr{D}) = \{h(x) : x \in \mathscr{D}\}$. Suppose that:*

a) *there is a collection* $D_1, D_2, \cdots, D_m$ *which is a finite measurable partition of* $\mathscr{D}$; *(These are Borel sets of* $\mathscr{B}$ *which are pairwise disjoint with union* $\mathscr{D}$.)

b) *on each* D_i, h *is one-to-one onto some subset* $\mathscr{H}_i$ *of* $\mathscr{H} = \cup_{i=1}^{m} \mathscr{H}_i$; *(That is,* $h(x_1) = h(x_2)$ *implies* $x_1 = x_2$; $h(D_i) = \mathscr{H}_i$.)

c) *for* $i = 1(1)m$, *the inverse function* h_i^{-1} *of* h_i *(h restricted to* D_i*) has a continuous derivative on* $\mathscr{H}_i$ *which is non-zero almost everywhere. Then,* $Y = h(X)$ *has a density given by*

$$f(y) = \sum_{i=1}^{m} f_X(h_i^{-1}(y)) \cdot \left| \frac{d}{dy} h_i^{-1}(y) \right| \cdot I\{y \in \mathscr{H}_i\} .$$

Proof: Let A be a Borel set of $\mathscr{H}$. Then,

$$P(Y \in A) = P(h(X) \in A) = P(X \in h^{-1}(A))$$
$$= \sum_{i=1}^{m} P(X \in D_i \cap h^{-1}(A)).$$

For $i = 1(1)m$, let $B_i = \{x : x \in D_i , h(x) \in A\}$ and $C_i = A \cap D_i$; then,

$$P(X \in D_i \cap h^{-1}(A)) = \int_{B_i} f_X(x)dx$$

$$= \int_{C_i} f_X(h_i^{-1}(y)) \cdot \left| \frac{d}{dy} h_i^{-1}(y) \right| \cdot dy$$

by the usual form for change of variables in integration. In turn, this is equal to

$$\int_A f_X(h_i^{-1}(y)) \cdot \left| \frac{d}{dy} h_i^{-1}(y) \right| \cdot I_{\mathscr{H}_i}(y) \cdot dy$$

so that

$$P(Y \in A) = \int_A \sum_{i=1}^{m} f_X(h_i^{-1}(y)) \cdot \left|\frac{d}{dy} h_i^{-1}(y)\right| \cdot I_{\mathscr{H}_i}(y) \cdot dy$$

and hence

$$f_Y(y) = \sum_{i=1}^{m} f_X(h_i^{-1}(y)) \cdot \left|\frac{d}{dy} h_i^{-1}(y)\right| \cdot I_{\mathscr{H}_i}(y) .$$

The following lemma contains further illustration of the "change of variable" technique.

Lemma. *For simplicity, assume* $m = 1$ *so that* $h : D \to \mathscr{H}$ *is one-to-one and onto with* $\frac{d}{dx} h(x) \neq 0$ *almost everywhere. Then,*

$$f_Y(y) = f_X(h^{-1}(y)) \cdot \left|\frac{d}{dy} h^{-1}(y)\right| .$$

Proof: a) If $h'(x) > 0$, then $\frac{d}{dy} h^{-1}(y) > 0$ on $\mathscr{H}$ and

$$P(Y \le y) = P(h(X) \le y) = P(X \le h^{-1}(y))$$

so that $f_Y(y) = f_Y(h^{-1}(y)) \cdot \frac{d}{dy} h^{-1}(y)$.

b) If $h'(x) < 0$, then $\frac{d}{dy} h^{-1}(y) < 0$ on $\mathscr{H}$ and

$$P(Y \le y) = P(h(X) \le y) = P[X \ge h^{-1}(y)] = 1 - P(X < h^{-1}(y)$$

so that $f_Y(y) = -f_X(h^{-1}(y)) \cdot \frac{d}{dy} h^{-1}(y)$. In either case,

$$f_Y(y) = f_X(h^{-1}(y)) \cdot \left|\frac{d}{dy} h^{-1}(y)\right| .$$

Exercise 3: Prove that when h is as in the Lemma,

$$\frac{d}{dx} h(x) < 0 \text{ if and only if } \frac{d}{dy} h^{-1}(y) < 0.$$

As a special case of the Theorem 1, we have

Corollary: *If* h *is one-to-one and onto from* $\mathscr{D}$ *to* $\mathscr{H}$, *and* h^{-1} *has a continuous non-zero derivative on* $\mathscr{H}$, *then*

$$f_Y(y) = f_X(h^{-1}(y)) \cdot \left|\frac{d}{dy} h^{-1}(y)\right| \cdot I_{\mathscr{H}}(y).$$

Exercise 4: Let X be a RV with $\mathscr{D} = [a,b]$; let $h : R \to R$ be measurable, continuous, strictly increasing (or strictly decreasing) on $[a,b]$ such that h^{-1} is continuously differentiable. Derive the result of the above corollary by first computing the CDF of $Y = h(X)$, then using differentiation.

Example: Let X have density $f_X(x) = e^{-x} \cdot I(0 < x < +\infty)$.

a) Let $Y = \log(X)$ (base e). Here

$$h : \mathscr{D} = (0, +\infty) \to \mathscr{H} = R$$

with $h(x) = \log(x)$ and the hypotheses of the corollary are satisfied.

For $y \in R$, $h^{-1}(y) = e^y$ so $f_Y(y) = e^y \cdot e^{-e^y}$.

b) Let $Z = e^{-X}$; then $\mathscr{H} = (0,1)$. $f_Z(z) = 1 \cdot I(0,1)$.

Example: Let X be a Cauchy RV with

$$f_X(x) = 1/\pi(1 + x^2)$$

for $x \in R$. Let $Y = X^2$. Now

$$h(x) = x^2,\ \mathscr{D} = R,\ \mathscr{H} = [0, +\infty].$$

Although h is not one–to–one, there is a partition for which h is one–to–one on each part:

$$\mathscr{D} = D_1 \cup D_2 \text{ where } D_1 = (-\infty, 0),\ D_2 = [0, +\infty).$$

On $\mathscr{H}_1 = h(D_1) = (0, \infty)$,

$$h_1^{-1}(y) = -\sqrt{y} \text{ and } \frac{d}{dy} h_1^{-1}(y) = -\frac{1}{2\sqrt{y}} \text{ is continuous.}$$

On $\mathscr{H}_2 = h(D_2) = [0,\infty)$, $h_2^{-1}(y) = \sqrt{y}$ but

$$\frac{d}{dy} h_2^{-1}(y) = \frac{1}{2\sqrt{y}} \text{ is continuous only on } (0, +\infty) = \mathscr{H}_2 - \{0\}.$$

Define $f_Y(0) = 0$; then

$$f_Y(y) = \frac{1}{2\sqrt{y}}[f_X(-\sqrt{y}) + f_X(\sqrt{y})] \cdot I(0 < y < +\infty)$$

$$= \frac{1}{\pi\sqrt{y}} \cdot \frac{1}{1+y} \cdot I(0 < y < +\infty) .$$

Exercise 5: Let X have density $f_X(x) = \frac{1}{\sigma\sqrt{2\pi}} e^{-(x-\mu)^2/2\sigma^2}$ for $x \in R$. Find the density of $Y = X^2$.

Exercise 6: Let X have density

$$f_X(x) = a\,e^{-a(x-b)} \cdot I(b \le x < +\infty) .$$

Find the density of $Y = \alpha X + \beta$.

Exercise 7: Let X be uniformly distributed over (–2,3); find the density of $Y = X^2$.

Exercise 8: Let U be uniformly distributed on [0,1]. Let F be the CDF of the RV X. Define $F^{-1}(y)$ as

$$\inf\{x \in R : F(x) \ge y\}.$$

Show that:

a) $F^{-1}(y) \le t$ iff $y \le F(t)$;

b) $Y = F^{-1}(U)$ is distributed as X;

c) if F is continuous, then $F(F^{-1}(y)) = y$ for all real y. Hint: there is a "decreasing" sequence $\{x_n\}$ for which

$\lim F(x_n) = y$ and $\lim x_n = x_o = \inf\{x : F(x) \ge y)$;

d) if F is continuous, then the RV $Z = F(X)$ is distributed as U.

The last result is referred to as the probability integral transformation.

LESSON 3. TRANSFORMATIONS OF RANDOM VECTORS

In Lesson 2, we studied the distribution of single transformations of one random variable; in this lesson we will extend the technique to more than one random variable.

Let $X = (X_1,X_2,\cdots,X_n)$ denote an n–dimensional random vector; let $Y = (Y_1,Y_2,\cdots,Y_m)$, $m \le n$, be defined by

$$Y_j = h_j(X_1,X_2,\cdots,X_n),\ j = 1(1)m,$$

with $h_j : R^n \to R$ all measurable. (We will assume, but ignore mentioning, measurability hereafter.) More compactly,

$$\text{for } h = (h_1,h_2,\cdots,h_m),\ Y = h(X) \text{ with } h : R^n \to R^m .$$

The extension mentioned above is a technique for finding the joint distribution of the Y_j's (i.e., the distribution of Y) from the joint distribution of the X_i's; note that we need the joint distribution of the X_i's and not only the individual distributions of the X_i's, except, of course, in the case where the X_i's are independent since then the joint distribution is obtained by multiplication. Once the joint distribution of the Y_j's is obtained, the distributions of any of the components Y_j are found by processes of marginalization; in particular, one finds a marginal density by "integrating out the excess" variables from a joint density.

When the joint density f_X is known, then it is possible to define the distribution of any function of the X_i's, say, $Y = h(X_1,X_2,\cdots,X_n)$, by first computing the CDF of Y. (The actual calculation may be quite a problem.) For example,

$$F_Y(y) = P(Y \le y) = P(h(X_1,X_2,\cdots,X_n) \le y)$$

$$= \int_A f_X(x_1,x_2,\cdots,x_n)dx_1dx_2\cdots dx_n$$

where $A = \{(x_1,x_2,\cdots,x_n) \in R^n : h(x_1,x_2,\cdots,x_n) \le y\}$. (Recall that we take $h \le y$ to mean $h_1 \le y_1, \cdots, h_m \le y_m$.)

Example: Let $X = (X_1, X_2)$ and $Y = X_1 + X_2$ be continuous type RVs .

a) $$F_Y(y) = \iint_{\{(x_1, x_2)\,:\, x_1 + x_2 \le y\}} f_X(x_1,x_2)\ dx_1\ dx_2$$

$$= \int_{-\infty}^{\infty}\int_{-\infty}^{w-x_2} f_X(x_1,x_2)dx_1dx_2$$

By differentiation,

$$f_W(w) = \int_{-\infty}^{\infty} f_X(w - x_2, x_2)\, dx_2 .$$

b) The transformation $Y_1 = X_1 + X_2$ and $Y_2 = X_2$ has inverse $X_1 = Y_1 - Y_2$, $X_2 = Y_2$ and Jacobian determinant 1 . (A review of "Jacobians" appears below.) The joint density of $Y = (Y_1,Y_2)$ is

$$f_X(y_1 - y_2, y_2) .$$

Integrating out y_2 yields the same result as in a).

c) In the particular case that X_1 and X_2 are real and independent, the density of $W = X_1 + X_2$, , namely,

$$f_W(w) = \int_{-\infty}^{\infty} f_{X_1}(w - x_2)\cdot f_{X_2}(x_2)\, dx_2$$

is called the *convolution* of f_{X_1} and f_{X_2} and is often abbreviated as $f_{X_1} * f_{X_2}(w)$.

In part a) of the example above, we looked at the CDF of a single RV $h(X) = X_1 + X_2$ so that $n = 2$ and $m = 1$. In part b), the distribution of $X_1 + X_2$ was obtained as the marginal distribution of the random vector $Y = (Y_1,Y_2)$. This suggests that in order to obtain the joint distribution of $(Y_1,Y_2,\cdots,Y_m)$ when $m < n$, we choose $n-m$ RVs $Y_j = h_j(X_1,X_2,\cdots,X_n)$, $j = m+1,\cdots,n$, to form $(Y_1,Y_2,\cdots,Y_m,Y_{m+1},\cdots,Y_n)$, then "marginalize". The choice is not totally arbitrary as the hypotheses of the theorem below make explicit. Actually, the same kind of restriction appears in the discrete case but the values of the Jacobian do not enter into those calculations; "invertibility" is assumed or sub–consciously checked for each point. Therefore, without loss of generality, we assume $m = n$.

Let us consider the discrete case first. Let

$X = (X_1,X_2,\cdots,X_n)$ be a discrete random vector

and $Y_j = h_j(X)$, $j = 1(1)m$. The mass (density) function of Y is obtained from f_X as follows.

$$f_Y(y_1,y_2,\cdots,y_m) = P[Y_1 = y_1, Y_2 = y_2, \cdots, Y_m = y_m]$$

$$= \sum_A f_X(x_1,x_2,\cdots,x_n),$$

$$A = h^{-1}(y_1,y_2,\cdots,y_m)$$

$$= \{(x_1,\cdots,x_n) : h_j(x_1,\cdots,x_n) = y_j, j = 1(1)m\}.$$

Example: The distribution of (X_1,X_2) is given by the table:

x_2 \ x_1	1	2	3	4
1	2/24	1/24	1/24	4/24
2	3/24	1/24	5/24	2/24
3	1/24	2/24	1/24	1/24

Let $Y_1 = X_1 + X_2$, $Y_2 = X_1 \cdot X_2$. The corresponding values of (Y_1,Y_2) are the pairs:

(2,1) (3,2) (4,3) (5,4)
(3,2) (4,4) (5,6) (6,8) .
(4,3) (5,6) (6,9) (7,12)

Then, $h^{-1}(2,1) = \{(1,1)\}$, $h^{-1}(3,2) = \{(1,2)\ (2,1)\}$,

$h^{-1}(4,4) = \{(2,2)\}$, $\quad h^{-1}(5,6) = \{(2,3)\ (3,2)\}$,

$h^{-1}(5,4) = \{(4,1)\}$, $\quad h^{-1}(6,8) = \{(4,2)\}$,

$h^{-1}(6,9) = \{(3,3)\}$, $\quad h^{-1}(7,12) = \{(4,3)\}$.

$$\begin{aligned} P(Y_1 = 3, Y_2 = 2) & \\ &= P(X_1 = 1, X_2 = 2) + P(X_1 = 2, X_2 = 1) \\ &= 3/24 + 1/24 = 4/\ 24 . \end{aligned}$$

$$\begin{aligned} P(Y_1 = 4, Y_2 = 3) & \\ &= P(X_1 = 3, X_2 = 1) + P(X_1 = 1, X_2 = 3) \\ &= 1/24 + 1/24 = 2/24 . \end{aligned}$$

$$\begin{aligned} P(Y_1 = 4) &= \textstyle\sum_{y_2} P(Y_1 = 4, Y_2 = y_2) \\ &= P(Y_1 = 4, Y_2 = 4) + P(Y_1 = 4, Y_2 = 3) \\ &= 1/24 + 2/14 = 3/24 . \end{aligned}$$

$$\begin{aligned} P(Y_2 = 4) &= \textstyle\sum_{y_1} P(Y_1 = y_1, Y_2 = 4) \\ &= P(Y_1 = 4, Y_2 = 4) + P(Y_1 = 5, Y_2 = 4) \\ &= 1/24 + 4/24 = 5/24 . \end{aligned}$$

Exercise 1: Complete the calculations to find the joint and marginal distributions of Y_1 and Y_2 in the example above.

In the rest of this lesson, we consider the continuous case with $m = n$, density f_X having support

$$\mathscr{D} = \{(x_1, \cdots, x_n) : f_X(x_1, \cdots, x_n) > 0\} .$$

The theorem below is a generalization of Theorem 1 in lesson 2. It is a form of the change of variable in a multiple integral where the derivative becomes the Jacobian. First, we review this concept.

Definition: *Suppose that* h_j *has first partial derivatives* $\dfrac{\partial h_j(x)}{\partial x_i}$ *for* $i = 1(1)n$ *and* $x = (x_1,\cdots,x_n)'$ *where the prime "′" denotes transpose and vectors are column-vectors. Then the gradient of* h_j *at* x *is the vector*

$$\nabla h_j(x) = \left(\frac{\partial h_j(x)}{\partial x_1}, \cdots, \frac{\partial h_j(x)}{\partial x_n} \right)'.$$

If $\nabla h_j(x)$ *exists for all* $j = 1(1)n$, *then the Jacobian of* $h = (h_1, \ldots, h_n)'$ *at* x *is defined to be the determinant (det) of the n by n matrix*

$$\left[\frac{\partial h_i(x)}{\partial x_j} \right]_{i,j = 1(1)n} = \frac{\partial h}{\partial x'}.$$

Among other notations for the Jacobian are the following:

$$J_h(x) = \left\| \begin{matrix} \dfrac{\partial h_1}{\partial x_1} & & \dfrac{\partial h_1}{\partial x_n} \\ & \cdots & \\ \dfrac{\partial h_n}{\partial x_1} & & \dfrac{\partial h_n}{\partial x_n} \end{matrix} \right\| = \frac{\partial(h_1,\cdots,h_n)}{\partial(x_1,\cdots,x_n)} = \left| \frac{\partial\, h}{\partial\, x'} \right| .$$

In the theorem below, we make a changes of variable using the transformations $h^{-1}(y) = g(y) = (g_1(y),\cdots,g_n(y))'$ with Jacobian

$$J_g(y) = J_{h^{-1}}(y) = \det\left[\frac{\partial g_i(y)}{\partial y_j} \right].$$

We are ready now to state the following results like those in Burrill, 1972 .

Theorem: $X = (X_1,\cdots,X_n)'$ *has density* f_X *with*

$$\mathscr{D} = \{x : f_X(x) > 0\};\ let\ y = h(x),$$

$$h = (h_1, \ldots, h_n),\ h_j : R^n \to R,\ j = 1(1)n.$$

Suppose:

a) *there is a finite measurable partition* $\{\mathscr{D}_1, \cdots, \mathscr{D}_m\}$ *of* $\mathscr{D}$ *such that the restriction of* h *to each* $\mathscr{D}_j$, *say* $h^j : \mathscr{D}_j \to \mathscr{H}_j$ *with* $\cup_{j=1}^m \mathscr{H}_j = \mathscr{H} = h(\mathscr{D})$, *is one-to-one and onto;*

b) *for the inverse* $(h^j)^{-1} = (g_{j1}, g_{j2}, \ldots, g_{jn})$, *the functions* g_{ji}, $j = 1(1)m$, $i = 1(1)n$, *all have continuous first partial derivatives on* $\mathscr{H}_j$ *and each Jacobian* $J_{(h^j)^{-1}}$ *is almost everywhere nonzero on* $\mathscr{H}_j$. *Then* Y *has joint density given by* $f_Y(y)$

$$= \Sigma_{j=1}^m f_X(g_{j1}(y), \cdots, g_{jn}(y)) \cdot |J_{(h^j)^{-1}}(y)| \cdot I_{\mathscr{H}}(y).$$

When m = 1, the theorem takes the form of the

Corollary: *If, in fact,*

a) *h is one-to-one with* $\mathscr{D} \to h(\mathscr{D})$, $h^{-1} = (g_1, \ldots, g_n)$;

b) *for each* $i = 1(1)n$, *all first partial derivatives of* g_i *exist and are continuous on* $h(\mathscr{D})$, *and* $J_{h^{-1}}$ *is almost everywhere nonzero on* $h(\mathscr{D})$; *then the joint density of* Y *is:*

$$f_Y(y) = f_X(g_1(y), \cdots, g_n(y)) \cdot |J_{h^{-1}}(y)| \cdot I_{h(\mathscr{D})}(y).$$

Example: Let $X = (X_1, X_2)$ have density

$$f_X(x_1, x_2) = \frac{1}{4} e^{-(|x_1| + |x_2|)}$$

with support $\mathscr{D} = \{x : f_X(x) > 0\} = R^2$. To find the density of $X_1 + X_2$, one might proceed as follows. Let

$$Y_1 = h_1(X_1,X_2) = X_1 + X_2 \text{ and } Y_2 = h_2(X_1,X_2) = X_1 .$$

Now $h : R^2 \to R^2$ is one–to–one and onto:

$$y_1 = x_1 + x_2,\ y_2 = x_1 \text{ if and only if } x_1 = y_2,\ x_2 = y_1 - y_2.$$

It follows that $\begin{array}{l} g_1(y_1,y_2) = y_2 \\ g_2(y_1,y_2) = y_1 - y_2 \end{array}$ and

$$J_{h^{-1}}(y_1,y_2) = \begin{vmatrix} 0 & 1 \\ 1 & -1 \end{vmatrix} = -1 .$$

Therefore, $f_Y(y_1,y_2) = f_X(y_2,\ y_1 - y_2)$, and so

$$f_{Y_1}(y_1) = \int_{-\infty}^{\infty} f_Y(y_1,y_2)\, dy_2 = \int_{-\infty}^{\infty} f_X(y_2,y_1 - y_2)\, dy_2$$

$$= \tfrac{1}{4}\int_{-\infty}^{\infty} \exp\{-(|y_2| + |y_1 - y_2|)\}\, dy_2 = \tfrac{1}{4}[1 + |y_1|]e^{-|y_1|} .$$

Example: Let X_1,X_2 be IID with common density f, support R; the joint density of (X_1, X_2) is

$$f(x_1)\cdot f(x_2) = f_{(X_1,X_2)}(x_1,x_2) .$$

Consider $Y_1 = \mathrm{Min}(X_1,X_2)$ and $Y_2 = \mathrm{Max}(X_1,X_2)$ .Now $(Y_1,\ Y_2) = h(X_1,X_2)$ is not one–to–one since, for example, $y_1 = 3,\ y_2 = 5$ has $x_1 = 3,\ x_2 = 5$ or $x_1 = 5,\ x_2 = 3$. But we can partition the support as

$$\mathcal{D}_1 = \{(x_1,x_2) : x_1 < x_2\} \text{ and } \mathcal{D}_2 = \{(x_1,x_2) : x_1 > x_2\}$$

on each of which h is 1:1 . This can be reflected in the density by taking

$$f_{X_1,X_2}(x_1,\ x_2) = f(x_1)f(x_2)\cdot I_{\mathcal{D}_1}(x) + f(x_1)f(x_2)\cdot I_{\mathcal{D}_2}(x) .$$

On $\mathscr{D}_1$, we have $\begin{bmatrix} y_1 \\ y_2 \end{bmatrix} = h^1(x_1,x_2) = \begin{bmatrix} x_1 \\ x_2 \end{bmatrix}$ with

$$\begin{bmatrix} x_1 \\ x_2 \end{bmatrix} = g_1(y_1,y_2) = \begin{bmatrix} y_1 \\ y_2 \end{bmatrix}$$

and

$$J_{[h^1]^{-1}} = \left|\frac{\partial x}{\partial y'}\right| = \begin{vmatrix} 1 & 0 \\ 0 & 1 \end{vmatrix} = 1 .$$

On $\mathscr{D}_2$, we have $\begin{bmatrix} y_1 \\ y_2 \end{bmatrix} = h^2(x_1,x_2) = \begin{bmatrix} x_2 \\ x_1 \end{bmatrix}$ with

$$\begin{bmatrix} x_1 \\ x_2 \end{bmatrix} = g_2(y_1,y_2) = \begin{bmatrix} y_2 \\ y_1 \end{bmatrix}$$

and

$$J_{[h^2]^{-1}} = \left|\frac{\partial x}{\partial y'}\right| = \begin{vmatrix} 0 & 1 \\ 1 & 0 \end{vmatrix} = -1 .$$

Then $f_{Y_1,Y_2}(y_1,y_2)$

$$\begin{aligned} = \{ & f_{X_1,X_2}(g_{11}(y_1,y_2),g_{12}(y_1,y_2)) \cdot |J_{[h^1]^{-1}}| \\ & + f_{X_1,X_2}(g_{21}(y_1,y_2),g_{22}(y_1,y_2)) \cdot |J_{[h^2]^{-1}}| \} \cdot \\ & I\{(y_1,y_2) : y_1 < y_2\} . \end{aligned}$$

This is often abbreviated as

$$f_Y(y) = \{ f(y_1)f(y_2) + f(y_2)f(y_1) \} \cdot I\{(y_1,y_2) : y_1 < y_2\}$$

or

$$\begin{aligned} f_Y(y) &= 2f(y_1)f(y_2) \quad \text{for } y_1 < y_2 \\ &= 0 \qquad \text{otherwise.} \end{aligned}$$

Example: Let X_1, X_2 be independent RVs with support R and densities

$$f_{X_1}(x) = \frac{1}{2} e^{-|x|}, \quad f_{X_2}(x) = \frac{1}{\sqrt{2\pi}} e^{-x^2/2}, \text{ respectively.}$$

Then $f_{X_1,X_2}(x_1,x_2) = \frac{1}{2\sqrt{2\pi}} e^{-|x_1|-x_2^2/2}$ on R^2. For the transformation

$$Y_1 = X_1 + X_2,\ Y_2 = X_2/X_1,$$

we need to restrict the domain to exclude $x_1 = 0$ (and correspondingly, $y_2 = -1$ below); of course, these have probability 0. Explicitly,

$$h_1(x_1,x_2) = x_1 + x_2 = y_1,\ h_2(x_1,x_2) = x_2/x_1 = y_2$$

and

$$x_1 = g_1(y_1,y_2) = \frac{y_1}{1 + y_2},\quad x_2 = g_2(y_1,y_2) = \frac{y_1 y_2}{1 + y_2}.$$

Then, $J_{h^{-1}}(y) = \begin{vmatrix} \frac{1}{1+y_2} & \frac{-y_1}{(1+y_2)^2} \\ \frac{y_2}{1+y_2} & \frac{y_1}{(1+y_2)^2} \end{vmatrix} = \frac{y_1}{(1 + y_2)^2}.$

Hence, $f_Y(y) = f_X(y_1/(1+y_2), y_1y_2/(1+y_2)) \cdot \frac{|y_1|}{(1 + y_2)^2}$

$$= \frac{1}{2\sqrt{2\pi}} \frac{|y_1|}{(1+y_2)^2} \exp\{-\left|\frac{y_1}{1+y_2}\right| - \frac{1}{2} \frac{y_1^2 y_2^2}{(1 + y_2)^2}\}$$

for $y = (y_1,y_2) \in R^2 - \{(y_1,y_2) : y_2 = -1\}$.

Exercise 2: Let X_1, X_2 be independent RVs with the same density

$$f(x) = \frac{1}{\sqrt{2\pi}} e^{-x^2/2},\ x \in R.$$

a) Find the joint density of $(X_1 - X_2, X_1)$.

b) Deduce the density of $X_1 - X_2$.

Exercise 3: Let X_1, X_2 be independent RVs with densities $f_{X_1}(x) = x\,e^{-x} \cdot I(x \geq 0)$, $f_{X_2}(x) = \frac{1}{2} x^2 e^{-x} \cdot I(x \geq 0)$, respectively.

a) Find the joint density of $(X_1 + X_2, \frac{X_1}{X_1 + X_2})$.

b) Deduce the marginal densities of $X_1 + X_2$ and $\frac{X_1}{X_1 + X_2}$.

c) Verify that $X_1 + X_2$ and $\frac{X_1}{X_1 + X_2}$ are independent.

Exercise 4: Let $X = (X_1, X_2, X_3)$ with support R^3 and joint density

$$f_X(x_1,x_2,x_3) = \frac{1}{8} e^{-(|x_1|+|x_2|+|x_3|)}$$

Find the joint density of $(X_1, X_1 + X_2, X_1 + X_2 + X_3)$.

LESSON 4. SAMPLING DISTRIBUTIONS IN NORMAL POPULATIONS–I

This lesson and the next contain applications of transformations and characteristic functions (Part III) to derive traditional distributions of functions of random samples from normal populations; these sampling distributions will be used later in confidence interval estimation and hypotheses testing.

Let $X_1, X_2, \cdots, X_n$ be a random sample from a (normal) population X with characteristic function

$$\varphi_X(t) = e^{i\mu t - \sigma^2 t^2/2} ;$$

for $\sigma > 0$, the density is $f(x) = \dfrac{1}{\sigma\sqrt{2\pi}} e^{-(x-\mu)^2/2\sigma^2}$, $x \in R$.

The sample mean and the sample variance are, respectively,

$$\bar{X} = \frac{1}{n}\sum_{i=1}^{n} X_i , \quad S^2 = \frac{1}{n}\sum_{i=1}^{n}(X_i - \bar{X})^2 .$$

1) Let $Y = \dfrac{\bar{X} - \mu}{\sigma/\sqrt{n}}$. As in lesson 15, Part III, the characteristic function of Y is

$$\varphi_Y(t) = \left[\varphi_{\frac{X-\mu}{\sigma\sqrt{n}}}(t)\right]^n = e^{-t^2/2} .$$

Therefore, Y is a standard normal RV; equivalently, $\bar{X}$ is normally distributed with mean μ and variance $\dfrac{\sigma^2}{n}$.

2) From exercise 5, Lesson 2, we see that when Z is $N(0,1)$, the density of Z^2 is

$$f(x) = (2\pi x)^{-1/2} e^{-x/2} \cdot I(x > 0) .$$

This is the density of a chisquare distribution with 1 degree of freedom; symbolically,

$$Z^2 \text{ is distributed as } \chi^2(1).$$

Example: a) A general form of the function f in 2) is the real "gamma" function: for $\alpha > 0$, $\Gamma(\alpha) = \int_0^\infty x^{\alpha-1} e^{-x}\, dx$. It can be shown (advanced calculus) that the integral is uniformly convergent for $\alpha > 0$ and divergent for $\alpha \le 0$.

b) For the special case $\alpha = \frac{1}{2}$, successive substitutions $x = t^2/2$ then $y = -t$ in the integral yield

$$\Gamma(\tfrac{1}{2}) = \int_0^\infty x^{-1/2} e^{-x}\, dx = \sqrt{2}\int_0^\infty e^{-t^2/2}\, dt$$

$$= \sqrt{2}\int_{-\infty}^0 e^{-y^2/2}\, dy .$$

Averaging the last two integrals yields

$$\Gamma(1/2) = \frac{\sqrt{2}}{2}\int_{-\infty}^\infty e^{-t^2/2}\, dt .$$

Then,

$$\Gamma^2(\tfrac{1}{2}) = \frac{1}{2}\int_{-\infty}^{+\infty}\int_{-\infty}^{+\infty} e^{-(t^2 + s^2)/2} dt\, ds .$$

Now switch to polar coordinates; then substitute $y = r^2/2$ to obtain:

$$\Gamma^2(1/2) = \frac{1}{2}\int_0^{2\pi}\int_0^\infty r\, e^{-r^2/2} dr\, d\theta$$

$$= \frac{1}{2}\int_0^{2\pi}\int_0^{+\infty} e^{-y} dy\, d\theta = \pi .$$

This establishes $\Gamma(\tfrac{1}{2}) = \sqrt{\pi}$ and hence the density of $\chi^2(1)$ can be written as:

$$f(x) = \frac{1}{2^{1/2}\,\Gamma(1/2)}\, x^{\frac{1}{2}-1}\, e^{-x/2} \cdot I(x > 0)\,.$$

Definition: *The family of chi-square random variables,* $\chi^2(k)$ *, (indexed by the positive real parameter k) has the density*

$$f(x;k) = x^{k/2\,-\,1}\, e^{-x/2} \cdot I(x > 0)\, /\, \Gamma(\tfrac{k}{2})2^{k/2}\,.$$

(Integral values of k are called degrees of freedom.)

Exercise 1: Show that $E[\chi^2(k)] = k$ and $Var[\chi^2(k)] = 2k$.

Example: The manipulations used to obtain the MGF are just a bit different than those used to find the moments. In

$$\int_0^\infty e^{tx}\, x^{(k/2)-1}\, e^{-x/2}\, dx = \int_0^\infty x^{(k/2)-1}\, e^{-x(1-2t)/2}\, dx\,,$$

substitute $x(1-2t) = y$ to obtain:

$$\int_0^\infty [y/(1-2t)]^{(k/2)-1}\, e^{-y/2}\, dy/(1-2t)$$

$$= \int_0^\infty y^{(k/2)-1}\, e^{-y/2}\, dy\, /(1-2t)^{k/2}.$$

It follows that the MGF $M_{\chi^2}(t) = E[e^{t\chi^2}] = 1/(1-2t)^{k/2}$. The CF is $\varphi_{\chi^2}(t) = M_{\chi^2}(it) = 1/(1-2it)^{k/2}$.

Exercise 2: Let $Z_1, \cdots, Z_k$ be IID N(0,1). Show that $\Sigma_{i=1}^{k} Z_i^2$ has the density of $\chi^2(k)$.

Note that the chi–square family is a particular case of the gamma family:

Definition: *The family of gamma distributions (indexed by the parameter* $\theta = (\theta_1, \theta_2)$ *with both components positive) has*

density $\quad f(x;\theta) = x^{\theta_1 - 1} e^{-x/\theta_2} \cdot I(x > 0) / \Gamma(\theta_1)\theta_2^{\theta_1}$.

Exercise 3: For the gamma density above, verify that

$$\int_{-\infty}^{\infty} f(x;\theta)\, dx = 1 .$$

3) We will show that nS^2/σ^2 can be written as $\sum_{i=1}^{n-1} Z_i^2$ where $Z_1, \cdots, Z_{n-1}$ are IID N(0,1) so that nS^2/σ^2 is distributed as $\chi^2(n-1)$.

In order to get the distribution of S^2, we need some information about Gaussian random vectors, i.e., random vectors having multivariate normal distributions (àla lesson 15, Part III). First recall that (a non-degenerate) $Y = (Y_1, \cdots, Y_n)$ is Gaussian if its CF is

$$\varphi_Y(t) = e^{it'\theta - t'\Sigma t/2}$$

where $t = (t_1, \cdots, t_n)'$; the mean is $\theta = (E[Y_1], \cdots, E[Y_n])'$ and the covariance matrix is

$$\Sigma = \Big[E[(Y_i - E[Y_i])(Y_j - E[Y_j])]\Big]_{i,j = 1(1)n} .$$

In particular, for the random sample $X_1, \cdots, X_n$ from $N(\mu, \sigma^2)$, the random vector $V = (X_1, \cdots, X_n)'$ is Gaussian with mean $\theta = \mu(1,1,\cdots,1)'$ and CF $\varphi_V(t) = e^{it'\theta - t'(\sigma^2 I)t/2}$.

For the Helmert matrix

$$A = \begin{bmatrix} \frac{1}{\sqrt{n}} & \frac{1}{\sqrt{n}} & \frac{1}{\sqrt{n}} & \cdots & \frac{1}{\sqrt{n}} \\ \frac{1}{\sqrt{2}} & -\frac{1}{\sqrt{2}} & 0 & \cdots & 0 \\ \frac{1}{\sqrt{2\cdot 3}} & \frac{1}{\sqrt{2\cdot 3}} & -\frac{2}{\sqrt{2\cdot 3}} & \cdots & 0 \\ \vdots & \vdots & \vdots & \cdots & 0 \\ \frac{1}{\sqrt{(n-1)n}} & \frac{1}{\sqrt{(n-1)n}} & \frac{1}{\sqrt{(n-1)n}} & \cdots & \frac{-(n-1)}{\sqrt{(n-1)n}} \end{bmatrix},$$

the random vector $U = (U_0, U_1, \cdots, U_{n-1})' = AV$ has components:

$U_0 = \sqrt{n}\,\overline{X}$,

$U_1 = (X_1 - X_2)/\sqrt{2}$,

$U_2 = (X_1 + X_2 - 2X_3)/\sqrt{2\cdot 3}$

$\cdots$

$U_{n-1} = (X_1 + X_2 + \cdots + X_{n-1} - (n-1)X_n)/\sqrt{(n-1)n}$.

In short form,

$$U_j = [\sum_{i=1}^{j} X_i - j\,X_{j+1}]/\sqrt{j(j+1)} \text{ for } j = 1,2,\cdots,(n-1).$$

The vector U is also multivariate Gaussian since it is obtained by a linear transformation of the multivariate Gaussian V.

Exercise 4: Continue with A, V, and U above.

a) Write out $U = AV$ when $n = 4$.

b) Show directly that for $j \geq 1$, U_j is $N(0, \sigma^2)$. Hint: find the characteristic function of U_j.

Exercise 5: In this exercise, only algebraic properties of S^2, A, V, U are used; the normal distribution is not involved.

a) Show that $A'A = I = AA'$.

b) Show that $A(1,1,\cdots,1)' = (\sqrt{n},0,\cdots,0)'$.

c) Show that $\Sigma_{i=1}^{n} X_i^2 = V'V = V'A'A\,V$

$$= U'U = n(\overline{X})^2 + \Sigma_{i=1}^{n-1} U_i^2 .$$

Hint: Try $n = 4$ first.

d) Show that $nS^2 = \Sigma_{i=1}^{n}(X_i - \overline{X})^2$

$$= \Sigma_{i=1}^{n} X_i^2 - n(\overline{X})^2 = \Sigma_{i=1}^{n-1} U_i^2 .$$

The derivation of the distribution of nS^2/σ^2 will be complete if we show that $U_1, U_2, \cdots, U_{n-1}$ are independent and then take $Z_j = U_j/\sigma$.

4) We can show a slightly stronger result which will be used later, namely, that all the U_j's, $j = 0,1,\cdots,(n-1)$, are independent. But then

$$\overline{X} = \frac{U_0}{\sqrt{n}} \quad \text{and} \quad S^2 = \frac{1}{n}\Sigma_{j=1}^{n-1} U_j^2$$

are functions of independent RV and are therefore independent of each other.

Using the results of exercise 5, we can evaluate the CF of U as $\varphi_U(t) = E[e^{it'U}] = E[e^{it'AV}] = \varphi_V(A't)$

$$= e^{it'A\theta - t'A(\sigma^2 I)A't/2}$$

$$= e^{it'A(1,1,\cdots,1)'\mu - t'AA't\sigma^2/2}$$

$$= e^{it'(\sqrt{n},0,\cdots,0)'\mu - t't\sigma^2/2}$$

$$= e^{it_1\sqrt{n}\mu - t'(\sigma^2 I)t/2} .$$

Since the covariance matrix of the normal random vector U, namely $\sigma^2 I$, is diagonal, the components of U are independent.

Exercise 6: a) Let $W_1, W_2, \cdots, W_n$ be IID as $\chi^2(1)$. Show that the sum $W_1 + W_2 + \cdots + W_n$ is distributed as $\chi^2(n)$.

b) Use the CLT to show that $(\chi^2(n) - n) / \sqrt{2n} \overset{D}{\to} N(0,1)$.

LESSON 5. SAMPLING DISTRIBUTIONS IN NORMAL POPULATIONS–II

Continuing as in Lesson 4, we now proceed to find the distribution of the random variable

$$T = \frac{\overline{X} - \mu}{S/\sqrt{(n-1)}}$$

where $\overline{X}$ and S^2 are the sample mean and sample variance of a random sample of size n from $N(\mu,\sigma^2)$. From Lesson 4, we already have: $\sqrt{n}\,(\overline{X} - \mu)$ is $N(0,\sigma^2)$;

nS^2/σ^2 is $\chi^2(n-1)$;

$\overline{X}$ and S are independent .

It follows that the random variable T can be written as

$$T = \frac{\sqrt{n}(\overline{X} - \mu)/\sigma}{\sqrt{nS^2/\sigma^2(n-1)}}$$

where the numerator is a standard normal random variable ; the denominator is the square–root of a chisquare random variable divided by its degrees of freedom ; and the numerator and denominator are independent . This is the original special case of the

Definition: *Let the RV U be N(0,1); let the RV V be $\chi^2(k)$; let U and V be independent. Then* $T = \dfrac{U}{\sqrt{V/k}}$ *is a Student-T random variable . The family has index parameter k > 0; when k is integral, it is called degrees of freedom.*

To find the density of T, we use a bivariate transformation of the joint density of U and V and integrate out the extra variable. Since U and V are independent, the non–zero part of their joint density is

$$f(u,v) = \frac{e^{-u^2/2}}{\sqrt{(2\pi)}} \cdot \frac{v^{k/2-1}\,e^{-v/2}}{\Gamma(k/2)\,2^{k/2}} .$$

The transformation is $t = u/\sqrt{v/k}$, $w = v$ with the inverse
$u = t\sqrt{w/k}$, $v = w$;

the Jacobian is $\partial(u, v)/\partial(t, w) = \sqrt{w/k}$. Then the joint density of (T, W) is

$$g(t,w) = f(t\sqrt{w/k}\,,w), \sqrt{w/k}\,) \text{ for } t \text{ real and } w > 0 .$$

$$g(t, w) = \frac{e^{-t^2/2k}\, w^{k/2 - 1}\, e^{-w/2}}{\sqrt{(2\pi)}\, 2^{k/2}\, \Gamma(k/2)} \sqrt{(w/k)}$$

$$= \frac{w^{(k+1)/2 - 1}\, e^{-w(1+t^2/k)/2}}{(\pi k)^{1/2}\, 2^{(k+1)/2}\, \Gamma(k/2)} .$$

The marginal density $f_T(t) = \int_0^\infty g(t,w)\, dw$. The integral is evaluated by use of the substitution $w(1+t^2/k)/2 = y$ to produce a gamma function. Then $f_T(t)$

$$= \frac{\int_0^\infty \left[\frac{2y}{1+t^2/k}\right]^{(k+1)/2 - 1} e^{-y}\, 2dy/(1+t^2/k)}{(\pi k)^{1/2}\, 2^{(k+1)/2}\, \Gamma(k/2)}$$

$$= \frac{\int_0^\infty y^{(k+1)/2 - 1}\, e^{-y}\, dy}{(\pi k)^{1/2}\, \Gamma(k/2)\, (1+t^2/k)^{(k+1)/2}}$$

$$= \frac{\Gamma((k+1)/2)}{(\pi k)^{1/2}\, \Gamma(k/2)\, (1+t^2/k)^{(k+1)/2}} .$$

Exercise 1: These questions all relate to this Student–T distribution.

a) Verify $\int_{-\infty}^{\infty} f_T(t)\, dt = 1$ directly.

b) Show that the mean does not exist for k = 1. (The Cauchy distribution.)
c) Show that for k > 2, the mean of these distributions is 0 and each variance is k/(k −2).
d) Argue that the MGF does not exist for any k .
e) Show that for each $x \in R$,

$$f_T(x) \to e^{-x^2/2}/\sqrt{(2\pi)}\ ,\ \text{as } k \to +\infty. .$$

As a consequence of exercise 1e, the CDF of T can be approximated by that of the standard normal (Scheffé, Lesson 15, Part III). In many methods books, this is done (tables are cut off) when the degrees of freedom is greater than 30.

Let $X_1, \cdots, X_n$ be a random sample from the population of X which is distributed as $N(\mu, \sigma^2)$; let $Y_1, \cdots, Y_m$ be a random sample from an independent population of Y distributed as $N(\nu, \tau^2)$. The corresponding sample variances are:

$$S_X^2 = \frac{1}{n}\Sigma_{i=1}^{n}(X_i - \overline{X}^2),\quad S_Y^2 = \frac{1}{m}\Sigma_{i=1}^{m}(Y_i - \overline{Y})^2 .$$

Then as for T above, $F = \dfrac{nS_X^2/\sigma^2(n-1)}{mS_Y^2/\tau^2(m-1)}$ is the original special case of

Definition: *Let U and V be independent chisquare random variables with positive parameters p and q respectively. Then* $F(p,q) = \dfrac{U/p}{V/q}$ *is the F(isher) random variable. When the parameters are integral, they are called degrees of freedom.*

Exercise 2: With the hypotheses in this definition, consider the transformation $R = \frac{U}{p} / \frac{V}{q}$, $W = \frac{V}{q}$. Following a procedure analogous to that for T, show that the density of R (= F(p,q)) is

$$\frac{\Gamma((p+q)/2)}{\Gamma(p/2)\ \Gamma(q/2)}(p/q)^{p/2}\frac{r^{p/2\ -\ 1}}{(1\ +\ pr/q)^{(p+2)/2}}I(0 < r) .$$

Exercise 3: Compute the mean and the variance of F(p,q) using the factored form $\frac{U}{p} / \frac{V}{q} = \frac{q}{p} U \cdot (1/V)$ to get the moments.

The Student and Fisher distributions introduced so far have been "central" because the underlying chisquare distributions have been "central", that is, involving squares of normal random variables with means 0. The following exercise and definition introduce the "non–central" case.

Exercise 4: Let U be a normal random variable with mean μ and variance $\sigma^2 > 0$. Find the MGF & PDF of $W = U/\sigma^2$.

Definition: *For i = 1(1)n, let U_i be distributed as $N(\mu_i, \sigma_i^2)$; suppose that all these RVs are independent. Then*

$W = \sum_{i=1}^{n} (U_i/\sigma_i)^2$ *is a non-central chisquare RV with*

n degrees of freedom and non-centrality parameter

$\lambda = \sum_{i=1}^{n} (\mu_i/\sigma_i)^2$. *We write* $W = \chi^2(n;\lambda)$.

We begin the investigation of W by computing its MGF:

$$M_W(t) = E[e^{tW}] = E[e^{t\Sigma(U_i/\sigma_i)^2}] = \Pi_{i=1}^{n} E[e^{t(U_i/\sigma_i)^2}] .$$

Each of these latter factors is of the form

$$\int_{-\infty}^{\infty} e^{t(x/\sigma)^2} e^{-(x-\mu)^2/2\sigma^2} dx / \sqrt{(2\pi)}\, \sigma$$

$$= \int_{-\infty}^{\infty} e^{-\mu^2/2\sigma^2} e^{\mu x/\sigma^2} e^{-x^2(1-2t)/2\sigma^2} dx / \sqrt{2\pi})\, \sigma$$

$$= e^{-\mu^2/2\sigma^2} \int_{-\infty}^{\infty} e^{(\mu/\sigma^2)(\sigma y/\sqrt{(1-2t)})} e^{-y^2/2} dy / \sqrt{(2\pi)}\sqrt{(1-2t)}$$

$$= \frac{e^{-\mu^2/2\sigma^2}}{\sqrt{(1-2t)}} \int_{-\infty}^{\infty} e^{(\mu/\sigma\sqrt{(1-2t)})y} e^{-y^2/2} dy / \sqrt{(2\pi)}$$

$$= \frac{e^{-\mu^2/2\sigma^2}}{\sqrt{(1-2t)}} e^{(\mu/\sigma\sqrt{(1-2t)})^2/2} = e^{\mu^2 t/\sigma^2(1-2t)}/\sqrt{(1-2t)} .$$

It follows that $M_W(t) = e^{(\Sigma(\mu_i/\sigma_i)^2 t/(1-2t)}/(1-2t)^{n/2}$.

From this, we see that W is a true generalization of chisquare since, when the means are all equal 0, we have the same MGFs. We see also that the the size of the non–centality parameter $\lambda = \sum_{i=1}^{n} (\mu_i/\sigma_i)^2$ suggests the depth of the departure from centrality. When the variances σ_i^2 are all equal (often 1), λ is essentially the square of the length of the vector of means $(\mu_1,\mu_2,\cdots,\mu_n)$. The density of W is

$$f_W(w) = e^{-\lambda} \sum_{i=0}^{\infty} \frac{\lambda^i \, w^{n/2+i-1} \, e^{-w/2}}{i! \, 2^{n/2+i} \, \Gamma(n/2+i)} \cdot I(0 < w) .$$

Exercise 5: Verfiy the formula for the density of $W = \chi^2(n;\lambda)$ by using it to calculate the MGF. Hint: interchange $\sum$ and $\int$.

Definition: *Let U be distributed as N(μ,1); let V be distributed as $\chi^2(n)$; let W be distributed as $\chi^2(p;\alpha)$; let Y be distributed as $\chi^2(q;\beta)$. Suppose that U, V, W, Y are independent. Then*

a) *The distribution of $T(n;\lambda) = U/\sqrt{V/n}$ is called the non-central T distribution with n degrees of freedom, non-centrality parameter $\lambda = \mu$.*

b) *The distribution of $F(p,q;\alpha,\beta) = \frac{W}{p} / \frac{Y}{q}$ is called the non-central F-distribution with degrees of freedom p and q and non-centrality parameters α and β.*

Let $X_1,\cdots,X_n$ be a random sample from $N(\mu, \sigma^2)$. For $\mu_0 \neq \mu$, $\sqrt{n-1}\,(\bar{X} - \mu_0)/S$ has the distribution $T((n-1);\lambda)$ with

$\lambda = \sqrt{n}(\mu - \mu_0)/\sigma$. Indeed, from the previous discussion, we have

$$\sqrt{n-1}\,(\bar{X} - \mu_0)/S = \frac{\bar{X} - \mu_0}{\sigma/\sqrt{n}} \Big/ \sqrt{nS^2/\sigma^2(n-1)}$$

which is of the form $U/\sqrt{V/(n-1)}$ where

$$U = (\bar{X} - \mu_0)/\sigma\sqrt{n}$$

is normally distributed with mean $(\mu - \mu_0)/\sigma\sqrt{n}$ and variance 1 and $V = n\,S^2/\sigma^2$ is distributed as $\chi^2(n-1)$, and U, V are independent .

Exercise 6: What is the distribution of $(n-1)(\bar{X} - \mu_0)^2/S^2$ above? Answer descriptively and by finding a density function.

It is more common to see the noncentral F defined as it is used in ANOVA, namely with a central chisquare in the denominator.

INE–Exercise. From the joint density of W and V as in the definition above, show that the (non–zero portion of the) density of $F_A = Wn/Vp$ is

$$e^{-\alpha}\sum_{i=1}^{\infty} \frac{\alpha^i\, x^{p/2 + i - 1}}{i!\,(1 + px/n)^{(n+p)/2 + i}} \cdot$$

$$\frac{\Gamma((n+p)/2 + i)\,(p/n)^{p/2 + i}}{\Gamma(n/2)\,\Gamma(p/2 + i)} .$$

LESSON 6. ORDER STATISTICS

This lesson contains a few results of that useful class of statistics named in the title above.

Let $X_1, \cdots, X_n$ be a random sample of real a RV X with CDF F. If we arrange these RVs in order of magnitude, we obtain a new set of RVs with $X_{(j)}$ the jth *order statistic* and $X_{(1)} \le X_{(2)} \le \cdots \le X_{(n)}$. In particular,

$$X_{(1)} = \text{Min}\{X_j, j = 1,\cdots,n\},\ X_{(n)} = \text{Max}\{X_j, j = 1,\cdots,n\}$$

are called the *extreme order statistics*. As usual, attention quickly focuses on functions of these order statistics such as:

the *sample range* $X_{(n)} - X_{(1)}$,

the *sample mean* $\frac{1}{n}\sum_{j=1}^{n} X_{(j)}$,

the *sample mid-range* $[X_{(1)} + X_{(n)}]/2$,

the *sample median*

$$\begin{aligned} m &= [X_{(n/2)} + X_{(n/2+1)}]/2 \quad \text{if n is even,} \\ &= X_{\left[\frac{n+1}{2}\right]} \quad \text{if n is odd .} \end{aligned}$$

And, as is to be expected, in order to be able to utilize order statistics in estimation and testing, we need to know their distributions.

Example: For min and max, we can write out joint distributions quite easily. Of course, $P(X_{(1)} \le X_{(n)}) = 1$. With $y < z$,

$$\begin{aligned} P(X_{(1)} \le y, X_{(n)} \le z) &= P(X_{(n)} \le z) - P(X_{(1)} > y, X_{(n)} \le z) \\ &= P(\text{all } X_j \le z) - P(y < X_1,\cdots,X_n \le z) \\ &= [F(z)]^n - [F(z) - F(y)]^n. \end{aligned}$$

As y increases, $P(X_{(1)} \leq y, X_{(n)} \leq z)$ increases to

$$P(X_{(1)} \leq X_{(n)} \leq z) = P(X_{(n)} \leq z) = [F(z)]^n.$$

As z increases to $+\infty$, $P(X_{(1)} \leq y, X_{(n)} \leq z)$ increases to

$$P(X_{(1)} \leq y) = 1 - (1 - F(y))^n.$$

Distributions of many order statistics are obtained on an ad–hoc basis, i.e., for each specific F. But when F is absolutely continuous (X is a continuous type RV), these new distribution functions are also differentiable and the corresponding densities can be obtained in terms of F and $F' = f$. We adopt this case in the remainder of this lesson. Note that since $P(X_{(i)} = X_{(j)}) = 0$,

$$P(X_{(1)} < X_{(2)} < \cdots < X_{(n)}) = 1.$$

Exercise 1: Perform appropriate differentiations of the CDFs in the example above to obtain the joint density of $X_{(1)}$ and $X_{(n)}$ and their marginal densities.

Theorem: *In the continuous case, the joint density of* $(X_{(1)}, \cdots, X_{(n)})$ *is given by*

$$g(y_1, \cdots, y_n) = n!\, \Pi_{j=1}^{n} f(y_j)\, I(y_1 < y_2 < \cdots < y_n)$$

(technically, almost surely).

Proof: We use the transformation technique of lesson 3. (In fact, this discussion is an n–dimensional version of the long min–max example in that lesson.) The joint density of $(x_1, \cdots, x_n)$ is

$$f(x_1, x_2, \cdots, x_n) = \Pi_{j=1}^{n} f(x_j).$$

There is a finite measurable partition $(\mathscr{D}_1, \mathscr{D}_2, \cdots, \mathscr{D}_{n!})$ of the support $\mathscr{D} = \{(x_1, \cdots, x_n) : f(x_1, \cdots, x_n) > 0\}$ such that the restriction of the transformation

$$h : (X_1, \cdots, X_n) \to (X_{(1)}, \cdots, X_{(n)}) \text{ to each } \mathscr{D}_i,\ i=1(1)n!,$$

is one–to–one; each $\mathscr{D}_i$ corresponds to one permutation of $(X_{(1)}, \cdots, X_{(n)})$. [For example, if $\mathscr{D}_1 = \{x_1 < \cdots < x_n\}$, then

$$h \mid \mathscr{D}_1 : (x_1, ..., x_n) \to y_1 = x_1, y_2 = x_2, \cdots, y_n = x_n$$

and the Jacobian of its inverse is $+1$.] The Jacobian of the inverse of $h \mid D_i$ is obviously ± 1. Hence the result follows.

Example: Let $f(x) = \theta e^{-\theta x} I(x > 0)$ for $\theta > 0$. For $n = 2$, the joint density of $(X_{(1)}, X_{(2)})$ is

$$g(y_1, y_2) = 2f(y_1)\cdot f(y_2)\cdot I(y_{(1)} < y_{(2)})$$

$$= 2\,\theta^2 e^{-\theta(y_1 + y_2)}\cdot I(y_1 < y_2).$$

Corollary: *The joint density of* $(X_{(r)}, X_{(s)})$, $1 \le r < s \le n$, *is given by*

$$\frac{n!}{(r-1)!(s-r-1)!(n-s)!}[F(x)]^{r-1} f(x)[F(y)-F(x)]^{s-r-1}$$

$$f(y)[1-F(y)]^{n-s} I(x < y).$$

Proof: This formula is obtained by integrating out the other variables in the joint density of $(X_{(1)}, \cdots, X_{(n)})$ given in the theorem above; technically one uses induction.

Exercise 2: Check the proof above for $n = 6, r = 2, s = 5$.

Exercise 3: For positive integers $1 < r_1 < r_2 \cdots < r_k < n$, generalize the corollary to obtain the joint density of

$$(X_{(r_1)}, \cdots, X_{(r_k)}).$$

Corollary: *The marginal density of* $X_{(r)}$ *is*

$$g_r(x) = \frac{n!}{(r-1)!(n-r)!}[F(x)]^{r-1}[1 - F(x)]^{n-r} f(x).$$

Proof: Integrating out the other variables in the joint density of $(X_{(1)}, \cdots, X_{(n)})$ makes

$$g(x_r) = n!\, f(x_r)\,.$$

$$\int_{-\infty}^{x_r}\int_{-\infty}^{x_{r-1}}\cdots\int_{-\infty}^{x_3}\int_{-\infty}^{x_2} f(x_1)dx_1 f(x_2)dx_2 \cdots f(x_{r-1})dx_{r-1}\,\cdot$$

$$\int_{x_r}^{\infty}\int_{x_{r+1}}^{\infty}\cdots\int_{x_{n-1}}^{\infty} f(x_n)dx_n f(x_{n-1})dx_{n-1}\cdots f(x_{r+1})dx_{r+1}$$

$$= n!\, f(x_r)\cdot\frac{[F(x_r)]^{r-1}}{(r-1)!}\cdot\frac{[1-F(x_r)]^{n-r}}{(n-r)!}\,.$$

Exercise 4: a) Show that the first corollary yields the same formula as Exercise 1a.

b) Show that the second corollary yields the same formula as Exercise 1b.

c) Write out all the details of the proof of the second corollary when $n = 4$, $r = 2$.

Theoretically, from the joint density of the order statistics, one can derive the distributions of functions of order statistics. For more details on order statistics, see David, 1970 .

Example: Consider the sample range $X_{(n)} - X_{(1)}$. Let

$$h : (X_{(1)}, X_{(n)}) \to (Y_1, Y_2)\,,$$

$$Y_1 = X_{(1)},\ Y_2 = X_{(n)} - Y_{(1)} \text{ or } X_{(1)} = Y_1,\ X_{(n)} = Y_2 + Y_1\,.$$

For $h = (h_1, h_2)$,

$$y_1 = h_1(x_{(1)}, x_{(n)}) = x_{(1)},\ y_2 = h_2(x_{(1)}, x_{(2)}) = x_{(n)} - x_{(1)}\,.$$

Then, $h^{-1}(y_1, y_2) = (g_1(y_1, y_2), g_2(y_1, y_2))$ where

$g_1(y_1, y_2) = y_1$, $g_2(y_1, y_2) = y_1 + y_2$. The Jacobian of h^{-1} is

$$J_{h^{-1}}(y_1,y_2) = \begin{vmatrix} \dfrac{\partial g_1}{\partial y_1} & \dfrac{\partial g_1}{\partial y_2} \\ \dfrac{\partial g_2}{\partial y_1} & \dfrac{\partial g_2}{\partial y_2} \end{vmatrix} = \begin{vmatrix} 1 & 0 \\ 1 & 1 \end{vmatrix} = 1.$$

By the first corollary, the joint density of $(X_{(1)}, X_{(2)})$ is

$$g_{12}(x_1,x_2) = n(n-1)f(x_1)f(x_2)[F(x_2) - F(x_1)]^{n-2} \cdot I(x_1 < x_2) .$$

It follows that the joint density of (Y_1, Y_2) is

$$\begin{aligned} \varphi(y_1,y_2) &= g_{12}(g_1(y_1,y_2),g_2(y_1,y_2)) \cdot |J_{h^{-1}}| \\ &= n(n-1)f(y_1)f(y_1 + y_2)[F(y_1 + y_2) - F(y_1)]^{n-2} . \end{aligned}$$

In the $(X_{(1)}, X_{(2)})$ plane, the support can be pictured as

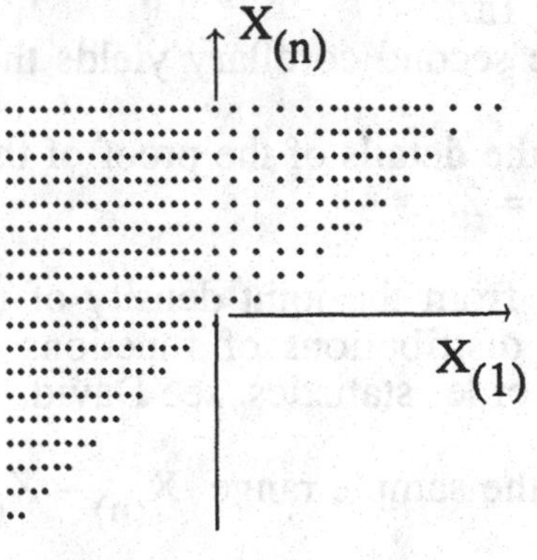

On the boundary, $X_{(1)} = X_{(n)}$, $y_2 = 0$. In the (y_1, y_2) plane, this half–plane region remains a half–plane:

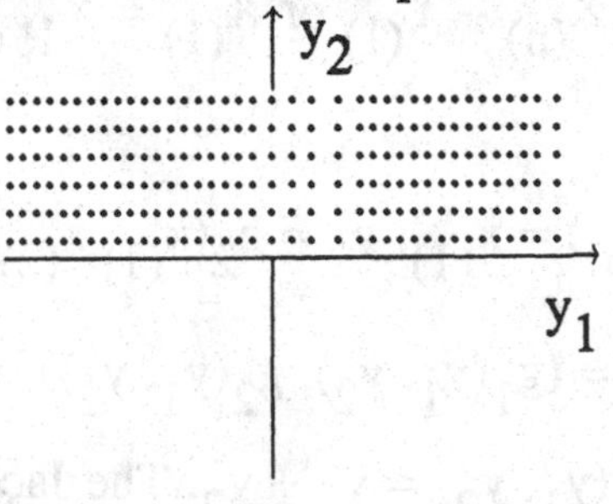

This emphasizes that y_1 can be any real number while y_2 is

any positive real number. From the joint density of Y_1, Y_2, we get the density of the range $Y_2 = X_{(n)} - X_{(1)}$ as

$$\int_R n(n-1)f(y_1)f(y_1 + y_2)[F(y_1 + y_2) - F(y_1)]^{n-2}dy_1 \qquad (*)$$

Exercise 5: Let $f(x) = I(0 \le x \le 1)$. Find the distribution of the sample range $X_{(n)} - X_{(1)}$; that is, substitute for f and F in (*) and perform the integration.

Exercise 6: For the random sample X_1, X_2, X_3, with $F' = f$, expand the integral $\int_{-\infty}^{+\infty}\int_{-\infty}^{x_3}\int_{-\infty}^{x_2} f(x_1)f(x_2)f(x_3)dx_1dx_2dx_3$.

Exercise 7: Let $X_1, \cdots, X_n$ be a RS from the distribution specified. Find the joint distribution of the n order statistics.

a) Uniform on (0,1)

b) Exponential with density $e^{-x} \cdot I(0 < x)$.

Exercise 8: Let $Y_1, \cdots, Y_n, Y_{n+1}$ be a RS from the exponential distribution with the density of exercise 7b). For $k = 1,2,\cdots,(n+1)$, let $W_k = Y_1 + \cdots + Y_k$.

a) Find the density of W_{n+1} .

b) Find the conditional density of $(W_1,\cdots,W_n)$ given $W_{n+1} = b$.

c) Show that the RVs $U_k = W_k/W_{n+1}$, $k = 1(1)n$, are distributed as the order statistics in exercise 7a).

Exercise 9: Consider again the distribution in exercise 7a). For $j = 1(1)n$, let $Y_j = -\log(X_{(j)})$ (base e). Show that

$$Y_1, \cdots, Y_n$$

are jointly distributed as the order statistics in exercise 7b).

LESSON 7. SUFFICIENT STATISTICS–I

In this lesson and the next, we investigate another useful class of statistics called sufficient statistics. They are "sufficient" in that they induce a significant "reduction" of the data without loss of the "information" contained in the sample; or, all the "information" in the sample is contained in the (generally smaller number of) "sufficient" statistics. Since "information" has a technical meaning which will be introduced later, our discussion here will involve this term only intuitively.

For simplicity, we present details for a real RV X but the ideas can be carried over (almost word for word) to random vectors. Thus, we begin with a (family of densities) model

$$\mathscr{F} = \{f(x;\theta) : x \in R, \theta \in \Theta \subset R^d\} .$$

When $X_1, \cdots, X_n$ is a random sample from $\mathscr{F}$, all "sampling information" about θ is contained in the sample. For ease of understanding, we try to summarize the data without losing information about θ. More formally, we ask, "Is there any function of the X_i's, a statistic $T(X_1,\cdots,X_n)$, which is simpler than $(X_1,\cdots,X_n)$, yet contains the same information about θ as the sample itself ?"

Definition: *A statistic $T(X_1,\cdots,X_n)$ is sufficient for θ (or for $\mathscr{F}$) if for each t, the conditional distribution of the sample $(X_1,\cdots,X_n)$ given the value $T = t$ is free of θ. (The phrase "free of θ" means that the domain as well as the range does not involve θ.)*

In this connection, the following principle is often used. If $A_i = \{X_i = x_i\}$ and $B = \{T(X_1,\cdots,X_n) = t\}$, then $\cap_{i=1}^{n} A_i \subset B$ so that $P(X_1 = x_1, \cdots, X_n) = P(X_1, \cdots, X_n, T = t)$. In the continuous cases, one deals with $\{X_1 \le x_1\}$, $\{T \le t\}$, etc.

Example: Let $\mathscr{F}$ be the Bernoulli model with density

$$f(x;\theta) = \theta^x(1-\theta)^{1-x}, \theta \in \Theta = (0, 1), x \in \{0,1\}.$$

The joint density of a random sample $X_1, \cdots, X_n$ is

$$L(x_1,\cdots,x_n;\theta) = \theta^{\Sigma x_i}(1-\theta)^{n-\Sigma x_i}, x_i \in \{0, 1\}.$$

Let $T = \Sigma_{i=1}^n X_i$. We know that T is a binomial RV with density $P(T = t) = \binom{n}{t} \theta^t(1-\theta)^{n-t}, t \in \{0, 1, \cdots, n\}$.

a) Then, $P(X_1 = x_1, \cdots, X_n = x_n \mid T = t)$

$= P(X_1 = x_1, \cdots, X_n = x_n, T = t) / P(T = t)$

$= P(X_1 = x_1, \cdots, X_n = t - \sum_{i=1}^{n-1} x_i) / P(T = t)$

$= \theta^t(1-\theta)^{n-t} / \binom{n}{t} \theta^t(1-\theta)^{n-t}$

$= 1/\binom{n}{t}$ which is free of θ.

b) Let $S = g(X_1,\cdots,X_n) \in R^k$ be any other statistic, a function of the sample, not a function of T alone. Let $A \in \mathscr{B}_k$. Then

$P(S \in A \mid T = t) = P(S \in A, T = t) / P(T = t)$

$= P[(X_1, \cdots, X_n) \in g^{-1}(A), T = t] / P(T = t)$

$= P((X_1, \cdots, X_n) \in g^{-1}(A) \mid T = t)$.

But this is just one "value" of the conditional density of $(X_1, \cdots, X_n)$ given $T = t$, so that this probability is free of θ.

Exercise 1: In the example just above, take $n = 4$; then

$$T = X_1 + X_2 + X_3 + X_4.$$

Let $S = X_1X_2X_3X_4$. Show that the distribution of S given T

is free of θ by direct calculations. That is, find
$P(S = 0 \mid t)$ and $P(S = 1 \mid t)$ for $t = 0,1,2,3,4$.

The result of part b) in the example above is completely general. If T is sufficient for θ, then the conditional distribution (given $T = t$) of any other statistic S (not a function of T alone) is also free of θ. The proof there is valid for any discrete model. It can be extended to other distributions by a "regularization" which includes the definition of

$$P((X_1, \cdots, X_n) \in A \mid T = t) \text{ when } P(T = t) = 0 .$$

It can be shown that $P(\cdot \mid T = t)$ is a probability measure on $\mathcal{B}_n$ when $P(T = t) \neq 0$. (See Lehmann, 1959.)

This result between S and T describes what is meant by "no loss of information about θ": since the conditional distribution, S given T, does not contain θ, knowledge of S after T does not relate to θ. And, a "reduction" is obtained when the dimension of T is less than n. The following theorem is often useful for finding or identifying sufficient statistics. We give the proof for discrete X; the general proof can be found in Lehmann, 1959.

Theorem: *(Neyman-Fisher Factorization) A statistic $T(X_1, \cdots, X_n)$ with values in R^m is sufficient for θ if and only if the joint density $f(x_1, \cdots, x_n; \theta)$ of $(X_1, \cdots, X_n)$ can be factored:*

$$f(x_1, \cdots, x_n; \theta) = g_\theta(T(x_1, \cdots, x_n)) \cdot h(x_1, \cdots, x_n) \qquad (*)$$

where for each $\theta \in \Theta$, g_θ is a non-negative function on R^m which depends on $(x_1, \cdots, x_n)$ only thru $T(x_1, \cdots, x_n)$ and h is a non-negative function on R^n free of θ.

Partial proof: Assume that the distribution is discrete. To simplfy notation, let $Y = (X_1, \cdots, X_n)$, $y = (x_1, \cdots, x_n)$; of course, $T^{-1}(t) = \{y : T(y) = t\}$.

First suppose that T is sufficient for θ. Then in view of the definition of sufficient statistics and the remark afterwards, we have $P_\theta(Y = y) = P_\theta(Y = y, T(Y) = T(y) = t)$

$$= P_\theta(Y = y \mid T = t)\cdot P_\theta(T = t) = g_\theta(t)\cdot h(y) .$$

Conversely, suppose (*). For each t such that $P_\theta(T = t) > 0$ for all θ, we have

$$P_\theta(Y = y \mid T = t) = P_\theta(Y = y, T = t)/P_\theta(T = t) .$$

Now $P_\theta(Y = y, T(Y) = t) \begin{array}{ll} = 0 & \text{when } T(y) \neq t \\ = P_\theta(Y = y) & \text{when } T(y) = t \end{array}$,

and $P_\theta(T(Y) = t) = \sum P_\theta(Y = y)$ (summation over $T^{-1}(t)$) .
It follows that

$$P_\theta(Y = y \mid T = t) \begin{array}{ll} = 0 & \text{when } T(y) \neq t \\ = g_\theta(t)\cdot h(y)/g_\theta(t)\sum h(z) & \text{when } T(y) = t , \end{array}$$

(summation again over $T^{-1}(t)$). Note that this function is now free of θ and hence T is sufficient for θ .

Example: Let the density of X (real) be

$$f(x ; \theta) = \frac{1}{\sqrt{2\pi}} e^{-(x-\theta)^2/2} .$$

The density of the sample is $f(x_1,\cdots,x_n;\theta)$

$$= \Pi_{i=1}^{n} f(x_i;\theta) = (2\pi)^{-n/2}\cdot e^{-\Sigma x_i^2/2 + \Sigma x_i\theta - n\theta^2/2}$$

$$= e^{\Sigma x_i\theta - n\theta^2/2} \cdot e^{-\Sigma x_i^2/2} (2\pi)^{-n/2} .$$

Now let $T(x) = \Sigma x_i$, $g_\theta(t;\theta) = e^{t\theta - n\theta^2/2}$,

$$h(x) = e^{-\Sigma x_i^2/2} (2\pi)^{-n/2} .$$

Then the factorization $f(x_1,\cdots,x_n) = g_\theta(T(x);\theta)\cdot h(x)$ shows that T is sufficient for θ .

Exercise 2: a) Let the density of X be

$$f(x;\theta) = \frac{1}{\theta\sqrt{2\pi}} e^{-(x-1)^2/2\theta^2} \quad \text{for } \theta > 0 \text{ , x real .}$$

Show that $T = \sum_{i=1}^{n} (X_i - 1)^2$ is sufficient for θ.

b) More generally, let the density be that of an arbitrary normal RV:

$$f(x;\theta) = \frac{1}{\sigma\sqrt{2\pi}} e^{-(x-\mu)^2/2\sigma^2}$$

with $\theta = (\mu , \sigma^2)$. Verify that:

i) when σ is known, ΣX_i is sufficient for μ ;

ii) when μ is known, $\Sigma(X_i - \mu)^2$ is sufficient for σ^2.

The result of the following algebraic exercise is a prototype for a large number of manipulations in normal distributions, in particular, in finding sufficient statistics.

Exercise 3: For real numbers $x_1, x_2, \cdots, x_n$, let $\bar{x} = \Sigma x_i/n$; let μ be a real number free of i. Show that

$$\Sigma(x_i - \mu)^2 = \Sigma(x_i - \bar{x})^2 + n(\bar{x} - \mu)^2 .$$

Example: Consider the model with $f(x;\theta)$ in Exercise 3b) above.

a) Then with the help of exercise 4, we can write

$$f(x_1,\cdots,x_n;\theta) =$$

$$\left[2\pi\sigma^2\right]^{-\frac{n}{2}} \exp\{-\frac{1}{2\sigma^2}\Sigma(x_i - \bar{x})^2 - \frac{n}{2\sigma^2}(\bar{x} - \mu)^2\} .$$

Let $T = (T_1, T_2)$, $T_1 = \bar{X}$, $T_2 = \Sigma(X_i - \bar{X})^2$. Then,

$f(x_1,\cdots,x_n;\theta) = g_\theta(t_1,t_2)\cdot h(x)$ where $g_\theta(t_1,t_2)$

$$= \left[2\pi\sigma^2\right]^{-n/2} \exp\{- t_2/2\sigma^2 - n(t_1 - \mu)^2/2\sigma^2\}$$

and $h(x) = h(x_1,\cdots,x_n) \equiv 1$.

This shows that T is sufficient for $\theta = (\mu , \sigma^2)$. Note that this does not say that T_1 is sufficient for μ and T_2 is sufficient for σ^2 (and they are not!) but that the vector (T_1, T_2) is sufficient for the vector (μ , σ^2) . To put it another way, T_1, T_2 are jointly sufficient for θ .

b) Incidentally, from Lesson 4, we know that:

$Y_1 = \sqrt{n}(T_1 - \mu)/\sigma$ is distributed as N(0,1);

$Y_2 = T_2/\sigma^2$ is distributed as $\chi^2(n-1)$;

the statistics T_1 and T_2 are independent .

It follows that Y_1 and Y_2 are independent RVs.

Exercise 4: Complete the proof of the following lemma by filling in the missing steps of the given outline.

Lemma: *Let T and S be as given in part a) of the proof of the factorization theorem. Let* $\eta : R^m \to R^m$ *be 1 to 1 so that for* $w = \eta(t)$, $t = \eta^{-1}(w)$ *and* $\| \partial\eta^{-1}/\partial w' \| \neq 0$. *Then,*

W = $\eta(T)$ *is sufficient for* θ *iff T is sufficient for* θ .

Partial proof: From the density of T and S , we can get the density of W and S , say $f_{W,S}(w,s;\theta) = \hat{g}_\theta(w)\cdot\hat{h}(w,s)$. Then,

$$f_{S|W}(s|w) = \hat{h}(w,s)/\int \hat{h}(w,s)\, ds$$

is free of θ and the conclusion follows.

Exercise 5: With respect to the last example, show that

$$\left(\frac{1}{n}\sum_{1}^{n} X_i\ ,\ \frac{1}{n-1}\sum_{1}^{n}(X_i - \bar{X})^2\right) \text{ is sufficient for } \theta .$$

This shows that there are many "sufficient statistics" and suggests that there must be other properties to be investigated; we will see some of these in the next several lessons.

Exercise 6: Let $X_1, \cdots, X_n$ be a random sample from the uniform distribution on $(0, \theta)$. Show that

$$X_{(n)} = \text{Max}(X_1, \cdots, X_n) \text{ is sufficient for } \theta.$$

Exercise 7: Let $X_1, \cdots, X_n$ be a random sample from the Poisson distribution with parameter $\lambda > 0$. Find a sufficient statistic for λ .

LESSON 8. SUFFICIENT STATISTICS–II

We continue the investigation of sufficient statistics with the same notation as in the Neyman–Fisher factorization theorem (NFT) of Lesson 7. We should perhaps reiterate that all probability results are "almost surely" but this phrase is not carried along.

Corollary: *A necessary and sufficient condition for a statistic* T *to be sufficient for* θ *is: for each* $\theta \neq \theta'$ *in* Θ, *the ratio* $f(x;\theta)/f(x;\theta')$ *is a function of* $T(x)$ *[depends on* x *only in terms of* $T(x)$*].*

Proof: a) Suppose that T is sufficient. Then by the NFT,

$$f(x;\theta) = g_\theta(T(x))\cdot h(x) .$$

It follows that $\dfrac{f(x;\theta)}{f(x;\theta')} = \dfrac{g_\theta(T(x))}{g_{\theta'}(T(x))} = \psi_{\theta,\theta'}(T(x))$ as required.

b) Conversely, if $\dfrac{f(x;\theta)}{f(x;\theta')} = \psi_{\theta,\theta'}(T(x))$, then

$$f(x;\theta) = f(x;\theta')\psi_{\theta,\theta'}(T(x)) .$$

Fixing θ' yields $f(x;\theta) = h(x)g_\theta(T(x))$, as required.

The following result may be used to show that a statistic is *not* sufficient.

Lemma: *For a model* $\mathcal{F}$, *let*

$$\mathcal{A}(\mathcal{F}) = \{z : f(z;\theta) > 0 \text{ for some } \theta \in \Theta\}.$$

Let T *be a statistic such that for some* $\theta \neq \theta' \in \Theta$, *and*

$$x \neq y \in \mathcal{A}(\mathcal{F}),\ T(x) = T(y) \text{ but } f(x;\theta)f(y;\theta') \neq f(x;\theta')f(y;\theta) ;$$

then T *is not sufficient for* θ.

Proof: Let S be a sufficient statistic; then,

$$f(x;\theta) = g_\theta(S(x))h(x) \text{ and } f(y;\theta) = g_\theta(S(y))h(y).$$

Hence, when $S(x) = S(y)$, $f(x;\theta)/f(y;\theta) = h(x)/h(y)$; that is, this ratio is free of θ. This means that for all $\theta, \theta' \in \Theta$

$$f(x;\theta)/f(y;\theta) = f(x;\theta')/f(y;\theta') . \qquad (*)$$

In other words, if S is sufficient, then $S(x) = S(y)$ implies the equality in (*). Therefore, T cannot be sufficient since $T(x) = T(y)$ but the equality in (*) does not hold.

In the proof of the lemma just above, we assumed the existence of a sufficient statistic S ; this is without loss of generality since the trivial statistic, the n observations themselves, is always sufficient.

The next example shows that if T is sufficient and S is a function of T, then S need not be sufficient; this is in contrast to the last lemma in lesson 7.

Example: For $X = (X_1, \cdots, X_n)$, let

$$f(x_1, \cdots, x_n;\theta) = f(x;\theta) = (2\pi)^{-n/2} e^{-\Sigma(x_i - \theta)^2/2} .$$

We know that $T = \Sigma X_i$ is sufficient for θ (lesson 7). We will show that $S = \varphi(T) = T^2$ is not sufficient by using the last lemma. To begin, we select x and y such that $S(x) = S(y)$ and consider the equality (*). Now,

$$\begin{aligned} &f(x;\theta)f(y;\theta') \\ &= (2\pi)^{-n}\exp\{-\tfrac{1}{2}[\Sigma\, x_i^2 + \Sigma\, y_i^2]\} \cdot \exp\{\theta T(x) \\ &\qquad + \theta' T(y) - \tfrac{n}{2}(\theta^2 + \theta'^2)\} \end{aligned}$$

and

$$\begin{aligned} &f(x;\theta')f(y;\theta) \\ &= (2\pi)^{-n}\exp\{-\tfrac{1}{2}[\Sigma\, x_i^2 + \Sigma\, y_i^2]\} \cdot \exp\{\theta' T(x) \\ &\qquad + \theta T(y) - \tfrac{n}{2}(\theta'^2 + \theta^2)\} \end{aligned}$$

so that

$$f(x;\theta)f(y;\theta') = f(x;\theta')f(y;\theta)$$

iff

$$\theta T(x) + \theta' T(y) = \theta' T(x) + \theta T(y)$$

iff

$$(\theta - \theta')[T(x) - T(y)] = 0 .$$

Since it is assumed that $\theta \neq \theta'$, it follows that $T(x) = T(y)$. But among those $x \neq y$ for which $S(x) = S(y)$, there are many (like $(-1)^2 = 1^2$) for which $T(x) \neq T(y)$. This contradiction means that S is not sufficient.

Exercise 1: Continue the example above. Take $n = 2$ and verify that when $x = (1,-3)$, $y = (1, 1)$, then $S(x) = S(y)$ but $T(x) \neq T(y)$.

We consider now that sufficient statistic which provides the greatest reduction of data without loss of information contained in the sample.

Definition: *A sufficient statistic T is minimal sufficient iff T is a function of any other sufficient statistic.*

More symbolically, T is minimal sufficient if whenever S is another sufficient statistic, $T = \varphi(S)$ for some function φ. The following is an elaboration of what this functional relationship means.

Lemma: *If T and S have a common domain $\mathscr{D}$, then T is a function of S iff*

for all $x, y \in \mathscr{D}$, $S(x) = S(y)$ implies $T(x) = T(y)$.

Proof: If $T = \varphi(S)$, then $T(x) = \varphi(S(x)) = \varphi(S(y)) = T(y)$ so that the conclusion is true. Conversely, define $\varphi : S(\mathscr{D}) \to$ range of T by

$$\varphi(S(x)) = T(x) .$$

Then φ is well-defined since if $s \in S(\mathscr{D})$, then

$$\varphi(s) = T(x) \text{ for all } x \in S^{-1}(s) .$$

The other way around, a sufficient statistic T is minimal if, when we reduce T further, say $S = \varphi(T) \neq T$, S is no longer sufficient. We will examine a "geometric" way of looking at minimal sufficient statistics after the next definition and exercise. We will use the notation of this definition whenever appropriate without specifically repeating it.

Definition: *Let a general RV X have range $\mathscr{X}$ and density*

$f(x;\theta)$ for $\theta \in \Theta$; let $X_1, X_2, \cdots, X_n$, be a random sample from this population so that the sample space is

$$\tilde{\mathscr{X}} = \mathscr{X}^n = \mathscr{X} \times \mathscr{X} \cdots \times \mathscr{X}.$$

Let T be a statistic mapping $\tilde{\mathscr{X}}$ into Θ. The subsets of $\tilde{\mathscr{X}}$ defined by

$$T^{-1}(\theta) = \{x \in \tilde{\mathscr{X}} : T(x) = \theta\} \text{ for each } \theta \in \Theta$$

are called the orbits (or contours) of T.

Exercise 2: Show that the collection of orbits of T (as above) makes up a partition of $\tilde{\mathscr{X}}$.

Now let S and T be sufficient statistics mapping $\tilde{\mathscr{X}}$ into Θ ; note that $S(\tilde{\mathscr{X}})$ and $T(\tilde{\mathscr{X}})$ are subsets of Θ. If S is minimal, then for each T , there is some function φ such that $S = \varphi(T)$ and $\varphi : T(\tilde{\mathscr{X}}) \to \Theta$. For each $\theta \in \Theta$,

$$S^{-1}(\theta) = T^{-1}(\varphi^{-1}(\theta)) = \cup \{T^{-1}(\theta') : \theta' \in \varphi^{-1}(\theta)\} .$$

It follows that for each $\theta \in \Theta$, there is a $\theta' \in \Theta$ such that

$$T^{-1}(\theta') \subseteq S^{-1}(\theta) .$$

This means that the partition induced by T is *finer* (has more subsets) than that induced by S. The following is perhaps the simplest example.

Example: Consider a sample of size 3 from the Bernoulli distribution with P("success") = θ. Let $S = X_1 + X_2 + X_3$ and $T = (X_1, X_2 + X_3)$. Then the sample space and the partitions induced by S and T can be represented as follows:

X_1	X_2	X_3	S		T	
0	0	0	0	I	0, 0	I
0	0	1	1	II	0, 1	II
0	1	0	1	II	0, 1	II
1	0	0	1	II	1, 0	III
0	1	1	2	III	0, 2	IV
1	0	1	2	III	1, 1	V
1	1	0	2	III	1, 1	V
1	1	1	3	IV	1, 2	VI

The conditional probabilities were computed in lesson 7:

$$P(X_1 = x_1, X_2 = x_2, X_3 = x_3 \mid S = s)$$

$$= P(X_1 = x_1, X_2 = x_2, X_3 = s - x_1 - x_2) \div P(S = s)$$

$$= \theta^s (1 - \theta)^{3-s} \div \begin{bmatrix} 3 \\ s \end{bmatrix} \theta^s (1 - \theta)^{3-s} = 1/\begin{bmatrix} 3 \\ s \end{bmatrix} ;$$

$$P(X_1 = x_1, X_2 = x_2, X_3 = x_3 \mid T = (t,r)) =$$

$$P(X_1 = t, X_2 = x_2, X_3 = r - x_2) \div P(X_1 = t, X_2 + X_3 = r)$$

$$= \theta^{t+r}(1 - \theta)^{3-t-r} \div \theta^t (1 - \theta)^{1-t} \begin{bmatrix} 2 \\ r \end{bmatrix} \theta^r (1 - \theta)^{2-r} = 1/\begin{bmatrix} 2 \\ r \end{bmatrix} .$$

Since both conditional densities are free of θ, both statistics S and T are sufficient. The partition for T has more subsets than that for S ; indeed, S is minimal (see exercise 4).

In passing, we note that all minimal sufficient statistics are equivalent in the sense that if T and S are both minimal sufficient, then there exists an one–to–one mapping ℓ (each is a function of the other) such that $T = \ell(S)$. Consequently, they each "contain" the same information.

Exercise 3: Let T and S be two sufficient statistics for which there exist x and y such that $S(x) = S(y)$ but $T(x) \neq T(y)$; show that T is not minimal.

The following gives a *sufficient* condition for minimality.

Lemma: *Let T be a sufficient statistic for* θ. *Let* $\mathscr{H}$ *be the set of* (x,y) *with* x,y *in* $\mathscr{A}(\mathscr{S})$ *for which there is a value* $h(x,y) > 0$ *such that* $f(x;\theta) = h(x,y) \cdot f(y;\theta)$ *for all* θ . *If for all* $(x,y) \in H$,

$T(x) = T(y)$, then T is minimal sufficient.

Proof: Let S be a sufficient statistic with $S(x) = S(y)$ for (x,y) in $\mathscr{H}$. Then, $f(x;\theta) = h(x)g_\theta(S(x))$ and $f(y;\theta) = h(y)g_\theta(S(y))$ imply

$$f(x;\theta) = [h(x)/h(y)]f(y;\theta) .$$

Since $h(x,y) = h(x)/h(y)$ satisfies the conditions for $\mathscr{H}$, $T(x) = T(y)$. The conclusion follows by application of the previous lemma.

Exercise 4: For the Bernoulli Binomial case with density $\binom{n}{x} \theta^x (1-\theta)^{n-x}$ for $x = 0(1)n$, show that $T(x) = x$ is minimal sufficient.

Example: Consider a random sample from the normal distribution with density $\frac{1}{\sqrt{2\pi}} e^{-(x-\theta^2)/2}$ for $\theta \in R$. We may write

$x = (x_1, \cdots, x_n)'$, Σx_i for $\sum_{i=1}^n x_i$, Σx_i^2 for $\sum_{i=1}^n x_i^2$; etc. From

$$f(x;\theta) = (2\pi)^{-n/2} e^{-\Sigma x_i^2} e^{\theta \Sigma x_i - n\theta^2/2} ,$$

we see that $T(x) = \Sigma X_i$ is sufficient. Now,

$$\frac{f(x;\theta)}{f(y;\theta)} = \exp\{-\tfrac{1}{2}[\Sigma x_i^2 - \Sigma y_i^2]\}\exp\{\theta[T(x) - T(y)]\}$$

$$= h(x,y) \cdot \exp\{\theta[T(x) - T(y)]$$

where $h(x,y) = \exp\{-[\Sigma x_i^2 - \Sigma y_i^2]\}$. This ratio is free of θ if and only if $T(x) = T(y)$. Then, $f(x;\theta) = h(x,y)f(y;\theta)$ and the provious lemma applies; that is, T is minimal sufficient.

Exercise 5: Let $f(x;\theta) = \theta^x(1-\theta)^{1-x}$ for $x \in \{0, 1\}$,

$\theta \in (0, 1)$. Let $T(X_1,\cdots,X_n) = \sum_1^n X_i$ and

$$S(X_1,\cdots,X_n) = (X_1,\cdots,X_n).$$

Show that:

a) T is a minimal sufficient statistic;
b) S is sufficient but not minimal.

The joint density for a random sample can be written as

$$f_n(x;\theta) = \prod f(x_i;\theta) .$$

For each x, the value of $\prod f(x_i;\theta)$ is just that of $\prod f(x_{(i)};\theta)$ where $x_{(1)} \le x_{(2)} \le \cdots \le x_{(n)}$ are the order statistics. Thus, for a given n, the value of the joint density depends on the observations x only thru these order statistics. This means that the order statistics are (jointly) sufficient for θ. The next example shows that, in some sense, this is the best that can be done.

Example: Let the population density be Cauchy:

$$1/\pi[1 + (x - \theta)^2] , \theta \text{ and } x \text{ real.}$$

The proof of the corollary at the beginning of this lesson contains the condition that, for a sufficient statistic to exist, ratios like

$$\frac{(1 + (x_1 - \theta)^2)\cdot(1 + (x_2 - \theta)^2)}{(1 + (x \;\; - \theta')^2)\cdot(1 + (x \;\; - \theta')^2}$$

must be free of θ and θ' ; this is obviously false! Therefore, the only non–trivial sufficient statistic for the Cauchy density is the order statistics.

LESSON 9. COMPLETE STATISTICS

This topic is taken up because, as will be seen in Parts V and VI, certain nice uniqueness properties of estimates and tests are based on "complete" sufficient statistics. In the following definition, "density" could be replaced by distribution function or probability measure function without effecting completeness.

Definition. *Let the model be*

$$\mathcal{F} = \{f(x;\theta),\ \theta \in \Theta \subseteq R^d,\ x \in \mathcal{X} \subseteq R^n\};$$

let $T(X_1,\cdots,X_n)$ *be a statistic.* T *is complete if the family of distributions of* T, $\{f_T(t;\theta) : \theta \in \Theta\}$, *is complete in the following sense:*

when g *is a measurable function and* $E_\theta[g(T)] = 0$ *for all* $\theta \in \Theta$, *then* $g = 0$ P_T*-a.s.*

In the popular cases, $E_\theta[g(T)] = \int g(t)f_T(t;\theta)\,dt$ or $\sum g(t)f_T(t;\theta)$.. "$g = 0$ P_T–a.s." means that $g(t) = 0$ except perhaps on a measurable set, say A (in the range of T), such that $P_\theta(T \in A) = 0$ for all $\theta \in \Theta$.

Lemma: *Completeness is preserved under a one-to-one transformation.*

Proof: Let $T = \varphi(S)$ where φ is an one–to–one transformation. Then

$$E_\theta[g(T)] = E_\theta[g(\varphi(S))] = E_\theta[(g \circ \varphi)(S)].$$

It follows that $g \circ \varphi = 0$ P_S–a.s. iff $g = 0$ P_T–a.s. Therefore T is complete iff S is complete.

Example: Consider a RS $X = (X_1,\cdots,X_n)$ from a Bernoulli population with $P(X_i = 1) = \theta \in (0,1)$; let $T = \sum_{i=1}^n X_i$.

A binomial density of T is

$$f(t;\theta) = \binom{n}{t}\theta^t(1-\theta)^{n-t},\ t \in \{0,1,\cdots,n\},\ \theta \in (0, 1).$$

Let $g: R \to R$ be measurable. If

$$E_\theta[g(X)] = \sum_{i=0}^{n} g(i)\binom{n}{i}\theta^i(1-\theta)^{n-i} = 0 \text{ for all } \theta \in (0,1),$$

then $\sum_{i=0}^{n} g(i)\binom{n}{i}p^i = 0$ for all $p = \dfrac{\theta}{1-\theta} > 0$.

This says that the polynomial in p of degree n has more than n roots; it follows from a fundamental theorem of algebra that this must be the zero polynomial: $g(i) = 0$ for all $i \in \{0,1,\cdots,n\}$. Therefore, the binomial T is complete.

If $n = 1$ in this example, T becomes the identity function and this gives the completeness of the Bernoulli model itself. It is of some interest to note that both these models will be complete if Θ is only a subset of (0,1) containing at least n+1 distinct points since that is all that are needed in the application of the algebra.

Example: Consider the uniform model with $n = 1$ and

$$f(x\,;\theta) = \frac{1}{\theta} I(0 < x < \theta),\ x \in R,\ \theta > 0.$$

Let $g : R \to R$ be measurable. If

$$E_\theta[g(X)] = \frac{1}{\theta}\int_0^\theta g(x)dx = 0 \quad \text{for all } \theta > 0,$$

then $\int_0^\theta g(x)dx = 0$ for all $\theta > 0$.

Since $\dfrac{d}{d\theta}\int_0^\theta g(x)dx = g(\theta)$ for almost all $\theta > 0$, $g(\theta) = 0$ for almost all θ and so this uniform model is complete.

Exercise 1: Consider the uniform model with $n = 1$ and

$$f(x;\theta) = \frac{1}{\theta} I\,(0 < x < \theta),\ x \in R, \text{and } \theta > 1.$$

Show that this model is not complete by constructing a function g on (0, 1) such that $g \neq 0$ a.e. and yet $\int_0^1 g(x)dx = 0$.

Hint: try $g(x) = \sin(2\pi x/\theta)$ for $0 < x < \theta$.

Example: Consider the model $N(1, \theta^2)$ with $\theta > 0$. If $g(x) = x - 1$, $g(x) \not\equiv 0$, but

$$E_\theta[g(X)] = \frac{1}{\theta\sqrt{2\pi}} \int_{-\infty}^{\infty} (x-1)\, e^{-(x-1)^2/2\theta^2}\, dx = 0$$

for all $\theta > 0$. Thus this normal model is not complete.

Exercise 2: Is the normal model $N(0, \theta^2)$ with $\theta > 0$ complete?

To verify the completeness of some other models, we need some theory of a slightly more general version of the MGF or CF called the Laplace transform. Herein μ takes the place of P in integration (Lessons 9 and 10, Part II).

Definition: *Let* μ *be a non-negative* σ*-finite measure for* $(R^k, \mathcal{B}_k)$. *For vectors* $t = (t_1, \cdots, t_k)'$ *and* $s = (s_1, \cdots, s_k)$, $s \cdot t = \sum_{i=1}^{k} s_i t_i$. *The Laplace transform of the measure* μ *is* $\varphi_\mu : R^k \to R^+$ *defined by* $\varphi_\mu(s) = \int_{R^k} e^{s \cdot t}\, d\mu(t)$. *The domain of* φ_μ *is* $\{s \in R^k : \varphi_\mu(s) < +\infty\}$. *For a Lebesgue-Stieltjes measure with* $d\mu(t) = f(t)dt$, φ_μ *is written as* φ_f *so that*

$$\varphi_f(s) = \int_{R^k} e^{s \cdot t} f(t)dt .$$

In classical analysis this is a Laplace transform of the function f.

We leave the proof of the uniqueness theorem for Laplace transforms to analysis but we use two forms:

if $\varphi_{\mu_1}(s) = \varphi_{\mu_2}(s)$ for all s in an open set of R, then $\mu_1 = \mu_2$;
if $d\mu_1 = f_1 d\nu$ and $d\mu_2 = f_2 d\nu$, then $f_1 = f_2$ ν–a.e.

The latter form has the corollary that if the Laplace transform of an integrable function is identically 0 , then the function must be zero almost everywhere.

Example: X is normal with $\theta = (\mu, \sigma^2) \in R \times (0,+\infty)$.
Then,

$$E_\theta[g(X)] = \int_{-\infty}^{\infty} g(x) \frac{1}{\sigma\sqrt{2\pi}} \exp\{-\frac{1}{2\sigma^2}(x-\mu)^2\}\, dx = 0 \text{ for all } \theta$$

iff

$$\int_{-\infty}^{\infty} e^{tx}\, g(x)\cdot e^{-x^2/2\sigma^2}\, dx = 0 \text{ for all } t \in R \text{ where } t = \mu/\sigma^2 .$$

But when the Laplace transform of $g(x)\cdot e^{-x^2/2\sigma^2}$ is identically zero, $g(x)\cdot e^{-x^2/2\sigma^2}$ must be 0 almost everywhere so that $g(x) = 0$ a.e. Thus this normal model is complete.

Example: Consider the normal model $N(0, \theta^2)$ with $\theta > 0$. If $T = X^2$, then the density of T is $\dfrac{e^{-t/2\theta}}{\theta\sqrt{2\pi t}}\cdot I(0 < t < \infty)$ and

$$E_\theta[g(T)] = \frac{1}{\theta\sqrt{2\pi}} \int_0^{\infty} g(t)\cdot t^{-1/2}\, e^{-t/2\theta}\, dt = 0 \text{ for all } \theta > 0$$

iff $\displaystyle\int_0^{\infty} g(t)\cdot t^{-1/2}\, e^{-t/2\theta}\, dt = 0$ for all $\theta > 0$;

that is, the Laplace transform of $g(t)\cdot t^{-1/2}$ is identically zero. Hence $g(t)\cdot t^{-1/2} \equiv 0$ a. e. so that $g = 0$, a. e. and $T = X^2$ is a complete statistic.

Exercise 4: a) Verify the density of $T = X^2$ above.

b) Show that the model $f(x;\theta) = \frac{1}{\theta\sqrt{2\pi}} e^{-x^2/2\theta^2}$, $\theta > 0$, is not complete. (The previous example shows only that there is a complete statistic.) Hint: $T = X$.

Exercise 5: Let $X_1, \cdots, X_n$ be a RS from the uniform model

$$f(x;\theta) = \frac{1}{\theta} I(0 < x < \theta) \text{ with } \theta > 0 .$$

Let $T = \text{Max}(X_1, \cdots, X_n)$. Show that T is a complete sufficient statistic for θ.

Example: (A sufficient statistic which is not complete).
Let $X_1, \cdots, X_n$ be a RS from the normal model $N(\theta, \theta^2)$ with $\theta > 0$; let $T = (T_1, T_2)$ where $T_1 = \sum_{i=1}^n X_i$ and $T_2 = \sum_{i=1}^n X_i^2$. Then,

$$f(x_1, \cdots, x_n;\theta) = (\theta\sqrt{2\pi})^{-n} \exp\{-\frac{1}{2\theta^2}[\Sigma x_i^2 - 2\theta\Sigma x_i + n\theta^2]\}$$

$$= (\theta\sqrt{2\pi})^{-n} \exp\{-T_2/2\theta^2 + T_1/2\theta - n/2\}$$

By the factorization theorem, T is seen to be sufficient for θ. But

$$E_\theta[T_1^2 - \frac{n+1}{2}T_2] = E_\theta[T_1^2] - \frac{n+1}{2}E_\theta[T_2]$$

$$= n^2[\theta^2/n + \theta^2] - \frac{n+1}{2} n[\theta^2 + \theta^2] = 0 \text{ for all } \theta > 0$$

while $g(x_1, \cdots, x_n) = T_1^2 - \frac{n+1}{2} T_2 \neq 0$. Therefore, T is not complete.

Exercise 6: Verify the expectations in the example above.

In the first example of this lesson, we proved that $T = \sum_{i=1}^{n} X_i$ is a complete sufficient statistic for the Bernoulli model; in Lesson 8, we also established its minimality. The example above shows that not all sufficient statistics are complete. But a complete sufficient statistic is minimal as we now proceed to show.

Theorem: *(Wackerly, 1976) If T is sufficient and complete, then T is minimal.*

Proof: We give an indirect proof (in pieces) by showing that a sufficient and non–minimal statistic is not a complete sufficient statistic.

Thus assume that T is sufficient but non–minimal. This implies that there is an $S = \varphi(T)$ which is minimal sufficient. Now there must be some s such that for

$$S^{-1}(s) = T^{-1}(\varphi^{-1}(s)) = \cup\{T^{-1}(t) : t \in \varphi^{-1}(s)\} ,$$

there is at least one pair α, β in $\varphi^{-1}(s)$ such that $P(T = \alpha)\cdot P(T = \beta) > 0$; otherwise, φ would be one–to–one and T would be minimal.

Also, for this α and β,

$$T^{-1}(\alpha) \cup T^{-1}(\beta) \subseteq S^{-1}(s). \qquad (*)$$

And, since $P(T = \beta) > 0$, so is $P(S = s)$. Define the function

$$g(x) = \begin{cases} 1 & \text{if } x = \alpha \\ -\dfrac{P(T = \alpha \mid S = s)}{P(T = \beta \mid S = s)} & \text{if } x = \beta . \\ 0 & \text{otherwise} \end{cases}$$

Since S is sufficient, $g(\cdot)$ does not depend on θ ; also, $g(x) \not\equiv 0$. Now

$$E_\theta[g(T)] = \Sigma g(t)P_\theta(T = t)$$

$$= P_\theta(T = \alpha) - \frac{P_\theta(t = \alpha \mid S = s)}{P_\theta(T = \beta \mid S = s)} P_\theta(t = \beta) . \qquad (**)$$

But by (*), $(T = \alpha) \subseteq (S = s)$ and $(T = \beta) \subseteq (S = s)$ so that

$$\frac{P_\theta(T = \alpha \mid S = s)}{P_\theta(T = \beta \mid S = s)} = \frac{P_\theta(T = \alpha, S = s)}{P_\theta(T = \beta, S = s)} = \frac{P_\theta(T = \alpha)}{P_\theta(T = \beta)}.$$

Substituting this in (**), yields $E_\theta[g(T)] = 0$

for all $\theta \in \Theta$ and it follows that T is not complete.

This theorem suggests that in the search for a complete sufficient statistic, we might look first for a minimal sufficient statistic and then check to see whether or not it is complete. Unfortunately, a minimal sufficient statistic need not be complete.

Example: Let $X = (X_1, X_2, \cdots, X_n)$ be a RS from the uniform distribution:

$$f(x;\theta) = I(\theta - 1/2 < x < \theta + 1/2) \text{ for } \theta \in R.$$

Let $X_{(1)} \le X_{(2)} \le \cdots \le X_{(n)}$ denote the order statistic.

a) The likelihood

$$L(x;\theta) = \prod_{i=1}^{n} I(\theta - 1/2 < x_i < \theta + 1/2)$$

$$= I(\theta - 1/2 < x_{(1)} < x_{(n)} < \theta + 1/2)$$

which shows immediately that $T = (X_{(1)}, X_{(n)})$ is a sufficient statistic. It is also minimal (see exercise 7).

b) Now let $g(x,y) = y - x - \dfrac{n-1}{n+1}$; then

$$E_\theta[g(X_{(1)}, X_{(n)})] = E_\theta[X_{(n)} - X_{(1)} - \frac{n-1}{n+1}]$$

$$= [\theta + \frac{n-1}{2(n+1)}] - [\theta - \frac{n-1}{2(n+1)}] - \frac{n-1}{n+1}$$

$= 0$ for all $\theta \in R$.

Thus T is not complete.

c) This example really shows more: for this model, there does not exist a complete sufficient statistic. Because, if there were such a statistic S, then by

Wackerly's theorem above, S would be minimal and hence equivalent to T. But then by the lemma at the beginning of this lesson, T would be complete which it is not.

Exercise 7: For part a) of the example just above, show that T is a minimal sufficient statistic for θ.

Exercise 8: A bottling line is allowed to continue operating so long as a randomly selected bottle is found to be acceptable. Let $0 < \theta < 1$ be the true proportion of defective bottles in this process. Suppose that counting doesn't start until time λ (> 0 minutes) . Then the corresponding probability density is

$$f(x;\theta,\lambda) = \theta(1-\theta)^{x-\lambda} \quad \text{for } x = \lambda, \lambda+1, \lambda+2, \cdots .$$

Let $X_{(1)} \le X_{(2)} \le \cdots \le X_{(n)}$ be the order statistic for a random sample from this population. Check sufficiency, completeness:

a) $T = (\Sigma X_{(i)}, X_{(1)})$ for (θ, λ) ;

b) $X_{(1)}$ for λ when θ is known;

c) $\Sigma X_{(i)}$ for θ when λ is known .

Here is another interesting example of a family with a sufficient statistic which is not complete. The density is

$$f(x:\theta) = 1 \cdot I\{\theta < x < \theta + 1\} .$$

The sufficient statistic is the pair $(X_{(1)}, X_{(n)})$. The statistic

$$h(X_{(1)}, X_{(n)}) = (X_{(n)} - X_{(1)}) - (n+2)(X_{(n)} - X_{(1)})^2/n$$

has mean 0 .

LESSON 10. EXPONENTIAL FAMILIES–I

In this lesson and the next, we will investigate a fairly general class of models admitting complete sufficient statistics.

Again, we begin with

$$\mathscr{F} = \{f(x;\theta),\ x \in \mathscr{X} \subseteq R^d,\ \theta \in \Theta \subseteq R^m\}$$

and a random sample $X_1, \cdots, X_n$ from $\mathscr{F}$. The trivial statistic, the identity $T(X_1, \cdots, X_n) = (X_1, \cdots, X_n)$, is sufficient for θ but has two "not nice" properties:

i) the dimension of T depends upon the sample size n;

ii) the conditional distribution of the sample given T is degenerate.

On the other hand, in the normal model with parameter $\theta = (\mu, \sigma^2)$, there is a two–dimensional sufficient statistic, namely $T = (\Sigma X_i, \Sigma X_i^2)$, which is invariant over (of the same form for) all sample sizes n.

As we will see, there is a larger class of models having such an invariance property; it is called the exponential family. Morever, for this class, the fixed dimensional sufficient statistic is also minimal, and under some mild conditions, complete.

Definition: *The model $\mathscr{F}$ is a k-parameter exponential family if the corresponding density is of the form*

$$f(x;\theta) = C(\theta)\, h(x)\, exp\{\Sigma_{i=1}^{k} Q_i(\theta) T_i(x)\} \qquad (*)$$

for some integer $k > 0$ and functions $C, h, Q_1, \cdots, Q_k$, $T_1, \cdots, T_k$.

The following discussion will clarify the way in which statisticians view this family.

i) $C(\theta)h(x)$ must be non–negative so we may as well take each of these functions to be non–negative. In addition, $C(\theta)$ cannot be zero for any θ since each integral of f must be 1.

ii) A priori, the form (*) is not unique; even the simple substitutions $\sigma \cdot Q_i$ for Q_i and T_i/σ for T_i, will determine another exponential family. However, this non–uniqueness will not confound the issue since all minimal sufficient statistics will be equivalent.

iii) Also, the integer k is not unique since the family $\{1, T_1, \cdots, T_k\}$ and/or the family $\{1, Q_1, \cdots, Q_k\}$ might be linearly dependent:

for some $\{\beta_j\text{'s}\}$ not all zero,

$$\beta_0 + \sum_{i=1}^{k} \beta_i T_i(x) = 0 \text{ for almost all } x \text{ in the support; } (**)$$

for some $\{\alpha_j\text{'s}\}$ not all zero,

$$\alpha_0 + \sum_{i=1}^{k} \alpha_i Q_i(\theta) = 0 \text{ for almost all } \theta \in \Theta\,. \ (***)$$

The k for which these families are linearly independent [(**) and (***) only hold for α's and β's all zero] is called the *order of the family*. The representation (*) with the smallest such k is called the *minimal exponential representation*. Unless otherwise stated, we always consider this type of representation.

iv) In general, the positive integer k need not be equal to the dimension of the parameter space Θ. (See exercise 1.)

v) The *support* is $\mathscr{S} = \{x : f(x;\theta) > 0\}$. Since $C(\theta) > 0$ and $f(x;\theta) > 0$ imply $h(x) > 0$, the support must be free of θ.

vi) The $\{Q_i\text{'s}\}$ cannot all be identically zero. For example, in the continuous case, when $f(x;\theta) = C(\theta)\, h(x)$, the support is $\mathscr{S} = \{x : h(x) > 0\}$ and

$$\int_{\mathscr{X}} f(x;\theta)\, dx = 1 \text{ implies } C(\theta) \int_{\mathscr{S}} h(x)\, dx = 1$$

so that $\mathscr{S}$ must contain θ.

vii) If in (*), one changes parameters to $\eta_i = Q_i(\theta)$, i=1(1)k, one obtains the exponential family with the *natural parameters* $\eta = (\eta_1, \cdots, \eta_k)$. Then we write the density as

$$f(x;\eta) = B(\eta) h(x) \exp\{\eta_1 T_1(x) + \cdots + \eta_k T_k(x)\} \qquad (****)$$

where (in the continuous case)

$$B(\eta) = 1 \Big/ \int_{\mathscr{X}} h(x) \exp\{\Sigma \eta_i T_i(x)\}\, dx$$

and the parameter space is $\mathscr{H} = \{\eta : B(\eta) < \infty\}$.

Exercise 1: For $\theta \in \Theta = R$ and $x \in R$, the density

$$f(x;\theta) = C(\theta) \exp\{-(x-\theta)^4\}$$

$$= C(\theta)\, e^{-\theta^4}\, e^{-x^4} \exp\{4\theta^3 x - 6\theta^2 x^2 + 4\theta x^3\}$$

has: $h(x) = e^{-x^4}$; $T_1(x) = x$, $T_2(x) = x^2$, $T_3(x) = x^3$;

$$Q_1(\theta) = 4\theta^3 , Q_2(\theta) = -6\theta^2 , Q_3(\theta) = 4\theta .$$

The parameter space has dimension $m = 1$ while the statistic T has dimension $k = 3$.

a) Show that 1, T_1, T_2, T_3 are linearly independent over R.

b) Show that 1, Q_1, Q_2, Q_3 are linearly independent over R.

c) Argue that $k = 3$ is the order of this exponential family.

Example: a) Consider the Bernouilli model with

$$f(x;\theta) = \theta^x(1-\theta)^{1-x} , x \in \{0,1\} \text{ and } 0 < \theta < 1.$$

Since $\theta^x = \exp\{x \log \theta\}$,

$$f(x;\theta) = (1-\theta)\cdot I_{\{0,1\}}(x)\cdot \exp\{[\log \frac{\theta}{1-\theta}]x\}$$

becomes a one parameter exponential family with

$$C(\theta) = 1 - \theta,\ h(x) = I_{\{0,1\}}(x) ,$$

$$Q(\theta) = \log \frac{\theta}{1-\theta},\ T(x) = x .$$

In terms of the natural parameter $\eta = \log \frac{\theta}{1-\theta}$,

$$f(x;\eta) = \frac{1}{1+e^{\eta}} e^{\eta x} \text{ for } x \in \{0,1\}\ \eta \in \mathcal{H} = R .$$

b) Consider the normal model with $\theta = (\mu, \sigma^2)$ and

$$f(x;\theta) = \frac{1}{\sigma\sqrt{2\pi}} \exp\{-(x-\mu)^2/2\sigma^2\} .$$

Since

$$f(x;\theta) = \frac{e^{-\mu^2/2\sigma^2}}{\sigma\sqrt{2\pi}} \cdot e^{\{(\mu/\sigma^2)x + (-1/2\sigma^2)x^2\}} ,$$

this is a two parameter exponential model with

$$C(\theta) = \frac{e^{-\mu^2/2\sigma^2}}{\sigma\sqrt{2\pi}} , \ h(x) \equiv 1 , \ \eta_1 = Q_1(\theta) = \mu/\sigma^2 ,$$

$$\eta_2 = Q_2(\theta) = -1/2\sigma^2 , \ T_1(x) = x , \ T_2(x) = x^2 .$$

c) (A change–point hazard rate model). If X is a positive continuous type RV with PDF f and CDF F, the hazard rate is $f(x)/(1 - F(x))$. The following is a density in which the hazard rate changes at an unknown point τ. We take $\Theta = \{\theta = (\alpha, \beta, \tau) : \alpha > 0, \beta > 0, \tau > 0\}$

$$f(x;\theta) \begin{aligned} &= 0 && \text{for } x < 0 \\ &= \alpha e^{-\alpha x} && \text{for } 0 \le x < \tau \, . \\ &= e^{-\alpha\tau}\ \beta e^{-(x-\tau)} && \text{for } \tau \le x < \infty \end{aligned}$$

It is obvious that the support $\mathscr{S} = [0, \infty)$ is free of θ but this model is not an exponential family because

$$f(x;\theta) = \left[\alpha e^{-\alpha x}\right]^{I(0 \le x < \tau)} \left[b e^{-\alpha\tau - \beta(x-\tau)}\right]^{I(\tau \le x < +\infty)}$$

is not of the form (*).

Exercise 2: Continuing in part c) of the example, show that when the hazard rate $f(x)/(1 - F(x))$ is a constant α,

$$f(x) = \alpha e^{-\alpha x} .$$

Exercise 3: Let $f(x;\theta) = \frac{1}{\theta} \cdot I(0 < x < \theta)$ for $\theta > 0$. Is this model an exponential family?

Exercise 4: Show that the following models are exponential families:

a) Binomial: $f(x;\theta) = \binom{n}{x} \theta^x (1 - \theta)^{n-x} \cdot I_{A_n}(x)$,

$\theta \in \Theta = (0, 1)$, $A_n = \{0, 1, \cdots, n\}$.

b) Multinomial: with density

$$\frac{x!}{x_1!x_2! \ \ldots \ x_m!} \theta_1^{x_1} \theta_2^{x_2} \cdots \theta_m^{x_m} I_A(x, \cdots, x),$$

$$W = \{\theta = (\theta_1, \cdots, \theta_m) : \theta_j \in (0,1) \text{ and } \Sigma_{j=i}^{m} \theta_j = 1\} ,$$

$$A = \{x = (x_1, \cdots, x_m) : x_j \in A_m \text{ and } \Sigma_{j=1}^{m} x_j = n\}.$$

c) Poisson: $f(x;\theta) = \dfrac{e^{-\theta} \theta^x}{x!} I_{A_\infty}(x)$, $\theta > 0$, A_∞ = the set of non–negative integers .

The following result shows that exponential families possess non–trivial sufficient statistics.

Theorem: *Let* $X_1, \cdots, X_n$ *be a random sample from the k-parameter exponential family (*) . Then there exists a non-trivial k-dimensional sufficient statistic for* θ.

Proof: The joint density of $(X_1, \cdots, X_n)$ is

$$\left[C(\theta)\right]^n \cdot \prod_{i=1}^{n} h(x_i) \cdot \exp\{\sum_{i=1}^{n} \sum_{j=1}^{k} Q_j(\theta) T_j(x_i)\}$$

or

$$\left[C(\theta)\right]^n \cdot \prod_{i=1}^{n} h(x_i) \cdot \exp\{\sum_{j=1}^{k} Q_j(\theta) [\sum_{i=1}^{n} T_j(x_i)] .$$

By the NFT, $T(X_1, \cdots, X_n) = (\Sigma_{i=1}^{n} T_1(x_i), \cdots, \Sigma_{i=1}^{n} T_k(x_i))$ is a k–dimensional sufficient statistic for θ.

This T statistic is obviously paired with the natural parameter $\eta = (Q_1(\theta), \cdots, Q_k(\theta))$ and so is called the *natural sufficient statistic* of the exponential family. "Reduction" of data will occur when $n > k$.

Corollary: *The natural sufficient statistic is minimal.*

Proof: We use the last lemma in Lesson 8. For $x = (x_1, \cdots, x_n)$ and $y = (y_1, \cdots, y_n)$ such that $f(x;\theta) > 0$ and $f(y;\theta) > 0$, we have $h^*(x) = \Pi_{i=1}^n h(x_i) > 0$ and $h^*(y) = \Pi_{i=1}^n h(y_i) > 0$.
For $j = 1(1)k$ let

$$S_j(x) = \Sigma_{i=1}^n T_j(x_i) \text{ and } S_j(y) = \Sigma_{i=1}^n T_j(y_i);$$

then, $\dfrac{f(x;\theta)}{f(y;\theta)} = \dfrac{h^*(x)}{h^*(y)} \exp\{\Sigma_{j=1}^k Q_j(\theta)\,[S_j(x) - S_j(y)]$.

This ratio is free of θ iff $\Sigma_{j=1}^k Q_j(\theta)\,[S_j(x) - S_j(y)]$ is free of θ. But if this happens for all $\theta \in \Theta$, then necessarily $S_j(x) = S_j(y)$ for all $j = 1(1)k$, since, otherwise, the $\{\eta_j = Q_j(\theta)\}$ would satisfy a linear constraint.

Example: a) For the normal model $N(\mu, \sigma^2)$,

$$T(X_1, \cdots, X_n) = (\Sigma_{i=1}^n X_i, \Sigma_{i=1}^n X_i^2).$$

b) For the "half–normal" model with $\theta \in \Theta = R^+$, the density is $f(x;\theta) = (\pi\theta/2)^{-1/2} \exp\{-x^2/2\theta\}\, I[0 < x < +\infty]$;

$$T(x_1, \cdots, x_n) = \Sigma_{i=1}^n X_i^2 .$$

Exercise 5: Find the natural sufficient statistic(s) of the following models.

a) $f(x;\theta) = \dfrac{1}{\Gamma(\alpha)\beta^\alpha} x^{\alpha - 1} e^{-x/\beta}\, I(0 < x < +\infty)$

where $x \in R$, $\theta = (\alpha, \beta) \in \Theta = (0, \infty)^2$,

$$\Gamma(\alpha) = \int_0^\infty x^{\alpha - 1} e^{-x}\, dx.$$

b) $f(x;\theta) = (x/\theta)\, e^{-x^2/2\theta} \cdot I(0 < x < +\infty)$ for $\theta > 0$.

Exercise 6: The Cauchy model has density

$$f(x;\theta) = \frac{\beta}{\pi}\frac{1}{\beta^2 + (x-\alpha)^2} \quad \beta/\pi[\beta^2 + (x-\alpha)^2]$$

for $x \in R$, $\theta = (\alpha, \beta) \in R \times (0, \infty)$. Is this model an exponential family?

It can happen that, in an exponential family, the natural minimal sufficient statistic is not complete. Then one can find (at least two) distinct functions of this statistic which have the same expected value. This situation can be contrasted with that in the Lehmann–Scheffe Theorem in Lesson 3, Part V.

Exercise 7: Let $X_1, \cdots, X_n$ be a RS from a normal population with mean = variance = θ.

a) Show that $S_1 = \Sigma X_i$ and $S_2 = \Sigma X_i^2$ are the (joint) sufficient statistics.

b) Let $T_1 = T_1(S_1, S_2) = S_1^2 + (n-1)S_1$,

$$T_2 = T_2(S_1, S_2) = nS_2 .$$

Show that T_1 and T_2 have the same expected value: $n^2\theta(1 + \theta)$.

LESSON *11. EXPONENTIAL FAMILIES–II

The exponential families have the property that if the model itself is complete, then its natural sufficient statistic is also complete; this is due to the fact that the family of distributions of the natural sufficient statistic is also an exponential family of the same type. We will give some details of the proof in special cases; as has been true before, extension to general distributions involves measure theory beyond what we have included.

First we recall some expressions from Lesson 10. If the exponential density for a RV is given as

$$f(y;\theta) = C(\theta)h(y)\exp\{\sum_{j=1}^{k} Q_j(\theta)T_j(y)\},$$

then the joint density for a RS $X = (X_1,\cdots,X_n)$ is

$$\tilde{C}(\theta)\tilde{h}(x)\exp\{\sum_{j=1}^{k} Q_j(\theta)S_j(x)\}$$

where for $j = 1(1)k$, $S_j(x) = \sum_{i=1}^{n} T_j(x_i)$. The natural parameter is $\eta = (\eta_1,\cdots,\eta_k)$; the natural statistic is $S = (S_1,\cdots,S_k)$.

In the proof of our first lemma, we will use an extension of the following exercise; part a) has appeared before.

Exercise 1: Let Y be a function mapping Ω_1 into Ω_2. Let A and B be sets in Ω_2. Show that:

a) $Y^{-1}(A \cup B) = Y^{-1}(A) \cup Y^{-1}(B)$;

b) when A and B are disjoint, so are $Y^{-1}(A)$ and $Y^{-1}(B)$.

Lemma: *Let $[\Omega_1,\mathcal{A}_1,P]$ be a probabilty space; let $(\Omega_2,\mathcal{A}_2)$ be another measurable space. Let Y be a $(\mathcal{A}_1,\mathcal{A}_2)$ measurable function on Ω_1 to Ω_2 ; that is, for each $A_2 \in \mathcal{A}_2$, $Y^{-1}(A_2) \in \mathcal{A}_1$. Then, $Q(A_2) = P(Y^{-1}(A_2))$ defines a probability* *measure on* $\mathcal{A}_2$.

Partial proof: The only property that really requires some work to be verified is "summability". Therefore, consider a union of pairwise disjoint sets $\{B_j\}$ in $\mathcal{A}_2$; then,

$$Y^{-1}(\cup B_j) = \cup Y^{-1}(B_j)$$

is also a union of pairwise disjoint sets in $\mathcal{A}_1$. Hence,

$$Q(\cup B_j) = P(Y^{-1}(B_j)) = P(\cup Y^{-1}(B_j))$$
$$= \sum P(Y^{-1}(B_j)) = \sum Q(B_j) .$$

Exercise 2: Fill–in the missing details of the proof of this lemma.

This result and that of the next lemma make up general versions of what is called integration by substitution in ordinary calculus courses.

Lemma: *Assume the hypotheses and conclusion of the previous lemma; let* X *be the identity RV on* Ω_1. *Let* $g : \Omega_2 \to R$ *be a non-negative measurable function. Then for any* $B \in \mathcal{A}_2$,

$$\int_B g(y)\, dQ(y) = \int_{Y^{-1}(B)} g(Y(x))\, dP(x)$$

Partial proof: For discrete type X and Y, the conclusion is

$$\sum_{y \in B} g(y) \cdot Q(y) = \sum_{x \in Y^{-1}(B)} g(Y(x)) \cdot P(x) .$$

This can be written as

$$\sum_y I_B(y) \cdot g(y) \cdot Q(y) = \sum_{x \in Y^{-1}(B)} I(x) \cdot g(Y(x)) \cdot P(x) . \qquad (*)$$

Case 1. $g = I_A$ for some $A \in \mathcal{A}_2$. Then (*) is

$$\sum_y I_B(y) \cdot I_A(y) \cdot Q(y) = \sum_x I_{Y^{-1}(B)}(x) \cdot I_A(Y(x)) \cdot P(X) . \quad (**)$$

The LHS of (**) is $\sum_y I_{B\cap A}(x)\cdot Q(y) = Q(A\cap B)$. The RHS of

(**) is $\sum_x I_{Y^{-1}(B)\cap Y^{-1}(A)}(x)\cdot P(X)$

$$= P(Y^{-1}(A)\cap Y^{-1}(B)) = P(Y^{-1}(A\cap B) = Q(A\cap B) .$$

Therefore, (**) is truly an equality.

Case 2. By linearity of summation (integration, Lesson 11, Part III), the equality (*) holds when g is a non-negative simple function.

Case 3. By the Lebesgue Monotone Convergence Theorem, the conclusion holds for non-negative measurable functions as limits of non-negative simple functions.

Case 4. Again by linearity, the conclusion holds for all measurable functions g .

General proofs of this lemma and the theorem below follow the case 1,2,3,4 format in the partial proof just completed. In the following, we retain all the notation used so far and $n > k$.

Theorem: *The family of distributions of the natural sufficient statistics of a k-parameter exponential family is a k-parameter exponential family.*

Partial proof: a) In the discrete case, start with

$$A = \{S_1 = s_1, \cdots, S_k = s_k\}$$

$$= \{x = (x_1,\cdots,x_n) : S_j(x_1,\cdots,x_n) = s_j \text{ for } j = 1(1)k\} .$$

Then, $P(S_j = s_j \text{ for } j = 1(1)k)$

$$= \sum_{x\in A} P(X_1 = x_1, \cdots, X_n = x_n)$$

$$= \sum_{x\in A} \tilde{C}(\theta)\cdot \tilde{h}(x) \exp\{\sum_{j=1}^{k} Q_j(\theta)\cdot s_j$$

$$= \tilde{C}(\theta)\cdot \sum_{x\in A} \tilde{h}(x)\cdot \exp\{\sum_{j=1}^{k} Q_j(\theta)s_j$$

which is precisely of the correct form.

b) In a (typical) continuous case, we can keep

$$S_j(X_1,\cdots,X_n) = \sum_{i=1}^{n} T_j(x_i) \quad \text{for } j = 1(1)k\,,$$

and take $S_j(X_1\cdots,X_n) = X_j$ for $j = k+1, k+2, \cdots, n$. The corresponding Jacobian determinant reduces to

$$\begin{vmatrix} \partial S_1/\partial x_1 & \cdots & \partial S_k/\partial x_1 \\ & \cdots & \\ \partial S_1/\partial x_k & \cdots & \partial S_k/\partial x_k \end{vmatrix} = \mathcal{J}$$

which is zero for at most a countable number of points (by the linear independence of the T_j's). The joint density of $(S_1,\cdots,S_k,S_{k+1},\cdots,S_n)$ is

$$\tilde{C}(\theta)\cdot\tilde{h}(s)\cdot\exp\sum_{j=1}^{k} Q_j(\theta)\cdot s_j$$

where $\tilde{h}(s) = |\mathcal{J}|\cdot\prod h(\tilde{x}_i)$ with $\{\tilde{x}_i\}$ values of $\{x_i\}$ in terms of $\{s_i\}$. Integrating out $s_{k+1}, \cdots, s_n$, reveals the density of $(S_1,\cdots, S_k)$ to be

$$\tilde{C}(\theta)\cdot h^*(s^*)\cdot\exp\sum_{j=1}^{k} Q_j(\theta)\cdot s_j$$

where $h^*(s^*)$ involves only $s_1, \cdots, s_k$.

With good imagination but without the minutiae of detailed proof (essentially from the uniqueness of the Laplace transform), we get the

Theorem: *The natural minimal sufficient statistic of the k-parameter exponential family is complete if* Θ *contains a non-degenerate open rectangle of* R^k.

Example: For the normal density

$$f(x;\theta) = C(\theta)\exp\{-\frac{1}{2\sigma_1^2}x_1^2 + \frac{\mu_1}{\sigma_1^2} - \frac{1}{2\sigma_2^2}x_2^2 + \frac{\mu_2}{\sigma_2^2}x_2\},$$

$$\theta = (\mu_1, \sigma_1^2, \mu_2, \sigma_2^2),\ x = (x_1, x_2),$$

$$T = (X_1, X_1^2, X_2, X_2^2)$$

$$\eta_1 = Q_1(\theta) = \frac{-1}{2\sigma_1^2},\ \eta_2 = \frac{\mu_1}{\sigma_1^2},\ \eta_3 = -\frac{1}{2\sigma_2^2},\ \eta_4 = \frac{\mu_2}{\sigma_2^2}.$$

The density is complete because

a) there are no linear relations among the components of T or $\eta = Q(\theta)$ and

b) $Q(\Theta)$ or Θ contains a 4 dimensional non–degenerate rectangle.

Example: For $f(x;\theta) = c\,e^{-(x-\theta)^4}$ with $\Theta = R$,

$$Q_1(\theta) = 4\theta^3,\ Q_2(\theta) = -6\theta^2,\ Q_3(\theta) = 4\theta.$$

The set of points $Q(\theta) = \{(4\theta^3, -6\theta^2, 4\theta) : \theta \in R]$ is a 1–parameter or 1–dimensional curve in three space; as such, it cannot contain a 3–dimensional rectangle so that this density is not complete.

Exercise 3: For the density

$$f(x;\theta) = C(\theta)\exp\{-\frac{1}{2\sigma_1^2}x_1^2 + \frac{\mu_1}{\sigma_1^2}x_1 - \frac{1}{2\sigma_2^2}x_2^2 + \frac{\mu_1}{\sigma_2^2}x_2\},$$

$$x = (x_1, x_2),\ \theta = (\mu_1, \sigma_1^2, \sigma_2^2) \in R \times (0, \infty)^2.$$

Show that $T = (X_1, X_1^2, X_2, X_2^2)$ is not complete. Hint: take $g(x_1,x_2) = x_1 - x_2$ in $E[g(X_1,X_2)]$.

We turn now to the converse problem in its simplest form: let $\mathcal{F} = \{f(x;\theta)\}$ be a model such that

a) the support of each $f(\ ;\theta)$ is free of θ;

b) there is a positive integer k such that for each sample

of size $n > k$, $\mathcal{F}$ admits a k–dimensional sufficient statistic .

Then, under some mild regularity conditions, $\mathcal{F}$ is an exponential family.

In this particular sense then, only exponential families have sufficient statistics. Of course, the situation is different when the support depends on θ ; for example, the maximum order statistic is sufficient for the uniform distribution on the interval $(0, \theta)$. First we need a result on the "linear algebraic structure" of exponential families.

Let $\mathcal{S}$ denote the common support assumed in a) above. For a fixed $\theta_o \in \Theta$, let $g_\theta : S \to R$ be defined by

$$g_\theta(x) = \log \frac{f(x;\theta)}{f(x;\theta_o)} .$$

Let $\mathcal{L}(\mathcal{S})$ be the linear space of functions on $\mathcal{S}$ generated by the constant function 1 and the functions g_θ for $\theta \in \Theta$; specifically, each function in $\mathcal{L}(\mathcal{S})$ is a finite linear combination of 1 and some of the g_θ's .

Lemma: *The model* $\mathcal{F}$ *is a k-parameter exponential family iff the dimension of* $\mathcal{L}(\mathcal{S})$ *is* $k + 1$.

Proof: a) Suppose that $\mathcal{F}$ is a k–parameter exponential family. Then, $g_\theta(x) = \alpha(\theta,\theta_o) + \sum_{j=1}^{k} [Q_j(\theta) - Q_j(\theta_o)]T_j(x)$

where $1, T_1, \cdots, T_k$ and $1, Q_1, \cdots, Q_k$ are linearly independent. This implies that all g_θ are linear combinations of 1 and $Q_1, \cdots, Q_k$ so that the dimension of $\mathcal{L}(\mathcal{S})$ is $k + 1$.

b) Suppose that the dimension of $\mathcal{L}(\mathcal{S})$ is $k + 1$. Let $\{1, \phi_1, \cdots, \phi_k\}$ be a basis for $\mathcal{L}(\mathcal{S})$ so that for some

$$\alpha_o, \alpha_1, \cdots, \alpha_k , \ \log \frac{f(x;\theta)}{f(x;\theta_0)}$$

$$= g_\theta(x) = \alpha_0(\theta) + \alpha_1(\theta)\phi_1(x) + \cdots + \alpha_k(\theta)\phi_k(x) .$$

Then, $f(x;\theta)$

$$= f(x;\theta_0)\cdot\exp\{\alpha_0(\theta) + \alpha_1(\theta)\phi_1(x) + \cdots + \alpha_k(\theta)\phi_k(x)\}$$

has the proper form. This part of the proof will be complete when we get $1, \alpha_1, \cdots, \alpha_k$ to be linearly independent on Θ. If they are not, then for some constants $b_0, \cdots, b_{k-1}$ (and relabeling if convenient),

$$\alpha_k(\theta) = b_0 + \sum_{j=1}^{k-1} b_j\alpha_j(\theta) .$$

It follows that $g_\theta(x) = g_\theta(x) - g_{\theta_0}(x)$

$$= \alpha_0(\theta) - \alpha_0(\theta_0)\Sigma_{i=1}^{k-1}[\alpha_j(\theta) - \alpha_j(\theta_0)]\phi_j^*(x)$$

where $\phi_j^*(x) = \phi_j(x) - b_j\phi_k(x)$; (see exercise 4). This implies that the dimension of $\mathscr{L}(\mathscr{A})$ is less than $k + 1$ which contradicts the hypothesis.

Exercise 4: For the proof above:

a) explain what is meant by the "relabeling" ;

b) verify the statement involving the ϕ^*'s .

Theorem: *(Barndorff-Nielsen and Pedersen (1968)). Let*

$$\mathscr{F} = \{f(x;\theta), x \in \mathscr{X} \subseteq \mathbb{R}, \theta \in \Theta\}$$

be a model such that $f(\cdot;\theta)$ is positive and continuous on $\mathscr{X}$. If there exists an one-dimensional sufficient and continuous statistic T for all samples of size $n > 1$, then $\mathscr{F}$ is a one-parameter exponential family.

Proof: In virtue of the previous lemma, it suffices to show that the dimension of $\mathscr{L}(\mathscr{A})$ is 2 but we must do this in pieces.

a) Under the regularity conditions of the theorem, let U be the linear space of continuous functions φ on $\mathscr{X}$ (of course, U depends on n and T) such that there exists ψ with

$$\Sigma_{i=1}^{n}\varphi(x_i) = \psi[T(x_1,\cdots,x_n)] . \qquad (*)$$

Since T is continuous and sufficient, we have $g_\theta \in U$ for all $\theta \in \Theta$. Indeed, sufficiency guarantees that

$$\tau = [\Pi_{j=1}^{n} f(x_j;\theta)]/[\Pi_{j=1}^{n} f(x_j;\theta_o)]$$

is a function of T alone so that $\Sigma_{j=1}^{n} g_\theta(x_j) = \log \tau$ is a function of T alone. Thus $\mathscr{L}(\mathscr{S}) \subseteq U$. Therefore, it will suffice to show that the dimension of U is finite.

b) Now observe that (*) can be written as

$$\varphi(x_1) + \varphi(x_2) = \psi\left[\overline{T}(x_1,x_2)\right] - \sum_{i=3}^{n} \varphi(x_i) \qquad (**)$$

by defining $\overline{T}(x_1,x_2)$ to be $T(x_1,x_2,x_3,\cdots,x_n)$; then, for each fixed $(x_3,\cdots,x_n)$, $\varphi(x_1) + \varphi(x_2)$ is a function of $\overline{T}(x_1,x_2)$.

c) If $\varphi_1, \varphi_2 \in U$ and φ_1 is not a constant, then there is a pair $x_1 \neq x_2$ such that $\varphi(x_1) \neq \varphi(x_2)$. This implies that $\varphi_1(x) + \varphi_1(y) = \psi[\overline{T}(x,y)]$ for all x and y and so $\overline{T}(x_0, x_1) \neq \overline{T}(x_0, x_2)$, for all $x_0 \in \mathscr{X}$.

d) Let $a = [\varphi_2(x_1) - \varphi(x_2)][\varphi_1(x_1) - \varphi_1(x_2)]^{-1}$; then,

$$\varphi_2(x_1) - a\,\varphi_1(x_1) = \varphi_2(x_2) - a\,\varphi_1(x_2).$$

The function $\varphi = \varphi_2 - a\,\varphi_1 \in U$ is such that when $\varphi(x_1) = \varphi(x_2)$, $\overline{T}(x_0,x_1) \neq \overline{T}(x_0,x_2)$ for all $x_0 \in \mathscr{X}$. By the result of the lemma below, φ is constant in a neighborhood of each $x_0 \in U$. Such a continuous function has to be a constant everywhere, i.e.,

$$\varphi_2 - a\,\varphi_1 = c \text{ or } \varphi_2 = c + a\,\varphi_1,$$

and hence the dimension of U is at most two.

INE–Exercise: Prove the lemma by filling in the details in the outline given below.

Lemma: *If* $x_0, x_1, x_2 \in \mathscr{X}$ *and* $\varphi \in U$ *are such that*

$\varphi(x_1) = \varphi(x_2)$ *and* $T(x_0,x_1) \neq T(x_0,x_2)$, *then* φ *is constant in a neighborhood of* x_0.

Outline of the proof: a) The argument for each case of the form $x_1 < x_2$, $\overline{T}(x_0,x_1) > \overline{T}(x_0,x_2)$ is similar, so that without loss of generality, assume $x_1 < x_2$ and $\overline{T}(x_0,x_1) < \overline{T}(x_0,x_2)$.

b) We can suppose that

$$t_1 = \overline{T}(x_0,x_1) < \overline{T}(x_0,x) < \overline{T}(x_0,x_2) = t_2 \qquad (***)$$

for all x in the interval (x_1, x_2) . Otherwise, let

$$x_1' = \sup\{x : x \le x_2 \text{ and } \overline{T}(x_0,x) = t_1\},$$

$$x_2' = \inf\{x : x \ge x_1' \text{ and } \overline{T}(x_0,x) = t_2\};$$

if x_1 and x_2 do not satisfy (***), replacing x_1, x_2 by x_1', x_2', respectively, will make x_1', x_2' satisfy (***) with $x_1' < x_2'$, $\overline{T}(x_0,x_1') < \overline{T}(x_0,x_2')$, $\varphi(x_1') = \varphi(x_2')$. For this end,

i) verify that $\overline{T}(x_0,x_1') = t_1$, $\overline{T}(x_0,x_2') = t_2$ by the continuity of $x \to \overline{T}(x_0,x)$;

ii) use the definitions of x_1',x_2' and i) to show that $x_1' < x_2'$;

iii) use (**) in the proof of the theorem and i) to show $\varphi_1(x_1) = \varphi_1(x_1')$, $\varphi_1(x_2) = \varphi_1(x_2')$ imply $\varphi_1(x_1') = \varphi_1(x_2')$.

iv) to complete (***) , use the intermediate value theorem applied to the function $x \to \overline{T}(x_0,x)$

and proof by negation.

c) Observe that $\varphi(x_1) = \varphi(x_2)$ implies that there is a $y_o \in (x_1,x_2)$ such that $\varphi(y_o)$ is an absolute minimum [maximum] of φ on $[x_1, x_2]$. Let $t_o = T(x_o,y_o)$. By using (***), verify that $\varphi(x_o) = \mathrm{Min}\{\psi(t), t \in [t_1, t_2]\}$ and

$$T(x_o,x_1) < t_o < T(x_o,x_2).$$

By the continuity of T, there is a neighborhood V of x_o such that

$$T(x,x_1) < t_o < T(x,x_2),\ t_1 < T(x,y_o) < t_2 \text{ for all } x \in V.$$

And for all $x \in V$,
there is an $\alpha(x) \in (x_1, x_2)$ with $T(x,\alpha(X)) = t_o$.
Finally, deduce that for all $x \in V$,

$$\psi(t_o) = \varphi(x) + \varphi(\alpha(x)) \geq \varphi(x) + \varphi(y_o) = \varphi[T(x,y_o)] \geq \psi(t_o)$$

implying that for all $x \in V$,

$$\varphi(x) + \varphi(y_o) = \psi(t_o) = \psi[T(x_o,y_o)] = \varphi(x_o) + \varphi(y_o)$$

and $\varphi(x) = \varphi(x_o)$.

PART V: STATISTICAL ESTIMATION

Overview

The distribution of a random quantity of interest, say X , is often assumed to be known ("for all practical purposes") except for one or more parameters. It is then required that these parameters be "estimated" using the information in a random sample of "X". Since, obviously, there are many different "estimators", one begins to search for the "best", indeed, to define what "best" is.

Lesson 1 contains some rigorous framework for estimating parameters. A popular measure of optimality, mean–square error with square–error loss function, is emphasized.

Conditional expectation is the essential tool used in subsequent lessons to find "good" estimators; the examination of this concept in lesson 2, though limited to purposes of estimation, is otherwise thorough.

The attainment of uniformly minimum variance unbiased estimators in lesson 3 is based on one theorem discovered by several authors but usually referred to as Rao–Blackwell and a second theorem by Lehmann and Scheffé .

Here, efficiency refers to the attainment of minimum variance and lesson 4 contains a discussion of this for "regular models"; it is seen that these models are fairly restrictive.

The title of lesson 5 describes its contents, extensions of some of the previous results to vectors of parameters.

The two methods of point estimation in lesson 6 are the method of moments (which is at least as old as Karl Pearson) and the method of maximum likelihood (MLE, emphasized by R.A. Fisher). The literature on the latter method is vast so more details are given in lessons 7, 11, 13.

At this point it is possible to deal with confidence intervals more generally and such material appears in lessons 8 and 9 .

Lessons 10 thru 14 contain asymptotic properties of estimators with special attention paid to MLE. These results are given with great detail (omitted from many other texts) under fairly general conditions.

LESSON 1. POINT ESTIMATION OF PARAMETERS

The model is $\mathcal{F} = \{f(x;\theta),\ x \in \mathcal{X} \subseteq R^m,\ \theta \in \Theta \subseteq R^d\}$. The observables $X_1, \cdots, X_n$ are a random sample from one of these distributions but its parameter θ_o, the "true" value, is unknown. We want to know how to use the sample (information) to provide a "really educated guess" about the unknown parameter θ or, more generally, some function $\varphi(\theta)$. Ideally, a (point) estimator of $\varphi(\theta)$ is a statistic $T(X_1, \cdots, X_n)$ taking values in the range $\varphi(\Theta)$, a subset of some R^p. However, requiring this of all statistics under consideration introduces some technicalities which we can avoid by requiring only that $T \in R^p$.

A word about phraseology: an *estimator* is a random variable $T(X_1, \cdots, X_n)$ under discussion before a sample has been observed while an *estimate* is a value $T(x_1, \cdots, x_n)$ after the sample has been observed.

The natural question is, "Which statistic should we use to estimate $\varphi(\theta)$?" Of course, the answer is, "The best one!" But then we have to spell out what we mean by "the best", and ask, "Do such estimators exist?"

If we use a statistic T as an estimator for $\varphi(\theta)$ when $\varphi(\theta)$ is real, then each value $(t - \varphi(\theta))$ is an "error" and one "overall error" is the *mean squared error* (MSE):

$$MSE_\theta(T) = E_\theta[(T - \varphi(\theta))^2].$$

Using only this measure of error, a best estimator might be defined to be an estimator T^* such that

for each estimator T, $MSE_\theta[T^*] \le MSE_\theta[T]$ for all $\theta \in \Theta$.

Note that some condition like for all $\theta \in \Theta$ is required since the true θ_o is unknown. The following example makes this point from another direction.

Example: Suppose that $\varphi(\theta_1) = \theta_1$ and $\varphi(\theta_2) = \theta_2 \neq \theta_1$. Then, $T_1(X_1,\cdots,X_n) \equiv \theta_1$ and $T_2(X_1,\cdots,X_n) \equiv \theta_2$ are such that

$$MSE_{\theta_1}[T_1] = MSE_{\theta_2}[T_2] = 0 .$$

These (degenerate) statistics T_1, T_2 both have the smallest possible MSE but at different points of Θ; therfore, there does not exist one T which has smallest possible MSE for all points of Θ. The other way around,

for each $T(X_1,\cdots,X_n)$ there is an estimator $T^* = \theta'$ which has smaller MSE than T at $\theta = \theta'$ but larger MSE than $T^{**} = \theta''$ at $\theta = \theta'' \neq \theta'$.

These "pathological" estimators are such that for $j = 1, 2$,

$$E_\theta[T_j(X_1,\cdots,X_n)] \neq \varphi(\theta) \text{ for some } \theta \in \Theta.$$

Such estimators are called biased estimators, where the *bias* for φ is defined to be

$$b(T,\theta) = E_\theta[T] - \varphi(\theta) .$$

This suggests that in order to define "best" estimators, we may want to restrict the class of possible estimators of $\varphi(\theta)$ to a smaller class, namely those which are unbiased.

Definition: *An estimator* $T(X_1,\cdots,X_n)$ *of* $\varphi(\theta)$ *(real) is unbiased if* $E_\theta[T(X_1,\cdots,X_n)]$ *exists and is equal to* $\varphi(\theta)$ *for all* $\theta \in \Theta \subseteq R$.

The following symbolism is common to many texts:
$X_1, \cdots, X_n$ is a RS from a distribution (population) with

finite *population mean* $\mu = E[X_i]$,

finite *population variance* $\sigma^2 = E[(X_i - \mu)^2]$.

The *sample mean* is $\bar{X} = \sum_{i=1}^{n} X_i/n$ and the *sample variance* is

$$S^2 = \sum_{i=1}^{n} (X_i - \overline{X})^2/(n-1) .$$

Exercise 1: Within the description in the paragraph above, show that:

a) $E[\overline{X}] = \mu$
b) $E[X_i^2] = \sigma^2 + \mu^2$
c) $E[\overline{X}^2] = \sigma^2/n + \mu^2$.

Example: a) Let $f(x;\theta) = \theta^x(1-\theta)^{1-x} I_{\{0,1\}}(x)$ for $\theta \in (0,1)$. Let $T(X_1, \cdots, X_n) = \overline{X}$ and $\varphi(\theta) = \theta$.

Then, $E_\theta[T] = \frac{1}{n} \Sigma_{i=1}^{n} E_\theta[X_i] = \theta$ for all $\theta \in (0,1)$.

b) For the model $\mathscr{F} = \{f(x;\theta), x \in R, \theta \in \Theta \subseteq R\}$, let $\varphi(\theta) = F_\theta(t) = P_\theta(X \le t) = E_\theta[I(X \le t)]$ for a fixed $t \in R$.

The *empirical (sample) distribution function* is defined by

$$F_n(s) = \frac{1}{n} \Sigma_{i=1}^{n} I(X_i \le s) \text{ for } s \in R .$$

Consider $T(X_1, \cdots, X_n) = F_n(t)$. Then,

$$E_\theta[F_n(t)] = \frac{1}{n} \Sigma_{i=1}^{n} E_\theta[I(X_i \le t)] = F_\theta(t) \text{ for all } \theta \in \Theta .$$

The empirical distribution function F_n is an unbiased estimator of the population distribution function F_θ .

c) For any distribution for which the population mean $\mu = E_\theta[X]$ is finite, let $T(X_1, \cdots, X_n) = \overline{X}$. Then

$E_\theta[T] = E_\theta[X]$; that is, the sample mean $\overline{X}$ is an unbiased estimator of the population mean μ .

d) For any distribution for which the population variance

$\sigma^2 = \varphi(\theta) = \mathrm{Var}_\theta(X) = E_\theta[(X - E_\theta(X))^2]$ is finite, let

$T(X_1,\cdots,X_n) = S^2$. Since

$$T = \frac{1}{n-1}[\Sigma_{i=1}^n X_i^2 - n\bar{X}^2],$$

$$E_\theta[T] = \frac{1}{n-1}\left[E_\theta[\Sigma_{i=1}^n X_i^2] - nE_\theta[\bar{X}^2]\right]$$

$$= \frac{1}{n-1}\{n[Var_\theta(X)+(E_\theta[X])^2] - n\left[\frac{Var_\theta(X)}{n}+(E_\theta[X])^2\right]\}$$

$$= \frac{1}{n-1}\{n\,Var_\theta(X) - Var_\theta(X)\} = Var_\theta(X) .$$

This shows that the sample variance S^2 is an unbiased estimator of the population variance σ^2 .

Exercise 2: Are the following estimators unbiased for the given parameter function $\varphi(\theta)$?

a) The density is $f(x;\theta) = \frac{1}{\sigma\sqrt{2\pi}}\exp\{-\frac{1}{2\sigma^2}(x-\mu)^2\}$ and $\theta = (\mu, \sigma^2)$.

$\varphi(\theta) = \sigma^4$; $T(X_1,\cdots,X_n) = \frac{n-1}{n+1}S^4$.

b) Consider the Bernouilli model and $\varphi(\theta) = \theta(1-\theta)$.

Let $T_1(X_1,\cdots,X_n) = \frac{\Sigma_{i=1}^n X_i[n - \Sigma_{i=1}^n X_i]}{n(n-1)}$;

$T_2(X_1,\cdots,X_n) = \bar{X}(1-\bar{X})$.

c) The density is $f(x;\theta) = \frac{1}{\theta}e^{-x/\theta}\cdot I(0 < x < +\infty)$ for $\theta > 0$ and $\varphi(\theta) = \theta^2$. $T_1 = S^2$; $T_2 = \bar{X}^2$.

If T is an unbiased estimator of $\varphi(\theta)$, it does not follow that $\psi(T)$ will be an unbiased estimator of $\psi(\varphi(\theta))$. For example, in exercise 1 , we saw that while

$$E[\bar{X}] = \mu \ , \ E[\bar{X}^2] \neq \mu^2 \ .$$

However, this will be true for "affine functions":

Lemma: *Let $\psi(y) = a + by$ where a and b are constants. Then when the expectations exist,*

$$E_\theta[T] = \varphi(\theta) \text{ implies } E[\psi(T)] = a + b\varphi(\theta) \ .$$

Exercise 3: Prove the previous lemma.

The following example shows one reason why the condition "$T \in \varphi(\Theta)$" might be included in the definition of estimator.

Example: Consider the Poisson model with density

$$f(x;\theta) = \frac{e^{-\theta}\theta^x}{x!} \quad \text{for } x \in \{0, 1, \cdots, \infty\} \text{ and } \theta > 0.$$

Let $\varphi(\theta) = e^{-2\theta}$ and $T(X_1) = (-1)^{X_1}$.

$$E_\theta[T(X_1)] = \sum_{x=0}^{\infty} (-1)^x \cdot e^{-\theta} \cdot \theta^x/x! = e^{-2\theta} \ .$$

But $e^{-2\theta} > 0$ while $T(X_1)$ is positive or negative according as X_1 is even or odd! Most people would consider that this *unbiased* estimator is absurd!

It can happen that there is no unbiased estimator for $\varphi(\theta)$. For example, consider the Bernouilli model (sample size 1) and $\varphi(\theta) = \theta^2$. If T is unbiased for θ^2 , then

$$E_\theta[T] = \theta\, T(1) + (1 - \theta)T(0) \text{ becomes}$$

$$\theta^2 = T(0) + [T(1) - T(0)]\theta \text{ for all } \theta \in (0,1) \ .$$

But this is impossible.

Exercise 4: Explain the "impossible" above .

Definition: *When a function $\varphi(\theta)$ has an [at least one] unbiased estimator, $\varphi(\theta)$ is said to be (unbiasedly) estimable.*

Given an estimable function $\varphi(\theta)$ (real) and two unbiased estimators T_1 and T_2 of $\varphi(\theta)$, we might compare these estimators by comparing their variances:

T_1 is preferred to T_2 if $Var_\theta(T_1) \le Var_\theta(T_2)$ for all $\theta \in \Theta$.

Certainly, the distribution with the smaller variance is more concentrated about the mean $\varphi(\theta)$; "on the average", T_1 is closer to $\varphi(\theta)$ than T_2 is . This is a special case of the following general notion of "quality".

A function $L : R^p \times \varphi(\Theta) \to R^+$ is called a *loss function.* The expected loss of an estimator T of $\varphi(\theta)$ is called the *risk*:

$$R(\theta,T) = E_\theta[L(T,\varphi(\theta))] .$$

The preferred estimators are those with the smaller risks.

Of course, the choice of L is rather arbitrary. For the quadratic loss function $L(x,y) = (x - y)^2$, the risk is MSE ; when the estimators are also unbiased, the risk is the variance. Note that "admissibility" in the next definition begins with a fixed loss function.

Definition: *Given a loss function L and a function $\varphi : \Theta \to R^p$, an estimator $T \in R^p$ is inadmissible if there exists another estimator $S \in R^p$ such that*

$$R(\theta,S) \le R(\theta,T) \text{ for all } \theta \in \Theta \text{ and}$$

$$R(\theta,S) < R(\theta,T) \text{ for at least one } \theta \in \Theta .$$

An estimator which is not inadmissible is called admissible.

Example: Let X be such that $E_\theta[X^2] < +\infty$; consider a random sample of size two from the distribution of X. If

$$S(X_1,X_2) = \frac{X_1 + X_2}{2} \text{ and } T(X_1,X_2) = X_1 ,$$

it is clear that T and S are both unbiased estimators of

$$\varphi(\theta) = E_\theta[X] .$$

But $Var_\theta(S) = \frac{1}{2} Var_\theta(X) \le Var_\theta(X_1) = Var_\theta(T)$ for all $\theta \in \Theta$;

(in general, the strict inequality actually holds). T is inadmissible.

Inconveniently, it can happen that for a given loss function, the most preferred (minimum risk) estimator among all possible *unbiased* estimators of $\varphi(\theta)$ might be inadmissible if it is considered among *all* possible estimators of $\varphi(\theta)$. Outcomes such as this have led to investigation of various classes of estimators and we shall see some of these in later lessons. Details of the next example are in Ibragimov and Has'minskii, 1981.

Example: Consider the normal model $N(\mu, \sigma^2)$ with σ known and $\varphi(\theta) = \theta = \mu \in \Theta = [0,1]$. For a random sample $X_1, \cdots, X_n$, let $T(X_1,\cdots,X_n) = \bar{X}$ and

$$S(X_1,\cdots,X_n) = \begin{cases} 0 & \text{if } \bar{X} < 0 \\ \bar{X} & \text{if } 0 \le \bar{X} < 1 \\ 1 & \text{if } 1 \le \bar{X} \end{cases}.$$

Then, $E_\theta[(S - \theta)^2] < E_\theta[(T - \theta)^2]$ for all $\theta \in \Theta$. Thus, for this model, the sample mean $T = \bar{X}$ is inadmissible.

LESSON 2. CONDITIONAL EXPECTATION

Conditional probability (Lesson 9, Part I) is just *probability* in a reduced or restricted sample space; conditional expectation is then expectation in that reduced sample space. This lesson will focus on expectation of one random variable with reduction imposed by values of another RV; some extensions will be noted without proofs. The following example is intended to suggest one origin of the notion.

Example: Suppose that the continuous type real RVs X and Y have joint PDF f and Y has marginal PDF g ; then the conditional PDF has values

$$f(x\,|\,y) = f(x,y)/g(y) \quad \text{(where defined).}$$

With the assumption that E[X] is finite, we can write, formally,

$$E[X] = \int\int x\, f(x,y)\, dx\, dy = \int \left[\int x\, f(x\,|\,y)\, dx\right] g(y)\, dy\ .$$

The integral in parentheses is a function of y called the conditional expectation of X given Y = y and denoted by E[X | y] . Note that we also get

$$E[X] = \int E[X\,|\,y]\, g(y)\, dy = E[\,E[X\,|\,Y]\,]\ . \qquad (*)$$

Inconveniently, justification of even this simple looking relation is somewhat involved. We will give some details, for the case when Y is discrete, which bring the role of conditional probability to the forefront. We use the notation and results of lessons 9 and 11, Part II but the proof of the major theorem will be left in analysis texts.

Recall that $X : [\Omega, \mathscr{A}, P] \to [R, \mathscr{B}]$ is a real–valued integrable RV iff

$$X^{-1}(B) \in \mathscr{A} \text{ for all } B \in \mathscr{B} \text{ and } E[X] \text{ is finite}\ .$$

First we consider the second RV Y to be discrete type so that $Y(\Omega) = \mathscr{Y}$ is at most countable; say

$$\mathscr{Y} = \{y_j,\ j = 0, 1, \cdots\} \subset R\ ,$$

with $P(Y = y_j) > 0$ for all those "j" . The σ–field in Ω

generated by Y is $\{Y^{-1}(B) : B \in \mathscr{B}\}$ and is denoted by $\sigma(Y)$ or $Y^{-1}(\mathscr{B})$. [Since Y is a RV, every $Y^{-1}(B)$ is in $\mathscr{A}$ so that $\sigma(Y)$ is actually a sub–σ–field of $\mathscr{A}$.] As Y is discrete, each element A of $\sigma(Y)$ is a union of some $A_j = \{\omega : Y(\omega) = y_j\}$. Since $A_j \cap A_i = \phi$ for $j \neq i$ (that is, $y_j \neq y_i$) and

$$\Omega = \cup_{j \geq 0} A_j ,$$

the collection $\{A_0, A_1, \cdots\}$ is a countable partition of Ω.

This allows us to make the

Definition: *Let X and Y be as in the paragraph above. The conditional expectation of X given Y is the function*

$$E[X|Y] : \Omega \to R \text{ with values}$$

$$E[X|Y](\omega) = \int_{A_j} X\, dP \,/\, P(A_j) \text{ for all } \omega \in A_j,\ j = 0, 1, \cdots .$$

Example: From Lesson 2, Part III, we take a density with support $\{0,1\}\times)\{,2\}\times\{2,3,4\}$ and values $f(x_1,x_2,x_3) = (x_1 + x_2x_3)/60$.

Of course, $X = X_1 + X_2$ is still a RV so that we can compute

$$E[(X_1 + X_2)|X_3].$$

ω	X_1	X_2	X_3	$f(x_1,x_2,x_3)$	$x_1 + x_2$
α	0	1	2	2/60	1
β	0	1	3	3/60	1
γ	0	1	4	4/60	1
δ	0	2	2	4/60	2
ε	0	2	3	6/60	2
η	0	2	4	8/60	2
θ	1	1	2	3/60	2
ι	1	1	3	4/60	2
κ	1	1	4	5/60	2
λ	1	2	2	5/60	3
μ	1	2	3	7/60	3
ν	1	2	4	9/60	3

There are four pairs of values of x_1, x_2 associated with each of the three values, 2, 3, 4 , of x_3:

for $x_3 = 2$, $\omega = \alpha, \delta, \theta, \lambda$ and $E[(X_1 + X_2)|X_3 = 2](\omega)$

$$= \frac{1\cdot 2/60 + 2\cdot 4/60 + 2\cdot 3/60 + 3\cdot 5/60}{2/30 + 4/60 + 3/60 + 5/60} = 31/14 ;$$

for $x_3 = 3$, $\omega = \beta, \varepsilon, \iota, \mu$ and $E[(X_1 + X_2)|X_3 = 3](\omega)$

$$= \frac{1\cdot 3/60 + 2\cdot 6/60 + 2\cdot 4/60 + 3\cdot 7/60}{3/60 + 6/60 + 4/60 + 7/60} = 44/20 ;$$

for $x_3 = 4$, $\omega = \gamma, \eta, \kappa, \nu$ and $E[(X_1 + X_2)|X_3 = 4](\omega)$

$$= \frac{1\cdot 4/60 + 2\cdot 8/60 + 2\cdot 5/60 + 3\cdot 9/60}{4/60 + 8/60 + 5/60 + 3/60} = 57/26 .$$

Exercise 1: Continue this example. By computing each side directly, show that $E[(X_1 + X_2)] = E[\,E[(X_1 + X_2)|X_3]\,]$.

The third part of the following theorem says that $E[X|Y]$ is unique up to the equivalence of RVs; this gives strict logical sense to the definition.

Theorem: *With the hypotheses in the definition above,*

a) *$E[X|Y]$ is a discrete type RV also measurable with respect to $\sigma(Y)$;*

b) *for each $A \in \sigma(Y)$, $\int_A x\, dP = \int_A E[X|Y]\, dP$;*

c) *if $Z = Z(Y)$ is a RV satisfying a) and b), then*

$$Z = E[X|Y] \quad P\text{-a.s.}$$

Proof: a) $E[X|Y]$ is constant on each $A_j \in \sigma(Y)$; hence, $E[X|Y]$ is in fact an elementary RV with respect to $\sigma(Y)$.

b) Each $A \in \sigma(Y)$ is the (at most countable) union of disjoint A_j ; say $A = A_{j1} \cup A_{j2} \cup A_{j3} \cup \cdots$. Let $E[X|Y](\omega) = a_{ji}$ for $\omega \in A_{ji}$ so that

$$\int_{A_{ji}} X\, dP = a_{ji} \cdot P(A_{ji}) .$$

Then, $\int_A X\, dP = \int_{\cup A_{ji}} X\, dP = \sum_i \int_{A_{ji}} X\, dP = \sum_i a_{ji}\, P(A_{ji})$

$$= \sum_i E[X|Y](\omega) \cdot P(A_{ji}) = \int_{\cup A_{ji}} E[X|Y]\, dP$$

$$= \int_A E[X|Y]\, dP .$$

c) If $B = \{Z > E[X|Y]\}$ and $C = \{Z < E[X|Y]\}$, then B and $C \in \sigma(Y)$ and $\{Z \neq E[X|Y]\} = B \cup C$.

Since Z satisfies a) and b),

$$\int_B X\, dP = \int_B Z(Y)\, dP = \int_B E[X|Y]\, dP$$

imply $\int_B (Z(Y) - E[X|Y])\, dP = 0$.

But the only way in which the integral of a positive function can be zero is if the corresponding set has probability zero; that is, $P(B) = 0$. Similarly, $P(C) = 0$ so that $Z = E[X|Y]$ P–a.s.

The following array of special cases contains important properties of conditional expectation which will be used with other RVs even though we will not repeat a proof from an analysis text.

Corollary: *a) If in part b) of the theorem,* $A = \Omega$ *, then the conclusion of that part is that*

$E[X|Y]$ is integrable and $E[X] = E[\, E[X|Y]\,]$ (extending (*) of the example and exercise 1). This also says that

$$E[(X - E[X|Y])] = 0 .$$

b) If Y is a constant, then $\sigma(Y) = \{\phi, \Omega\}$ and

$$E[X|Y] = E[X] .$$

c) *Let* $X = I_A$ *for* $A \in \mathcal{A}$; $A_j = \{\omega : Y(\omega) = y_j\}$. *Then,*

for $\omega \in A_j$, $E[I_A|Y](\omega) = \int_{A_j} I_A \, dP / P(A_j)$

$$= P(A \cap A_j) / P(A_j) = P(A|Y = y_j) ;$$

for $\omega \notin A_j$, $E[I_A|Y](\omega) = 0$.

(This emphasizes the fact that $E[I_A|Y]$ *is a random variable; it is only when* $\omega \in A_j$ *that the value of this RV is the conditional probability of* A *given* Y *.)*

d) *From the results of iii), we get*

$$E[P(A|Y)] = \sum_{j \geq 0} P(A|Y = y_j) \cdot P(Y = y_j)$$

$$= \sum_{j \geq 0} P(A \cap A_j) = P(A) = E[I_A] .$$

Exercise 2: Let $X_1, \cdots, X_n$ be a RS from a Bernoulli population with parameter $\theta = P(X_i = 1)$; let

$$Y = X_1 + \cdots + X_n .$$

Find $E[X_1 | Y]$. Hint: try n = 2, 3 first.

The following is a rearrangement of the powerful Radon–Nikodym Theorem applied to our situation; proofs may be found in texts like Halmos (1950) and Loève (1963). This also makes the next definition easy.

Theorem: *On the probability space* $[\Omega, \mathcal{A}, P]$, X *is a real integrable RV and* Y *is an arbitrary RV with a corresponding sub-σ-field* $\sigma(Y)$ *; a measure* ν *is defined by*

$$\nu(A) = \int_A X \, dP \text{ for each } A \in \sigma(Y) .$$

Then there is a RV denoted by $E[X|Y]$ *which*

a) *is unique up to equivalence of RVs;*

b) *is measurable with respect to* $\sigma(Y)$;

c) *satisfies* $\int_A X\,dP = \int_A E[X|Y]\,dP$ *for all* $A \in \sigma(Y)$.

Definition: *The random variable* $E[X|Y]$ *in the theorem above is called the conditional expectation of* X *given* Y .

One of the useful properties of conditional expectation is that it leads back to the notion of conditional distribution (Lessons 2,4, Part III). For example, in part iii) of the corollary above,

$$E[I_A|Y](\omega) = E[I_A|Y(\omega)] = E[I_A|Y = y_j] = P(A|Y = y_j)$$

defines a probability measure on $\mathscr{A}$ for each y_j . With only a bit more care, our first example in this lesson follows from the formal conditional probability

$$P(X \in B|Y = y) = \int_B (f(x,y)/g(y))\,dx .$$

Exercise 3: Let X and Y have the joint density

$$f(x,y) = e^{-x-y} \cdot I(0 < x, 0 < y) .$$

Compute: $E[X|Y = 2]$, $P(X \le 3|Y = 1)$, $P(X > x|Y \le y)$.

Loosely speaking, conditional expectations have, almost surely, all the properties of expectation (Lesson 11, Part II) . For example,

Exercise 4: Prove the linearity of conditional expectation: on $[\Omega, \mathscr{A}, P]$, let X_1 , X_2, Y be real RVs with X_1, X_2 having finite expectation; let a_1, a_2 be constants. Then a.s. ,

$$E[a_1X_1 + a_2X_2|Y] = a_1E[X_1|Y] + a_2E[X_2|Y] .$$

Hint: use the definiton and the fact that for all $A \in \sigma(Y)$,

$$\int_A (a_1X_1 + a_2X_2)\,dP = a_1 \cdot \int_A X_1\,dP + a_2 \cdot \int_A X_2\,dP .$$

The following is another special property of conditional expectation which is in the same spirit as those enunciated in the earlier corollary .

Corollary: *Let X and Z be real random variables such that* $E[|X|]$ *and* $E[|Z\cdot X|]$ *are both finite. Suppose that for the RV* Y, $Z \in \sigma(Y)$ *. Then a.s.,* $E[Z\cdot X|Y] = Z\cdot E[X|Y]$.

Partial proof: Since $E[X|Y]$ and Z both belong to $\sigma(Y)$, so does their product. It remains to verify that

$$\text{for all } A \in \sigma(Y), \int_A Z\cdot X\,dP = \int_A Z\cdot E[X|Y]\,dP . \qquad (**)$$

Case 1. Let $Z = I_B$ for $B \in \sigma(Y)$. Then $A\cap B \in \sigma(Y)$ and by definition, $\int_{A\cap B} X\,dP = \int_{A\cap B} E[X|Y]\,dP$. It follows that

$$\int_A Z\cdot X\,dP = \int_A I_B\cdot X\,dP = \int_{A\cap B} E[X|Y]\,dP$$

$$= \int_A I_B\cdot E[X|Y]\,dP = \int_A Z\cdot E[X|Y]\,dP .$$

Case 2. From the validity for indicator functions and linearity, we get validity for non-negative simple functions.

Case 3. From the validity of non-negative simple functions, we get validity for non-negative measurable functions by a limit process.

Case 4. With $Z = Z^+ - Z^-$, we obtain the last justification by linearity.

In effect, this corollary says that, in the conditional distribution given Y , $Z = E[X|Y]$ behaves like a constant. Finally, we note that one can justify analogous formulas for multidimensional RVs and functions $H = h(X)$ when the corresponding expectations exist. This was done already in the particular case of exercise 1. A general form is:

$$E[Z(Y)h(X)|Y] = Z(Y)\cdot E[h(X)|Y] .$$

Example: Let X and Y be jointly distributed; assume that $E[X^2] < \infty$.

a) Since $Z = E[X|Y]$ given Y is constant, the variance of the conditional distribution of X given Y is:

$$\begin{aligned}
\mathrm{Var}(X|Y) &= E[(X - E[X|Y])^2|Y] \\
&= E[\{X^2 + (E[X|Y])^2 - 2X\cdot E[X|Y]\}|Y] \\
&= E[X^2|Y] + (E[X|Y])^2 - 2E[X\cdot E[X|Y]|Y] \\
&= E[X^2|Y] + (E[X|Y])2 - 2E[X|Y]\cdot E[X|Y] \\
&= E[X^2|Y] - (E[X|Y])^2
\end{aligned}$$

just as in ordinary expectation.

b) $E[(X - E[X|Y])\cdot E[X|Y]]$

$$\begin{aligned}
&= E[\ E[\{(X - E[X|Y])\cdot E[X|Y]\}|Y]\ ] \\
&= E[\ E[(X - E[X|Y])|Y]\cdot E[X|Y]\ ] \\
&= E[\ 0\cdot E[X|Y]\ ] = 0\ .
\end{aligned}$$

c) $E[(X - E[X|Y])\cdot E[X]]$

$$= E[(X - E[X|Y])]\cdot E[X] = 0\cdot E[X] = 0\ .$$

d) From b) and c), it follows that

$$\begin{aligned}
\mathrm{Var}(X) &= E[(X - E[X|Y] + E[X|Y] - E[X])^2] \\
&= E[(X - E[X|Y])^2] + E[(E[X|Y] - E[X])^2] \\
&\quad + 2E[(X - E[X|Y])\cdot(E[X|Y] - E[X])] \\
&= E[\ E[(X - E[X|Y])^2|Y] + E[\ (E[X|Y] - E[X])^2\ ] \\
&= E[\ \mathrm{Var}(X|Y)\ ] + \mathrm{Var}(E[X|Y])\ .
\end{aligned}$$

Exercise 5: For the density $f(x,y) = x + y$ with support $(0,1)\times(0,1)$, compute: $Z(y) = E[X|y] = \left[\dfrac{2 + 3y}{3 + 6y}\right]$;

$W(y) = \mathrm{Var}(X|y) = \int x^2\, f(x|y)\, dx - \left[\int x\, f(x|y)\, dx\right]^2$;

$\mathrm{Var}Z(Y)$; $E[W(Y)]$.

LESSON 3. UNIFORMLY MINIMUM VARIANCE UNBIASED ESTIMATORS

In this lesson, we consider only the quadratic loss function and use variance as a measure of the quality of various unbiased estimators. In this sense, the "best" estimator is given by:

Definition: *Let* $\varphi(\theta)$ *be an estimable function and* $\mathscr{W}$ *be the set of all its unbiased estimators (which also have variance). The statistic* $T \in \mathscr{W}$ *is called a uniformly minimum variance unbiased estimator (UMVUE) of* $\varphi(\theta)$ *if:*

a) $E_\theta[T^2] < \infty$ *for all* $\theta \in \Theta$;

b) *for any* $U \in \mathscr{W}$, $Var_\theta(T) \le Var_\theta(U)$ *for all* $\theta \in \Theta$.

To identify an UMVUE, it is helpful to have other criteria such as the following:

Theorem: *Let* $\mathscr{N}$ *be the set of unbiased estimators of zero, that is, statistics* S *with* $E_\theta[S] = 0$ *for all* $\theta \in \Theta$. *An unbiased estimator* T *of* $\varphi(\theta)$ *is an UMVUE if and only if*

$$\text{for each } S \in \mathscr{N},\ E_\theta[TS] = 0 \text{ for all } \theta \in \Theta . \qquad (*)$$

Proof: a) If $E_\theta[T^2] = 0$ for all $\theta \in \Theta$, then $T = 0$ a.s. It follows that $E_\theta[TS] = 0$ and so does $\varphi(\theta)$.

b) (Sufficiency: assume (*)) Let T' be another unbiased estimator of $\varphi(\theta)$. Since $E_\theta[T - T'] = 0$ for all $\theta \in \Theta$, $T - T' \in \mathscr{N}$ so that $E_\theta[T(T - T')] = 0$ or

$$E_\theta[T^2] = E_\theta[T \cdot T'] .$$

Then application of the CBS inequality (Lesson 11, Part II), yields

$$E_\theta[T^2] = E_\theta[T \cdot T'] \le (E_\theta[T^2])^{1/2} \cdot (E_\theta[(T')^2])^{1/2} .$$

This is equaivalent to

$$(E_\theta[T^2])^{1/2} \le (E_\theta[(T')^2])^{1/2}$$

and consequently, $\text{Var}_\theta(T) \le \text{Var}_\theta(T')$.

c) (Necessity: assume T is an UMVUE) Let $S \in \mathcal{N}$. For each $\lambda \in R$, $T + \lambda \cdot S$ is also an unbaiased estimator of $\varphi(\theta)$ with $\text{Var}_\theta[T + \lambda \cdot S]$

$$= \text{Var}_\theta(T) + \lambda^2 \text{Var}_\theta(S) + 2\lambda \cdot E_\theta[S \cdot (T - \varphi(\theta))] .$$

Since T is an UMVUE, $\text{Var}_\theta[T] \le \text{Var}_\theta[T + \lambda \cdot S]$. Together these imply

$$\lambda^2 \text{Var}_\theta(S) + 2\lambda \cdot E_\theta[S \cdot (T - \varphi(\theta))] \ge 0 \text{ for all } \lambda \in R .$$

By completeing the square in λ , this can be written as

$$\left[\lambda + E[\theta S \cdot (T - \varphi(\theta))]/\text{Var}_\theta(S)\right]^2$$

$$\ge \left[E_\theta[S \cdot (T - \varphi(\theta))]/\text{Var}_\theta(S)\right]^2 ,$$

say $(\lambda + r)^2 \ge r^2$ for all real λ . But, this inequality is false for $-r < \lambda < -r/2$ when $r > 0$ and for $-r/2 < \lambda < -r$ when $r < 0$.

It follows that $E[S \cdot (T - \varphi(\theta))]$ must be 0 whence

$$E_\theta[S \cdot T] = E_\theta[S \cdot \varphi(\theta)] = E_\theta[S]\varphi(\theta) = 0 .$$

Exercise 1: Verify the statements about λ and r in the proof above.

Example: Recall that the binomial model with parameters n and θ is complete. This means that the only unbiased estimator of 0 is $S(X) = 0$. Hence $E[g(X) \cdot S(X)] = 0$ for all g , in particular, $g(X) = X/n$. By virtue of the theorem above, X/n is the UMVUE of $E[X/n] = \theta$.

The following lemma and exercise arrange the CBS inequality into a form needed for the next theorem.

Lemma: *Let X and Y be joint random variables with finite second moments. Then their covariance exists and equals*

$$Cov(X,Y) = E[(X - E[X])\cdot(Y - E[Y])] .$$

If both variances are positive, their correlation exists:

$$\rho(X,Y) = Cov(X,Y)[Var\ X]^{-1/2}[Var\ Y]^{-1/2} .$$

Proof: Via the next to the last lemma in Lesson 11, Part II, $E[X^2]$ and $E[Y^2]$ both finite imply $E[X]$ and $E[Y]$ both finite. Hence their variances are finite and the CBS inequality yields

$$\left[E[(X - E[X])\cdot(Y - E[Y])]\right]^2 \le E[(X - E[X])^2]\cdot E[(Y - E[Y])^2].$$

It follows that the covariance is finite and equals

$$E[X\cdot Y] - E[X]\cdot E[Y] .$$

The final sentence follows by the definiton of correlation.

Exercise 2: Continuing the lemma, show that:

a) $|\rho(X,Y)| \le 1$;

b) $\rho(X,Y) = 1$ if and only if there exist (non-random) real α, β such that $P[X = \alpha\cdot Y + \beta] = 1$. Hint: first find a non-random λ such that

$$0 \le Var(X + \lambda\cdot Y) = Var\ X - [Cov(X,Y)]^2 / Var\ Y .$$

Theorem: *There is at most one UMVUE for an estimable function $\varphi(\theta)$.*

Proof: Suppose that T and S are both UMVUE of $\varphi(\theta)$; then

$$E_\theta[T - S] = 0 \text{ and } Var_\theta(T) = Var_\theta(S) .$$

Now $Cov(T,S) = E_\theta[(T - \varphi(\theta))\cdot(S - \varphi(\theta))] = E_\theta[TS] - \varphi^2(\theta)$.

By the first theorem, $E_\theta[T(T - S)] = 0$ or $E_\theta[T^2] = E_\theta[TS]$.

It follows that $Cov(T,S) = E_\theta[T^2] - \varphi^2(\theta) = Var_\theta(T)$ and so $\rho_\theta(T,S) = 1$ for all $\theta \in \Theta$. By exercise 2,

$$T = \alpha\cdot S + \beta ,\ P_\theta\text{–a.s. for } \alpha = Cov(T,S)/Var_\theta(S) = 1 .$$

That is, $T = S + \beta$ P_θ–a.s. But since $E_\theta[T] = E_\theta[S] = \varphi(\theta)$,

β must be 0 and so $P_\theta[T = S] = 1$ for all $\theta \in \Theta$.

The following theorem tells us that in the search for UMVUE, we may focus our attention on sufficient statistics.

Theorem: *(Rao-Blackwell) Let*

$$\mathcal{F} = \{f(x;\theta),\ x \in \mathcal{X} \subseteq R,\ \theta \in \Theta \subseteq R\}$$

and $T(X_1, \cdots, X_n)$ *be a sufficient statistic for* $\varphi(\theta)$ *where* $\varphi : \Theta \to R$. *Furthermore, assume that* $S(X_1, \cdots, X_n)$ *is an unbiased estimator of* $\varphi(\theta)$ *with* $E_\theta[S^2] < \infty$ *for all* $\theta \in \Theta$ *and that* S *is not a function of* T *alone. Consider the function defined by* $U(t) = E_\theta[S \mid T = t]$. *Then,*

a) $U(T) = E[S \mid T]$ *is a statistic and a function of* T;

b) $U(T)$ *is an unbiased estimator of* $\varphi(\theta)$;

c) $Var_\theta[U(T)] \le Var_\theta(S)$ *for all* $\theta \in \Theta$.

Proof: a) Since T is sufficient, $U(T)$ does not depend on θ so that $U(T)$ is indeed a statistic and a function of T.

b) Since (Lesson 2) $E_\theta[\, E[S \mid T)]\,] = E_\theta[S]$,

$$E_\theta[U(T)] = E[\, E[S \mid T]\,] = \varphi(\theta) \text{ for all } \theta \in \Theta.$$

c) By exercise 3 below, for all $\theta \in \Theta$,

$$Var_\theta[U(T)] = E_\theta[U^2(T)] - \varphi^2(\theta)$$
$$\le E_\theta[S^2] - \varphi^2(\theta) = Var_\theta(S). \qquad (**)$$

Exercise 3: Prove the relations (**). Hint: first show that for

$U = U(T)$, $E_\theta[S^2] = E_\theta[(S - U + U)^2] = E_\theta[(S - U)^2] + E_\theta[U^2]$.

Example: Consider the exponential model

$$f(x\,;\,\theta) = \frac{1}{\theta} e^{-x/\theta} \cdot I(0 < x < \infty) \text{ with } \varphi(\theta) = \theta > 0.$$

For a RS of size m, let $T = \sum_{i=1}^{m} X_i$. By the factorization theorem (lesson 7, Part IV), it can be seen that T is sufficient for θ. Now $S(X_1, \cdots, X_m) = X_1$ is an unbiased estimator for θ and is not a function of T. By the R–B theorem,

$$U(T) = E[X_1 | T]$$

is an unbiased estimator of θ with smaller variance than X_1.

Exercise 4: For a direct check on the variance inequality, compute $\mathrm{Var}_\theta(X_1)$, $\mathrm{Var}_\theta(U(T))$ in the example above. Hint:

$E[T|T] = T = \sum_{i=1}^{n} E[X_i|T]$ implies $E[X_1|T] = \frac{T}{n}$.

Exercise 5: Let $X_1, \cdots, X_n$ be a random sample from the Bernouilli distribution with parameter θ. Show that

$$E[X_1 | \sum_{i=1}^{n} X_i]$$

is an unbiased estimator of θ with a smaller variance than that of X_1.

Under the hypotheses of the Rao–Blackwell theorem, $U(T) = E[S|T]$ is an unbiased estimator with variance not larger than that of S; however, this does not say that U(T) is an UMVUE for $\varphi(\theta)$. But, the following theorem says that if, in addition, the sufficient statistic T is complete, then $E[S|T]$ is indeed an UMVUE.

Theorem: *(Lehmann-Scheffé) Let $X_1, \cdots, X_n$ be a random sample from $\mathcal{F}$, and let $\varphi(\theta)$ be an estimable function. If $T = T(X_1, \cdots, X_n)$ is a complete sufficient statistic and for all $\theta \in \Theta$, $U = U(T)$ is an unbiased estimator of $\varphi(\theta)$ with finite variance, then U is the unique UMVUE of $\varphi(\theta)$.*

Proof: By the Rao–Blackwell theorem, the search for a UMVUE of $\varphi(\theta)$ may be restricted to the class of unbiased estimators of $\varphi(\theta)$ which are functions of the sufficient statistic T. But completeness implies that this class of estimators has only one element in the a.s. sense: if $V(T)$ is another unbiased estimator of $\varphi(\theta)$, then for all $\theta \in \Theta$,

$E_\theta[U(T) - V(T)] = 0$ implies that $U(T) - V(T) = 0$ $P_{\theta,T}$–a.s. .

Note that the Lehmann–Scheffé theorem provides *sufficient* conditions for an UMVUE but not necessary conditions; that is, the theorem provides a method of construction of an UMVUE when a complete sufficient statistic exists. Since a complete sufficent statistic is also minimal (Lesson 9, Part IV), after obtaining $E[S] = \varphi(\theta)$, the search can proceed by looking for sufficient statistics which are minimal; then, the conditioning can be employed. But, it can happen that even if a complete sufficient statistic does not exist, a UMVUE does; searching for an UMVUE under such conditions will be discussed in the next lesson.

Example: For $\theta \in \Theta = \{2, 3, \cdots\}$ and $A = \{1, 2, \cdots, \theta\}$, let

$$f(x;\theta) = \frac{1}{\theta}\cdot I_A(x) .$$

Let $g : A \to R$ be defined by $g(x) = \begin{cases} 1/2 & \text{for } x = 1 \\ -1/2 & \text{for } x = 2 \\ 0 & \text{for } x \geq 3 \end{cases}$.

Then (exercise 1), $E_\theta[g(X)] = 0$ for all $\theta \in \Theta$, but $g \not\equiv 0$ P_θ–a.s., i.e. the model is not complete.

Exercise 6: Continue the example. Show that:

a) $E_\theta[g(X)] = 0$; b) $E_\theta[X] = \dfrac{\theta + 1}{2}$.

Example: (Continuing from above) Consider the statistic given by

$$T(x) = \begin{cases} 2x - 1 & \text{for } x \geq 3 \\ 2 & \text{for } x = 1, 2 \end{cases}.$$

Since $E_\theta[T(X)] = \frac{1}{\theta}\sum_{x=1}^{\theta} T(x) = \frac{1}{\theta}[2 + 2 + \sum_{x=1}^{\theta} (2x - 1)]$

$$= \frac{1}{\theta}[1 + 3 + \sum_{x=1}^{\theta} (2x - 1)] = \frac{1}{\theta}\sum_{x=1}^{\theta} (2k - 1)$$

$$= E_\theta[2X - 1] = \theta ,$$

$T(X)$ is an unbiased estimator of θ. Now, let $S(X)$ be such that $E_\theta[S(X)] = 0$ for all $\theta \in \Theta$; that is,

$$\frac{1}{\theta}\sum_{x=1}^{\theta} S(x) = 0 \text{ for } \theta = 2, 3, \cdots .$$

For $\theta = 2 : \frac{1}{2}[S(1) + S(2)] = 0$ implies $S(1) = -S(2)$.

For $\theta = 3 : \frac{1}{3}[S(1) + S(2) + S(3)] = 0$ implies $S(3) = 0$.
In fact, it follows that for all $\theta \geq 3$, $S(x) = 0$, for all $x \geq 3$.

Hence, for some $\alpha > 0$, $S(x) = \begin{cases} \alpha & \text{if } x = 1 \\ -\alpha & \text{if } x = 2 \\ 0 & \text{if } x \geq 3 \end{cases}$.

By the first theorem in Lesson 3,

$$E_\theta[T \cdot S] = \frac{1}{\theta}\sum_{x=1}^{\theta} T(x) \cdot S(x)$$

$$= T(1) \cdot S(1) + T(2) \cdot S(2) = 0 \text{ for all } \theta \in \Theta$$

implies that T is an UMVUE of θ .

Example: Let $X_1, X_2, \cdots, X_n$ be a RS from the Bernoulli distribution with parameter $\theta = P(X_i = 1)$. Recall that

$T = \sum_{i=1}^{n} X_i = \Sigma X_i$ is a complete sufficient statistic for θ . (All the sums following have this same index.)

a) Since $\bar{X} = \frac{1}{n}\Sigma X_i$ is an unbiased estimator of θ and a function of the complete sufficient statistic T , $\bar{X}$ is the UMVUE of θ .

b) Let $\varphi(\theta) = \theta(1 - \theta)$ and

$$S^2(X_1, \cdots, X_n) = \frac{1}{n-1} \Sigma (X_i - \bar{X})^2 .$$

Since $Var_\theta(X) = \theta(1-\theta)$, we see that S^2 is an unbiased estimator of $\varphi(\theta)$. By the (L–S) theorem, the UMVUE of $\varphi(\theta)$ is $U(T) = E_\theta[S^2 | T]$.

c) Let us compute this conditional expectation. Since $X_i \in \{0, 1\}$, we have $X_i^2 = X_i$, $i = 1(1)n$, and

$$\Sigma (X_i - \bar{X})^2 = \Sigma X_i - \left[\Sigma X_i\right]^2 /n.$$

$$\text{Then } E_\theta[S^2 | T] = E_\theta[\frac{1}{n-1} \Sigma (X_i - \bar{X})^2 | \Sigma X_i]$$
$$= \frac{1}{n-1} [\Sigma X_i - \left[\Sigma X_i\right]^2/n]$$
$$= (T - T^2/n)/(n-1) .$$

Exercise 7: Let $X_1, X_2, \cdots, X_n$ be a RS from the normal distribution with mean $\theta \in R$ and variance σ^2 known.

a) Show that $E[X_1 | \bar{X}]$ is an UMVUE of θ.

b) Show that $E[X_1 | \bar{X}] = \bar{X}$ a.s. Hint: use a technique paralleling that of the exponential case before exercise 4.

Exercise 8: Consider the Poisson model with parameter $\theta > 0$ and $\varphi(\theta) = \theta \cdot e^{-\theta}$. For a random sample $X_1, \cdots, X_n$, show

a) that $T = \begin{cases} 1 & \text{for } X_1 = 1 \\ 0 & \text{for } X_1 \neq 1 \end{cases}$ is an unbiased estimator for θ;

b) that $\sum_{j=1}^{n} X_j$ is a complete sufficient statistic for θ;

c) derive the UMVUE of $\varphi(\theta)$.

LESSON 4. EFFICIENT ESTIMATORS

When complete sufficient statistics are not available, one must search for "best estimators" using other techniques; the following imposes some additional conditions. Our notation here is that for X a continuous type RV; if X is discrete, integration is replaced by summation. It is possible to have only one list of conditions by using the Lebesgue–Stieltjes or Riemann–Stieltjes format.

Definition: *A model* $\mathscr{F} = \{f(x;\theta),\ x \in R,\ \theta \in \Theta \subset R\}$ *is called regular if it satisfies the following conditions:*

(R_1) *the support of* $f(\cdot\,;\theta)$ *is a set* $\mathscr{X}$ *independent of* θ ;

(R_2) Θ *is an open interval (finite or infinite) of* R ;

(R_3) $f(x;\cdot)$ *is differentiable in* Θ *for* $x \in \mathscr{X}$ *except possibly on a set* $\mathscr{A}$ *with* $P_\theta(\mathscr{A}) = 0$ *for all* $\theta \in \Theta$;

(R_4) *Fisher Information* $I(\theta) = E_\theta[\left[\frac{\partial}{\partial\theta}\log f(X;\theta)\right]^2]$ *satisfies* $0 < I(\theta) < +\infty$;

(R_5) $0 = \frac{\partial}{\partial\theta}(1) = \frac{\partial}{\partial\theta}\int_{\mathscr{X}} f(x;\theta)\,dx = \int_{\mathscr{X}} [\frac{\partial}{\partial\theta}f(x;\theta)]\,dx$.

The following are simple consequences of the definiton . First, for $\tilde{X} = \{X_1, \cdots, X_n\}$ a random sample from $\mathscr{F}$, let

$$f(\tilde{x};\theta) = \prod_{i=1}^{n} f(x_i;\theta) .$$

a) Since $\frac{\partial}{\partial\theta}\log f(\tilde{x};\theta) = \sum_{i=1}^{n} \frac{\partial}{\partial\theta}\log f(x_i;\theta)$,

$$I_n(\theta) = E_\theta[\left[\frac{\partial}{\partial\theta}\log f(\tilde{X};\theta)\right]^2] < +\infty \text{ by } (R_4) .$$

$I_n(\theta)$ is called the Fisher Information in the sample $X_1, \cdots, X_n$.

b) The random variables $\{ \frac{\partial}{\partial\theta}\log f(X_i;\theta) \}$ are IID for they are disjoint functions of the IID RVs $\{X_i\}$ and have zero mean by (R_5):

$$E_\theta[\frac{\partial}{\partial\theta}\log f(X_i;\theta)] = \int_{\mathscr{X}} \frac{1}{f(x;\theta)}[\frac{\partial}{\partial\theta} f(x;\theta)]\cdot f(x;\theta)\,dx$$

$$= \int_{\mathscr{X}} [\frac{\partial}{\partial\theta} f(x;\theta)]\,dx = \frac{\partial}{\partial\theta}\int_{\mathscr{X}} f(x;\theta)\,dx = 0 \text{ for all } \theta \in \Theta .$$

c) It follows that the variance of each $\frac{\partial}{\partial\theta}\log f(X_i;\theta)$ is $I(\theta)$ and hence $I_n(\theta) = n\cdot I(\theta)$ is the variance of their sum. Or, $\frac{\partial}{\partial\theta}\log f(\tilde{X};\theta)$ is a RV with mean 0 and variance $n\cdot I(\theta)$.

Exercise 1: Assume that, in addition, $f(x;\cdot)$ is twice differentiable with respect to θ and that

$$\frac{\partial^2}{\partial\theta^2}\int_{\mathscr{X}} f(x;\theta)\,dx = \int_{\mathscr{X}} \frac{\partial^2}{\partial\theta^2} f(x;\theta)\,dx .$$

Show that $I(\theta) = -E_\theta[\frac{\partial^2}{\partial\theta^2}\log f(X;\theta)]$.

The following theorem contains the *Information Inequality* or *lower bound* developed in different forms by Cramér, Rao, Dini, Fréchet . We use the notation $\tilde{\mathscr{X}} = \mathscr{X}\times\cdots\times\mathscr{X}$ (n times) for the n–fold cartesian product of a set $\mathscr{X}$ with itself.

Theorem: *Let $X_1, \cdots, X_n$ be a random sample from a regular $\mathscr{F}$ so that $\tilde{x} = (x_1,\cdots,x_n) \in \tilde{\mathscr{X}}$. Let $\varphi(\theta)$ define an estimable function of θ ; let $T(\tilde{X}) = T(X_1,\cdots X_n)$ be an unbiased estimator of $\varphi(\theta)$ with finite variance for all $\theta \in \Theta$. If,*

$$\varphi'(\theta) = \frac{\partial}{\partial\theta}E_\theta[T(X_1,\cdots X_n)] = \int_{\tilde{\mathscr{X}}} T(\tilde{x})\frac{\partial}{\partial\theta}f(\tilde{x};\theta)\,d\tilde{x} ,$$

then $Var_\theta(T(\tilde{X})) \geq [\varphi'(\theta)]^2 / n\cdot I(\theta)$.

Proof: By the CBS inequality,

$$\left[E_\theta[\{T(\tilde{X}) - \varphi(\theta)\}\cdot\frac{\partial}{\partial\theta}\log f(\tilde{X};\theta)]\right]^2 \le \mathrm{Var}_\theta T(\tilde{X})\cdot I_n(\theta) . \quad (*)$$

Since $E_\theta[\frac{\partial}{\partial\theta}\log f(\tilde{X};\theta)] = 0$,

$$E_\theta[\{T(\tilde{X}) - \varphi(\theta)\}\cdot\frac{\partial}{\partial\theta}\log f(\tilde{X};\theta]$$

$$= E_\theta[T(\tilde{X})\cdot\frac{\partial}{\partial\theta}\log f(\tilde{X};\theta)]$$

$$= \int_{\mathscr{X}} T(\tilde{x})\cdot\frac{\partial}{\partial\theta} f(\tilde{x};\theta)\, d\tilde{x}$$

$$= \frac{\partial}{\partial\theta} E_\theta[T(\tilde{X})] = \varphi'(\theta) .$$

Thus the left–hand side of (*) is $(\varphi'(\theta))^2$ and the conclusion follows.

When the equality actually holds, the variance of the unbiased estimator T is as small as the variance of any unbiased estimator can be so that then T is an UMVUE .

Exercise 2: Let X be a discrete RV with support $\mathscr{X} = \{a_1, \cdots, a_m\}$ and probability function

$$p_i(\theta) = P_\theta(X = a_i),\ i = 1(1)m,\ \theta \in [\alpha, \beta] .$$

Suppose that all p_i are differentiable on $[\alpha,\beta]$. Let $\lambda_1, \cdots, \lambda_m$ be real numbers and $H(\theta) = \sum_{i=1}^{m} \lambda_i p_i(\theta)$. Show that the lower bound of $\sum_{i=1}^{m} \{\lambda_i - H(\theta)\}^2 p_i(\theta)$ is

$$[H'(\theta)]^2 \div \{\sum_{i=1}^{m} p_i(\theta)\cdot[\frac{\partial}{\partial\theta}\log p_i(\theta)]^2\} .$$

Exercise 3: Consider the Bernoulli model with density

$$f(x;\theta) = \theta^x(1-\theta)^{1-x} \text{ for } x \in \{0, 1\} \text{ and } \theta \in (0,1) .$$

Find $E_\theta[X]$, $E_\theta[X^2]$, $E_\theta[(1-X)^2]$.

Example: Continue with the Bernoulli model; take $\varphi(\theta) = \theta$ and a RS $X_1, \cdots, X_n$. Note that Θ is an open set (actually an interval) of R. Obviously, $\frac{\partial}{\partial\theta} f(x;\theta)$ exists for each x and sliding derivatives under any finite summation sign is valid. Let $T(X_1,\cdots,X_n) = \bar{X} = \frac{1}{n}\sum_{i=1}^{n} X_i$.

a) T is an unbiased estimator of φ : $E_\theta[T] = \theta$ for all $\theta \in (0,1)$.

b) Since $\frac{\partial}{\partial\theta} \log f(x;\theta) = \frac{x}{\theta} - \frac{1-x}{1-\theta}$,

$$I(\theta) = E_\theta[\left[\frac{X}{\theta} - \frac{1-X}{1-\theta}\right]^2] = \frac{1}{\theta(1-\theta)}.$$

It follows that "Information" lower bound of the variance is $\theta(1-\theta)/n$.

c) The variance of T is $\theta(1-\theta)/n$ which equals this lower bound so T is an UMVUE.

Example: Let $\mathscr{F}$ be the normal model with mean $\theta \in R$, variance σ^2 known. Let $\varphi(\theta) = \theta$ and $T = \bar{X}$. Since

$$\frac{\partial}{\partial\theta} \log f(x;\theta) = \frac{(x-\theta)}{\sigma^2}, \quad I(\theta) = E_\theta[\frac{1}{\sigma^4}(X-\theta)^2] = \frac{1}{\sigma^2}.$$

The "information" lower bound is σ^2/n which is precisely $\text{Var}_\theta(\bar{X})$; hence $\bar{X}$ is an UMVUE of θ.

Definition: *An unbiased estimator T of $\varphi(\theta)$ such that*

$$Var_\theta(T) = [\varphi'(\theta)]^2/nI(\theta) \text{ for all } \theta \in \Theta$$

is called an efficient estimator of $\varphi(\theta)$.

When the assumptions of the theorem above are satisfied, an efficient estimator is necessarily an UMVUE; you might also have noticed that the statistics in the examples were also complete and sufficient. However, an UMVUE need not be an efficient estimator. The next exercise is preparation for such an example.

Exercise 4: Consider the Poisson model with density $\theta^x e^{-\theta}/x!$ and a random sample $X_1, \cdots, X_n$. Show that:

a) the distribution of $S = \sum_{i=1}^{n} X_i$ is Poisson with parameter $n\theta$;

b) S is a complete sufficient statistic for θ;

c) $E_\theta\left[t^{X_i}\right] = e^{-\theta(1-t)}$ for all real t;

d) $I_n(\theta) = n/\theta$.

Example: Continue with the exercise and take $\varphi(\theta) = e^{-\theta}$. Then

$$\frac{\partial}{\partial\theta}\log f(x;\theta) = \frac{x}{\theta} - 1 \text{ and } I(\theta) = E_\theta[(\frac{X}{\theta} - 1)^2] = \frac{1}{\theta}.$$

a) Let $T(X_1,\cdots,X_n) = \begin{bmatrix} 1 & \text{if } X_1 = 0 \\ 0 & \text{if } X_1 \geq 1 \end{bmatrix}$. Then,

$$E_\theta[T] = P_\theta(X_1 = 0) = e^{-\theta} = \varphi(\theta).$$

Since $S(X_1,\cdots,X_n) = \sum_{i=1}^{n} X_i$ is a complete sufficient statistic, the UMVUE of $\varphi(\theta)$ is

$$U = E[T|S] = P_\theta(X_1 = 0|S)$$

(by Lehmann–Scheffé).

b) We need only this conditional probability but it is just as easy to find the conditional distribution of X_1 given S. The joint distribution of X_1 and S is

$$P_\theta(X_1 = k, S = m)$$

$$= P_\theta(X_1 = k, \sum_{j=2}^{n} X_j = m - k)$$

$$= P_\theta(X_1 = k)\cdot P_\theta[\sum_{j=2}^{n} X_j = m - k] .$$

By part a) of exercise 4, this becomes

$$\frac{\theta^k\cdot e^{-\theta}}{k!}\cdot\frac{[(n-1)\theta]^{m-k}\cdot e^{-(n-1)\theta}}{(m-k)!}$$

$$= \frac{\theta^m\cdot e^{-n\theta}}{k!\,(m-k)!}(n-1)^{m-k} .$$

Since $P(S = m-k) = (n\theta)^{m-k}\cdot e^{-n\theta}/(m-k)!$,

$$P_\theta(X_1 = k\,|\,S = m) = \binom{m}{k}\left(\frac{1}{n}\right)^k\left(1-\frac{1}{n}\right)^{m-k} .$$

This says that the conditional distribution of X_1 given $S = m$ is binomial with parameters m & $1/n$. In particular,

$$P_\theta(X_1 = 0\,|\,S) = (1 - 1/n)^S = U .$$

c) Now we can compute the variance of U .

$$Var_\theta(\{1 - 1/n\}^S) = E_\theta[U^2] - \{E_\theta[U]\}^2$$

$$= E_\theta[(1 - 1/n)^{2S}] - e^{-2\theta} .$$

By part c) of exercise 4,

$$E_\theta[(1 - 1/n)^{2S}] = e^{-n\theta(1 - (1 - 1/n)^2)} = e^{-2\theta + \theta/n} .$$

Therefore,

$$Var_\theta(U) = e^{-2\theta + \theta/n} - e^{-2\theta} = e^{-2\theta}(e^{\theta/n} - 1) .$$

d) But the Information lower bound is

$$\frac{[\varphi'(\theta)]^2}{n\,I(\theta)} = \frac{e^{-2\theta}}{n/\theta} = \theta\cdot e^{-2\theta}/n .$$

Since for $x > 0$, $e^x - 1 > x$,

$$e^{-2\theta}(e^{\theta/n} - 1) > \theta \cdot e^{-2\theta}/n \text{ for all } \theta > 0.$$

Thus $U = (1 - 1/n)^S$ is the UMVUE of $\varphi(\theta) = e^{-\theta}$ yet its variance does not achieve the information lower bound; U is not efficient.

Exercise 5: (One direct version of the lower bound).

a) For each $\theta \in [\alpha,\beta]$, $f(\cdot;\theta)$ is a probability density (with respect to Lebesgue measure Λ) on R^n, $n \geq 1$.

b) $\frac{\partial}{\partial\theta} f(x;\theta)$ exists for all x and all θ.

c) There is a function q such that $|\frac{\partial}{\partial\theta} f(x;\theta)| \leq q(x)$ for all x and θ.

d) All the functions q, A, $A \cdot f$, $A \cdot q$, $A \cdot \frac{\partial f}{\partial\theta}$, $(\frac{\partial f}{\partial\theta})^2/f$ are integrable with respect to Λ.

e) $\psi(\theta) = \int_{-\infty}^{\infty} A(x) \cdot f(x;\theta)\, d\Lambda(x)$; $\varphi(\theta)$ is an arbitrary function of θ. Then

$$\int_{-\infty}^{\infty} [A(x) - \varphi(\theta)]^2 \cdot f(x;\theta)\, d\Lambda(x) \geq$$

$$[\psi'(\theta)]^2 / \int_{-\infty}^{\infty} [\tfrac{\partial}{\partial\theta} f(x;\theta)]^2 / f(x;\theta)\, d\Lambda(x) .$$

Hint: these integrals are essentially Riemann with "$d\Lambda(x) = dx$".

Exercise 6: Let the distribution of X be binomial $B(n, \theta)$. Show that $\frac{X}{n}$ is the UMVUE of θ and is also efficient.

If a model does not satisfy all the regularity conditions (R_1) to (R_5), then an UMVUE may have a variance that is actually less than the information lower bound (when it exists.)

Exercise 7: Consider the uniform model

$$\mathcal{F} = \{f(x;\theta) = \frac{1}{\theta} \cdot I(0 < x < \theta),\ x \in R,\ \theta > 0\} .$$

a) Does $\mathscr{F}$ satisfy the conditions of regularity?
b) Compute $I(\theta)$.
c) Let $X_1, \cdots, X_n$ be a random sample from $\mathscr{F}$ with the maximum order statistic $T = \max\{X_1, \cdots, X_n\}$.

Show that $U = \frac{n+1}{n} T$ is an unbiased estimator of $\varphi(\theta) = \theta$.
d) Compute $\mathrm{Var}_\theta(U)$.
e) Use the Lehmann–Scheffé theorem to show that U is the UMVUE of θ .
f) Compare $\mathrm{Var}_\theta U$ with the "information" lower bound. Explain .

We close this lesson on unbiased estimation of one–dimensional real parameters by showing a relation between efficient estimation and exponential families.

Under the regularity conditions, we see that when

$$\mathrm{Var}_\theta T(\tilde{X}) = [\varphi'(\theta)]^2/n \cdot I(\theta) \text{ (and finite),}$$

the CBS inequality (*) is actually an equality so that

$$T(\tilde{X}) - \varphi(\theta) = k_\theta \frac{\partial}{\partial\theta} \log f(\tilde{x};\theta)$$

where k_θ is some function of θ but not of $\tilde{x}$. Integrating

$$(T(\tilde{x}) - \varphi(\theta))/k_\theta = \frac{\partial}{\partial\theta} \log f(\tilde{x};\theta) \text{ wrt to } \theta$$

yields:

$Q(\theta)T(\tilde{x}) + S(\theta) = \log f(\tilde{x};\theta) + A(\tilde{x})$ for some functions Q, S, A.

It follows that

$$f(\tilde{x};\theta) = \exp\{Q(\theta)T(\tilde{x}) + S(\theta) + A(\tilde{x})\}$$

which is a one–parameter exponential family and hence $T(\tilde{X})$ is a sufficent statistic. In this sense, only exponential families have efficient (sufficient) statistics.

LESSON 5. UNBIASED ESTIMATION: THE VECTOR CASE

In this lesson, we extend the results of some preceeding lessons to the case where the parameter space Θ is a subset of R^d for $d > 1$. We have not introduced special symbols for vectors of parameters; we let the context determine their dimensionality. Recall first that a real–valued random variable X is said to be integrable (with respect to P) iff $E[|X|] < \infty$.

Definition: *For $x \in R^m$, the column vector $x = (x_1, \cdots, x_m)'$ where the prime denotes transpose. The norm of x is*

$$\| x \| = \left[\sum_{i=1}^{m} x_i^2\right]^{1/2} = (x'x)^{1/2} .$$

A random vector $X = (X_1, \cdots, X_m)'$ is a measurable map from $[\Omega, \mathcal{A}, P]$ to $(R^m, \mathcal{B}_m)$ where each X_j is a real-valued RV. $\| X \|$ is the real valued RV $(X'X)^{1/2}$. If each component X_j in X is integrable, X is said to be integrable with mean (the m by 1 vector) $E[X] = (E[X_1], \cdots, E[X_m])'$.

Exercise 1: Prove the following corollaries.

a) $\| x \|^2 = x'x = tr(x'x) = tr(xx')$ where
$$tr(B) = \sum_{i=1}^{m} b_{ii}$$
is the *trace* of the m by m matrix $B = [b_{ij}]$.

b) $E[\| X \|] < \infty$ iff $E[|X_j|] < \infty$ for $j = 1(1)m$.
Hint: first show that
$$\| X \| \le |X_1| + \cdots + |X_m| \le m \cdot \| X \| .$$

c) When $E[\| X \|^2] < \infty$ so is $E[\| X \|]$.
Hint: use the CBS inequality.

d) When $E[\| X \|^2] < \infty$, $E[\| X - E[X] \|^2] < \infty$ and

$$E[(X_j - E[X_j])^2] < \infty \text{ for } j = 1(1)m .$$

Definition: *The covariance matrix of the random vector X is*

$$Cov(X) = \Sigma(X) = E[(X - E[X])(X - E[X])']$$
$$= \left[\Sigma_{ij}(X)\right] \text{ for } i, j = 1, \cdots, m ,$$

when $\sigma_{ij}(X) = \Sigma_{ij}(X) = E[(X_i - E[X_i])(X_j - E[X_j])]$ *are all finite. If X is "understood",* $\Sigma(X)$ *may be written as* $\Sigma = (\sigma_{ij})$.

Exercise 2: For the above , show that

$$\Sigma(X) = E[XX'] - (E[X])(E[X])' .$$

Lemma: *The covariance matrix* $\Sigma(X)$ *is symmetric. Moreover,* $\Sigma(X)$ *is non-negative: for all* $y \in R^m$, $y'\Sigma(X)y \geq 0$.

Proof: Since $E[X_iX_j] = E[X_jX_i]$, symmetry is obvious. Now,

$$\begin{aligned} y'\Sigma(X)y &= y'E[(X - E[X])(X - E[X])']y \\ &= E[y'(X - E[X])(X - E[X])'y] \\ &= E[y'(X - E[X])(y'(X - E[X])'] \\ &= E[(y'(X - E[X]))^2] \geq 0. \end{aligned}$$

Non–negativity may be symbolized by $\Sigma(X) \geq 0$. This also allows us to write,

for random vectors X_1, X_2, $\Sigma(X_1) \leq \Sigma(X_2)$

if

$$y'(\Sigma(X_2) - \Sigma(X_1))y \geq 0 \text{ for all } y \in R^m .$$

Exercise 3: Show that this "partial ordering" satisfies the transitive property:

$$\Sigma(X_1) \leq \Sigma(X_2) \text{ and } \Sigma(X_2) \leq \Sigma(X_3) \text{ imply } \Sigma(X_1) \leq \Sigma(X_3) .$$

Definition: *Let the model be*

$$\mathcal{F} = \{f(x;\theta), x \in R^m, \theta \in \Theta \subseteq R^d\}$$

with a random sample $X^1, X^2, \cdots, X^n$. *Let* $\varphi : \Theta \longrightarrow \varphi(\Theta)$; *then* $T = t(X^1, \cdots, X^n)$ *is an estimator of* $\varphi(\theta)$ *iff* $t(x^1, \cdots, x^n) \in \varphi(\Theta)$ *for all (a.s.) sample points* $(x^1, \cdots, x^n)$. T *is also unbiased iff* $E_\theta[T] = \varphi(\theta)$ *for all* $\theta \in \Theta$.

Unlike the one dimensional case, here we restrict the range of T to $\varphi(\Theta)$ so as to eliminate "absurd" estimators. The following example shows that the basic properties of the sample mean and variance carry over to the multi–dimensional case.

Example: Continue from the definition; write $\sum_{i=1}^{n}$ as $\sum$.

a) The sample mean is an unbiased estimator of the population mean. Suppose that $\varphi(\theta) = E_\theta[X]$ is finite in R^m. Then, $T(X^1, \cdots, X^m) = \bar{X} = \frac{1}{n}\sum X^i$ is in R^m and $E_\theta[T] = \sum E_\theta[X^i]/n = E_\theta[X]$.

b) Suppose that the covariance matrix exists:

$$\varphi(\theta) = \Sigma_\theta(X) .$$

Let $S = S(X^1, \cdots, X^n)$

$= \frac{1}{n-1}\sum[(X^i - T)(X^i - T)']$ and

$$E_\theta[X] = E[X^i] = \mu .$$

Now, $\sum(X^i - \mu)(X^i - \mu)'$

$$= \sum(X^i - T + T - \mu)(X^i - T + T - \mu)'$$

$$= \sum(X^i - T)(X^i - T)' + \sum(X^i - T)(T - \mu)$$

$$+ \sum(T - \mu)(X^i - T)' + \sum(T - \mu)(T - \mu)'$$

$$= \sum(X^i - T)(X^i - T)' + \sum(T - \mu)(T - \mu)'$$

since $\sum(X^i - T)(T - \mu)' = \left[\sum(X^i - T)\right](T - \mu)' = 0$.

This can be rewritten as

$$(n-1)S = \sum (X^i - \mu)(X^i - \mu)' - n(T - \mu)(T - \mu)' .$$

Taking expectation on both sides of this last equality yields

$$(n-1)E_\theta[S] = nCov(X) - nCov(T) . \qquad (*)$$

c) The (co)variance of the sample mean is 1/n times that of the population. Since

$$T - M = \sum X^i /n - \mu = \sum (X^i - v)/n ,$$

$$Cov(T) = E_\theta[(T - \mu)(T - \mu)']$$

$$= E_\theta\left[\left[\sum (X^i - \mu)/n\right]\left[\sum (X^j - \mu)/n\right]'\right]$$

$$= \sum_{i=1}^{n} \sum_{j=1}^{n} E_\theta[X^i - \mu)(X^j - \mu)']/n^2 .$$

When $i \neq j$, X^i and X^j are independent so that

$$E_\theta[(X^i - \mu)(X^j - \mu)'] = \begin{cases} 0 & \text{for } i \neq j \\ \Sigma_\theta(X) & \text{for } i = j \end{cases} .$$

Therefore, $Cov(T) = Cov(X)/n = \Sigma_\theta(X)/n$. (**)

d) The sample covariance S is an unbiased estimator of the population covariance. Putting (**) in (*), yields

$$(n-1)E_\theta[S] = n\Sigma_\theta(X) - n\Sigma_\theta(X)/n$$

or $E_\theta[S] = \Sigma_\theta(X)$.

Exercise 4: Let A be a p by m matrix of constants; let X by an m dimensional random vector. Show that:

a) $E[AX] = AE[X]$ when $E[X]$ exists;

b) $Cov(AX) = ACov(X)A'$ when $Cov(X)$ exists .

Unless stated otherwise, we will take the *quadratic loss function* in the vector case to be

$$L(\theta,a) = \|\theta - a\|^2 = (\theta - a)'(\theta - a) \text{ for } \theta, a \in R^d.$$

In general, one might consider

$$(\theta - a)'A(\theta)(\theta - a) \text{ for any } A(\theta) \geq 0 ,$$

so in particular, for the identity matirx I in R^d .
In any case, for a given L , the *risk* of an estimator T is

$$R(\theta,T) = E_\theta[L(\theta,T)] .$$

This is $E_\theta[\|\theta - T\|^2]$ when $\varphi(\theta) = \theta$ and $A(\theta) = I$.

Lemma: *Let T and S be unbiased estimators of $\varphi(\theta)$ such that $\Sigma_\theta(T) \leq \Sigma_\theta(S)$. Then $R(\theta,T) = E_\theta[\|\varphi(\theta) - T\|^2] \leq R(\theta,S)$.*

Proof: a) By hypothesis,

$$y'(\Sigma_\theta(S) - \Sigma_\theta(T))y \geq 0 \text{ for all } y \in R^d .$$

In particular, for $y = (1,0,\cdots,0)'$, this yields

$$\sigma_{11}(S) - \sigma_{11}(T) \geq 0 .$$

For $y = (0,1,0,\cdots,0)$, we get $\sigma_{22}(S) - \sigma_{22}(T) \geq 0$; etc. It follows that

$$tr(\Sigma_\theta(S) - \Sigma_\theta(T)) \geq 0 .$$

b) Now $E_\theta[\|\varphi(\theta) - T\|^2] = E_\theta[(\varphi(\theta) - T)'(\varphi(\theta) - T)]$

$$= E_\theta[tr(\varphi(\theta) - T)'(\varphi(\theta) - T))]$$

$$= E_\theta[tr(\varphi(\theta) - T)(\varphi(\theta) - T)')]$$

$$= trE_\theta[(\varphi(\theta) - T)(\varphi(\theta) - T)']$$

$$= tr\left[\Sigma_\theta(T)\right] .$$

Then $R(\theta,S) - R(\theta,T) = tr\left[\Sigma_\theta(S)\right] - tr\left[\Sigma_\theta(T)\right]$

$$= tr\left[\Sigma_\theta(S) - \Sigma_\theta(T)\right] .$$

This is non–negative by a) .

This suggests that we might restrict our choice of

estimators T to those which are unbiased with a "minimum" covariance matrix: for each unbiased estimator S,

$$\Sigma_\theta(T) \le \Sigma_\theta(S) \text{ for } \theta \in \Theta .$$

Exercise 5: Let $\varphi(\theta) = (\varphi_1(\theta),\cdots,\varphi_d(\theta))' \in R^d$ and $T(X^1,X^2,\cdots,X^n) = (T_1(X^1,\cdots,X^n),\cdots,T_d(X^1,\cdots,X^n))'$ be an unbiased estimator of $\varphi(\theta)$. Show that if T has a minimum covariance matrix among all unbiased estimators of $\varphi(\theta)$, then for each $j = 1, \cdots, d$, T_j is an unbiased estimator of $\varphi_j(\theta)$ with minimum variance. Hint: use a technique like that in part a) of the lemma above.

The theorem below (see for example Rao, 1965) contains extensions to the vector case of the main results for real parameters. Their proofs are "straightforward" and in some cases, one has merely to read the original statements as if they were vectors/matrices to begin with. Usually, the function φ has the same dimensionality as θ, but that is not absolutely necessary.

Theorem: *Let T be an unbiased estimator of $\varphi(\theta) \in R^p$. Let S be a sufficient statistic for $\theta \in R^d$. If φ is differentiable, let $D(\theta)$ be the $p \times d$ matrix of partial derivatives $\partial\varphi_i/\partial\theta_j$, $i = 1(1)p$, $j = 1(1)d$.*

a) *$E[T|S]$ is an unbiased estimator of $\varphi(\theta)$ with a covariance matrix "less than" that of T.*

b) *If S is also complete, then $E[T|S]$ is an unbiased estimator of $\varphi(\theta)$ with minimum covariance matrix.*

c) *Under appropriate extensions of the regularity conditions, the Fisher Information matrix is the d by d matrix*

$$I(\theta) = \left[E_\theta[\psi\frac{\partial}{\partial\theta_i} \log f(X;\theta)\varphi\cdot\psi\frac{\partial}{\partial\theta_j} \log f(X;\theta)\varphi]\right] .$$

If it exists, $I(\theta)$ is positive definite and non-singular.

d) *It follows from c) that $\Sigma_\theta(T) \ge D(\theta)\left[I(\theta)\right]^{-1}D(\theta)'$.*

LESSON 6. TWO METHODS OF POINT ESTIMATION

So far in Part V, we have looked at some desirable properties of estimators of parameters of a model. In this lesson, we introduce two popular methods for finding estimators: the method of moments and the method of maximum likelihood. The latter can be used in many situations and will be examined more thoroughly in Lessons 7, 11, 13.

Since we can trace the method of moments back past Karl Pearson, this technique is probably the oldest general method of estimation. It is also a simple technique based upon the intuitive idea that the natural estimators of population moments are sample moments. More precisely, let $X_1, \cdots, X_n$ be a random sample of a univariate population "of X" . Suppose that the *population moments* $\mu_j = E[X^j]$, $j = 1, \cdots, k$ exist. Define the *sample moments* (statistics) as $m_j = \frac{1}{n}\sum_{i=1}^{n} X_i^j$, $j = 1, \cdots, k$. (To be consistent, we should use M_j but historically it has been m_j .) Then the *method of moments estimator* (MME) of μ_j is m_j .

Exercise 1: Show that the sample moments m_j are unbiased estimators of the population moments μ_j .

If a new parameter of interest is a known function of the μ_j's, say $\alpha = g(\mu_1, \cdots, \mu_k)$, then we can take as an estimator $\hat{\alpha} = g(m_1, \cdots, m_k)$ simply substituting m_j for μ_j in the known function g . If a system

$$\alpha_i = g_i(\mu_1, \cdots, \mu_k)\ , i = 1(1)k\ ,$$

has a unique solution $\mu_j = \varphi_j(\alpha_1, \cdots, \alpha_k)$, $j = 1(1)k$, then the *estimating equations* are $m_j = \varphi_j(\alpha_1, \cdots, \alpha_k)$, $j = 1(1)k$, with the solution

$$\hat{\alpha}_i = g_i(m_1, \cdots, m_k)\ , i = 1(1)k\ .$$

Example: Let X be any RV with finite mean μ and variance σ^2. Let $\alpha_1 = \mu$, $\alpha_2 = \sigma^2$ so that

$$E[X] = \mu_1 = \alpha_1 \text{ and } E[X^2] = \mu_2 = \alpha_2 + \alpha_1{}^2 .$$

Then the estimating equations are

$$m_1 = \frac{1}{n}\sum_{i=1}^{n} X_i = \alpha_1 \text{ and } m_2 = \frac{1}{n}\sum_{i=1}^{n} X_i{}^2 = \alpha_2 + \alpha_1{}^2$$

and the moment estimators of α_1, α_2 are

$$\hat{\alpha}_1 = m_1 = \frac{1}{n}\sum_{i=1}^{n} X_i = \bar{X} \text{ and}$$

$$\hat{\alpha}_2 = m_2 - \hat{\alpha}_1{}^2 = m_2 - m_1{}^2$$

$$= \frac{1}{n}\sum_{i=1}^{n} X_i{}^2 - \left[\sum_{i=1}^{n} X_i/n\right]^2 = \frac{1}{n}\sum_{i=1}^{n} (X_i - \bar{X})^2 .$$

Exercise 2: Let $X_1, X_2, \cdots, X_n$ be IID Bernoulli RVs with $P(X_i = 1) = \theta$. Find MMEs of θ, $\theta(1 - \theta)$, $\cos\theta$.

Exercise 3: Suppose that θ can be considered the constant probability that a light switch "fails" in each independent trial. Let X be the number of "successes" before a failure. Find a MME estimator of θ based on X. Hint: to find $E[X]$, first find the MGF of this geometric RV.

It is seen that the method of moments technique can be simple and quick. Although moment estimators can be biased, it is often possible to remove the bias by a minor correction. In the example above, $\hat{\alpha}_2$ is biased, but

$$\frac{n}{n-1}\hat{\alpha}_2 = \frac{1}{n-1}\sum_{i=1}^{n} (X_i - \bar{X})^2$$

is unbiased. Of course, moment estimators are, in general, not uniquely determined.

Example: Let X be a Poisson RV with parameter $\lambda > 0$. Since

$E[X] = Var(X) = \lambda$, the method of moments solution of the first example will lead to two different estimators of λ, namely,

$$\hat{\lambda} = m_1 \text{ and } \hat{\lambda} = m_2 - m_1^2 .$$

Moreover, when the parameter of interest is a function of the population moments, there are at least two ways to obtain estimators based on sample moments. For example, we can estimate $g(\mu_1)$ by $g(m_1)$ or we (might be able to) find $H = h(X_1, \cdots, X_n)$ with $E[H] = g(\mu_1)$ so that

$$\hat{g}(\mu_1) = h(x_1, \cdots, x_n) .$$

The following exercise and example also illustrate the non–uniqueness of the MMEs.

Exercise 4: Let X have a lognormal distribution with density

$$f(x) = \frac{1}{x\sigma\sqrt{2\pi}} \cdot e^{-(\log x - \mu)^2/2\sigma^2} \cdot I(0 < x < \infty) .$$

a) Show that $E[X]$

$$= \int_{-\infty}^{\infty} e^y \cdot e^{-(y-\mu)^2/2\sigma^2} \, dy / \sigma\sqrt{2\pi} = e^{\mu + \sigma^2/2} .$$

b) Show that $Var(X) = e^{2\mu + \sigma^2}(e^{\sigma^2} - 1)$.

Example: Let $X_1, \cdots, X_n$ be a RS from the log–normal distribution above.

a) For $\mu_1 = E[X] = e^{\mu + \sigma^2/2}$ and

$$\mu_2 = E[X^2] = \left[e^{\mu + \sigma^2/2}\right]^2 \cdot e^{\sigma^2} ,$$

the estimating equations $m_1 = e^{\mu + \sigma^2/2}$ and

$$m_2 = \left[e^{\mu + \sigma^2/2}\right]^2 \cdot e^{\sigma^2}$$

yield $\hat{\sigma}^2 = \log(m_2/m_1^2)$ and

$$\hat{\mu} = \log m_1 - (1/2)\log m_2/m_1^2 .$$

b) On the other hand, $Y_1 = \log X_1, \cdots, Y_n = \log X_n$ is a RS from the normal distribution with $E[Y_i] = \mu$ and $E[Y_i^2] = \sigma^2 + \mu^2$. The estimating equations

$$\sum_{i=1}^{n} Y_i/n = \mu \text{ and } \sum_{i=1}^{n} Y_i^2/n = \sigma^2 + \mu^2$$

yield $\tilde{\sigma}^2 = \sum_{i=1}^{n} (\log X_i - \tilde{\mu})^2/n$ and

$$\tilde{\mu} = \sum_{i=1}^{n} (\log X_i)/n .$$

Obviously these differ from $\hat{\sigma}$ and $\hat{\mu}$.

In general, moment estimators are not functions of sufficient statistics and so they cannot be efficient either. For example, consider a RS of X uniformly distributed over the interval $[\alpha,\beta]$; the sufficient statistic is the pair

$$X_{(1)} = \min\{X_1, \cdots, X_n\}, X_{(n)} = \max\{X_1, \cdots, X_n\} .$$

Since $E[X] = \dfrac{\alpha + \beta}{2}$, $Var(X) = \dfrac{(\beta - \alpha)^2}{12}$, the estimating equations

$$\bar{X} = (\hat{\alpha} + \hat{\beta})/2 , \frac{1}{n}\sum_{i=1}^{n} (X_i - \bar{X})^2 = \frac{(\hat{\beta} - \hat{\alpha})^2}{12} \text{ yield}$$

$$\hat{\alpha} = \bar{X} - [\tfrac{3}{n}\sum_{i=1}^{n} (X_i - \bar{X})^2]^{1/2} \text{ and}$$

$$\hat{\beta} = \bar{X} + [\tfrac{3}{n}\sum_{i=1}^{n} (X_i - \bar{X})^2]^{1/2}$$

which are not functions of the sufficient statistic $(X_{(1)}, X_{(n)})$ alone.

We turn now to the method of maximum likelihood used so effectively by Fisher. The following simple example illustrates the principle.

Example: Consider the problem of estimating the probability $\theta = P(X = 1)$ as in n independent tosses of a coin. Specifically, suppose it is known that $\theta \in \Theta = \{\frac{1}{4}, \frac{2}{4}, \frac{3}{4}\}$ and $n = 4$. The density of X is

$$f(x;\theta) = \theta^x(1-\theta)^{1-x} I_{\{0,1\}}(x).$$

Given a random sample x_1, x_2, x_3, x_4, one would like to find the true value of θ. For a given θ, the joint density of X_1, X_2, X_3, X_4 is

$$g(x_1,x_2,x_3,x_4 | \theta) = \prod_{i=1}^{n} f(x_i;\theta) = \theta^{\Sigma x_i}(1-\theta)^{4-\Sigma x_i}.$$

Let's tabulate this probability for each $\theta \in \Theta$:

θ	1/4	2/4	3/4	$\Sigma x_i =$
	81/256	16/256	1/256	0
	27/256	16/256	3/256	1
	9/256	16/256	9/256	2
$g(x_1,x_2,x_3,x_4 \| \theta)$	3/256	16/256	27/256	3
	1/256	16/256	81/256	4

We see that the probability that we will observe $\Sigma x_i = 0$ is greatest when $\theta = \frac{1}{4}$; hence, in a kind of reverse logic, when $\Sigma x_i = 0$ is observed, $\theta = \frac{1}{4}$ is a "most likely" value. Thus, we estimate θ to be $\frac{1}{4}$ when we observe $\Sigma x_i = 0$. By similar reasoning, we estimate $\theta = \frac{1}{4}$ when $\Sigma x_i = 1$. Etc. In summary,

$$\hat{\theta} = \frac{1}{4} \text{ if } \Sigma x_i = 0 \text{ or } \Sigma x_i = 1$$

$$\hat{\theta} = \frac{1}{2} \text{ if } \Sigma x_i = 2$$

$$\hat{\theta} = \frac{3}{4} \text{ if } \Sigma x_i = 3 \text{ or } \Sigma x_i = 4$$

For each (x_1,x_2,x_3,x_4), the value $\hat{\theta}$ is the value θ^* in Θ such that

$$g(x_1,x_2,x_3,x_4|\theta^*) = \underset{\theta \in \Theta}{\text{Max}}\ g(x_1,x_2,x_3,x_4|\theta) .$$

The estimator $\hat{\theta}$ is called the maximum likelihood estimator (MLE) of θ .

Exercise 5: Consider a similar Bernoulli experiment with n = 15 and $\theta = .6$ or $.3$. If $\Sigma x_i = 6$, what is the MLE of θ ? Hint: it is necessary to calculate only two probabilities.

For discrete cases like the Bernoulli above, the likelihood is the probability and one might call for maximum probability estimators; but the term likelihood has historical precedence and, today, "maximum probability estimators" has a different meaning. (See Weiss and Wolfowitz, 1970.) We can now formulate the method of maximum likelihood in general.

Definition: *Let X have density $f(x;\theta)$ for $x \in R^m$, $\theta \in \Theta \subseteq R^d$; let $X_1, \cdots, X_n$ be a random sample "of X " . The joint density of the X_i's, as a function of θ , is called the likelihood function:*

$$L_n(x_1,\cdots,x_n;\theta) = \prod_{i=1}^{n} f(x_i;\theta) .$$

The maximum likelihood estimate (MLE) of θ is defined to be a value $\hat{\theta}_n(x_1,\cdots,x_n)$ (if it exists) for which

$$L_n(x_1,\cdots,x_n;\hat{\theta}_n) = \underset{\theta \in \Theta}{Max}\ L_n(x_1,\cdots,x_n;\theta).$$

Note: although the practice abuses notation, it customary to write $\hat{\theta}_n$ for both the RV $\hat{\theta}_n(X_1,\cdots,X_n)$ and its value; in the same vein, we may write $L_n(\theta)$ or even L_n for the likelihood.

Example: Let $f(x;\theta) = \frac{1}{\theta} I(0 \le x \le \theta)$ for $\theta > 0$. Then,

$$L_n(x_1,\cdots,x_n;\theta)=\theta^{-n}\prod_{i=1}^{n} I(0\le x_i\le\theta)\,.$$

As a function of θ, L_n is maximal when θ^n is minimal; for a given $x_1,\cdots,x_n$, this occurs when

$$\theta = x_{(n)} = \max\{x_1,\cdots,x_n\}\,.$$

Then, $\hat{\theta}_n = X_{(n)} = \mathrm{Max}\{X_1,\cdots,X_n\}$ is unique and is a sufficient statistic.

The *log-likelihood function* $\log L_n(x_1,\cdots,x_n;\theta)$ can be used to determine a MLE $\hat{\theta}_n$ since $\hat{\theta}_n$ maximizes L_n iff $\hat{\theta}_n$ maximizes $\log L_n$. Unfortunately, it can happen that a MLE does not exist.

Example: (This estimation problem in survival analysis has been discussed by a number of authors including Nguyen, Rogers, and Walker, 1984.) Let X have density

$$f(x;\theta) = ae^{-ax}I(0\le x\le\tau)$$
$$+\,be^{-a\tau-b(x-\tau)}I(x>\tau)$$

where $x \in R$, $\theta = (a,b,\tau) \in \Theta = (0,\infty)^3$. Let the RV $W(\tau)$ be the number of $X_i \le \tau$. The log likelihood

$$\log L_n = \sum_{i=1}^{n} I(0\le X_i\le\tau)\log a - a\sum_{i=1}^{n} X_iI(0\le X_i\le\tau)$$
$$+\sum_{i=1}^{n}[1-I(0\le X_i\le\tau)][\log b-(a-b)\tau]$$
$$-\,b\sum_{i=1}^{n} X_i[1-I(0\le X_i\le\tau)]\,.$$

can be put in terms of $W = W(\tau)$ and $X_{(1)},\cdots,X_{(n)}$ as

$$W \log a - a\sum_{i=1}^{W} X_{(i)} + (n - W)[\log b - (a\text{-}b)\tau]$$
$$- b\sum_{i=W+1}^{n} X_{(i)}]/n!\,. \qquad (*)$$

For $X_{(n-1)} \le \tau < X_{(n)}$, the right hand side of (*) is, effectively,

$$(n-1)\log a - a\sum_{i=1}^{n-1} X_{(i)} - a\,\tau + \log b - b(X_{(n)} - \tau)\,.$$

If we take $\hat{b} = 1/(X_{(n)} - \hat{\tau}) > 0$ and let $\hat{\tau}$ get close to $X_{(n)}$, $\log L_n$ will tend to ∞. Obviously, an unbounded function cannot have a max.

If the parameter space is changed to have $a > b$, then $\log L_n$ will be bounded but since Θ is not bounded, $\sup L_n$ might not be achieved.

The following example shows that the MLE might not be unique.

Example: Let X be uniformly distributed on $[\theta, \theta + 2]$ for $\theta > 0$. The likelihood function is

$$L_n = \begin{cases} 2^{-n} & \text{if } \theta \le x_{(1)} \le x_{(n)} \le \theta + 2 \\ 0 & \text{otherwise} \end{cases}$$

and the support is equivalent to $x_{(n)} - 2 \le \theta \le x_{(1)}$. From this, it can be seen that any value of θ between $x_{(n)} - 2$ and $x_{(1)}$ will give the same value 2^{-n} of L_n; in other words, $\hat{\theta}_n$ can be any number in $(x_{(n)} - 2, x_{(1)})$.

The reader might be tempted to start "differentiating" to find max/min. The last two examples should suggest that this is not always possible; one needs to examine boundary points as well. On the other hand, for "interior stationary points", one can use the classical test based on second or higher order derivatives. (See e.g. Fleming, 1977.) *The* max can then be obtained from all the local maxima.

Example: Consider the Bernouilli model with $\theta \in [0,1]$.

$$L_n(\theta) = \begin{cases} 0 & \text{if } \theta = 0 \text{ or } 1 \\ \theta^n & \text{if } 0 < \theta < 1 \ , \ \Sigma x_i = 0 \\ \theta^{\Sigma x_i}(1-0)^{n-\Sigma x_i} & \text{if } 0 < \theta < 1 \ , \\ & \qquad 1 \le \Sigma x_i < n \\ (1-\theta)^n & \text{if } 0 < \theta < 1 \ , \ \Sigma x_i = n \end{cases}$$

For the third part of this L_n ,

$$\log L_n(\theta) = (\Sigma x_i)\cdot \log \theta + (n - \Sigma x_i)\cdot \log(1-\theta) \ ,$$

$$\frac{\partial}{\partial\theta} \log L_n(\theta) = \frac{\Sigma x_i}{\theta} - \frac{n - \Sigma x_i}{1-\theta} \ ,$$

and the log–likelihood equation is

$$\frac{\Sigma x_i}{\theta} - \frac{n - \Sigma x_i}{1-\theta} = 0 \ . \qquad (*)$$

Its solution is $\hat{\theta}_n = \frac{1}{n}\sum_{i=1}^{n} x_i$.

Here $\hat{\theta}_n$ is indeed the point of maximum since it is the only candidate and at the boundaries

$$L_n(0) = L_n(1) = 0 < L_n(\hat{\theta}_n) \ .$$

Moreover, except at the boundaries,

$$\frac{\partial^2}{\partial\theta^2} \log L_n(\theta) = -\frac{\Sigma x_i}{\theta^2} - \frac{n - \Sigma x_i}{(1-\theta)^2} < 0 \ .$$

Exercise 6: Continue this example. Find the log–likelihood equation and discuss its solution for the other values of L_n .

Exercise 7: Find the MLE of θ in the following models.

a) $f(x;\theta) = \theta e^{-\theta x} I(x > 0)$ for $\theta > 0$.

b) $f(x;\theta)$ in "survival analysis" above with τ known.

c) $f(x;\theta)$ is Poisson.

LESSON 7. MAXIMUM LIKELIHOOD ESTIMATION

In this lesson, we continue the discussion of MLE introduced in Lesson 6; in particular, we examine its relations with sufficient statistics and invariance. General conditions for existence and uniqueness of MLE can be found in Makelainen, Schmidt and Styan, 1981. We begin with an

Example: The order statistic $(X_{(1)}, X_{(n)})$ is a sufficient statistic for θ in the model $f(x;\theta) = 1/2 \cdot I(-\theta \leq x \leq \theta + 2)$. Any value in the interval $[X_{(n)} - 2, X_{(1)}]$ is a MLE (Lesson 6). In particular, $\hat{\theta} = X_{(1)}$ is a MLE but is not sufficient by itself. However, this $\hat{\theta}$ is a function of the sufficient statistic.

Loosely speaking, the functional relation between MLE and sufficient statistics noted in this example is common but the precise relation is as follows.

Theorem: *Let* $X_1, \cdots, X_n$ *be a RS from a model having a sufficient statistic. If the MLE exists and is unique, then it is a function of any sufficient statistic.*

Proof. Let T be a sufficient statistic; then by the factorization theorem (with $x = (x_1, \cdots, x_n)$), $f(x;\theta) = h(x)g_\theta(T(x))$.

First observe that the MLE $\hat{\theta}(x)$ is the value of θ which maximizes $g_\theta(T(x))$. For each value t in the range of T, consider the function $\theta \to g_\theta(t)$. If there is a value t in the range of T such that this function has, say, two maxima, at $\theta_1 \neq \theta_2$, then, for $x \in T^{-1}(t)$, the values $\hat{\theta}(x) = \theta_1$ and $\tilde{\theta}(x) = \theta_2$ would be two different MLEs contray to hypothesis.

To show that $\hat{\theta}$ is necessarily a function of T, it suffices to verify that for all x and y,

$$\text{if } T(x) = T(y) = t, \text{ then } \hat{\theta}(x) = \hat{\theta}(y).$$

Now, $x, y \in T^{-1}(t)$, and $\hat{\theta}(x), \hat{\theta}(y)$ both maximize the function $\theta \longrightarrow g(\theta,t)$. So by the previous observation, this function has a unique maximum and $\hat{\theta}(x) = \hat{\theta}(y)$.

The following exercise and example (Martin, 1985) show that the requirement of the uniqueness of the MLE in this theorem is essential.

Exercise 1: Consider a RS of size one for the discrete RV X with density

$$f(x;\theta) = \begin{cases} 1/4 & \text{if } x = 1, 2 \\ (1+\theta)/4 & \text{if } x = 3 \\ (1-\theta)/4 & \text{if } x = 4 \end{cases} \qquad \text{for } \theta \in [0,1] .$$

Let a and b be arbitrary numbers in [0, 1]. Complete the table to show that the function

$$\delta(1) = a, \delta(2) = b, \delta(3) = 1, \delta(4) = 0$$

is a MLE.

x	f(x ; a)	f(x; b)	f(x; 1)	f(x;0)
1	1/4		1/4	
2	1/4	1/4		
3	(1+a)/4		1/2	
4	(1−a)/4		0	
max	(1+a)/4	(1+b)/4	1/2	1/4

Thus, there are infinitely many MLE $\hat{\theta}$ = some δ .

Example: Continuing exercise 1, let T be a statistic with three distinct values: $T(1) = T(2) = t_1$, $T(3) = t_3$, $T(4) = t_4$.

T will be sufficient for θ if $f(x;\theta) = h(x)g_\theta(T(x))$ (Lesson 7, Part IV). In fact, for $h(x) \equiv 1$,

$$f(x;\theta) = g_\theta(T(x))$$

where $g_\theta(t) = 1/4$ for $t = t_1$ $(x = 1, 2)$
$= (1+\theta)/4$ for $t = t_3$ $(x = 3)$
$= (1-\theta)/4$ for $t = t_4$ $(x = 4)$.

To show that T is minimal sufficient, we use the last lemma in Lesson 8, Part IV:
if T is sufficient and for each x and y $\in \{1, 2, 3, 4\}$,

$$f(x;\theta)= k(x,y)f(y;\theta) \text{ for some function } k(x,y) > 0 \qquad (*)$$

implies $T(x) = T(y)$, then T is minimal.

Here we have $f(x;\theta) = f(y;\theta) = 1/4$ for x and y $\in \{1, 2\}$

$$f(x;\theta) = f(y;\theta) = (1+\theta)/4 \text{ for } x = y = 3$$

$$f(x;\theta) = f(y;\theta) = (1-\theta)/4 \text{ for } x = y = 4 .$$

There are no other pairs satisfying (*) . Obviously, $T(x) = T(y)$ for these six pairs. Therefore, T is minimal.

Finally, if δ is a function of T , then

$$T(x) = T(y) \text{ must imply } \delta(x) = \delta(y) .$$

But for $a \neq b$, this is impossible.

Under the conditions of the last theorem, if a MLE exists, it will also maximize $g_\theta(T(x))$ and hence it will depend on T in some fashion. If only the existence and *not* the uniqueness is assumed, then there is only a *relation* between $\hat{\theta}(x)$ and $T(x)$ described as follows:

for each t , $R(\theta,t) = 1$ if $\theta \in M_t$, the set of maximum points of the function $\theta \to g_\theta(t)$;

in general, $\hat{\theta}(x) \in M_{T(x)}$. When uniqueness is assumed, R is a function.

For ease of reference here (or see Fleming, 1977), we state a second derivative test for max/min.

f is a real valued function on an open subset ϕ of R^d . The gradient $\nabla f = \frac{\partial f}{\partial \varphi}$ is the column vector of partial derivatives $\partial f/\partial\varphi_1, \partial f/\partial\varphi_2, \cdots, \partial f/\partial\varphi_d$. The Hessian is the matrix of second order partial derivatives $H(\varphi) = \dfrac{\partial^2 f}{\partial\varphi\, \partial\varphi'} = \left[\dfrac{\partial^2 f}{\partial\varphi_i\, \partial\varphi_j}\right]$.

Suppose $f \in C^2$, that is, the second order partial derivatives are continuous. Let $\partial f/\partial\varphi = 0$ at some point φ^0 interior to ϕ. Then, $f(\varphi^0)$ is a local max [min] when $H(\varphi^0)$ is negative [positive] definite: $\alpha' H(\varphi^0)\alpha < 0$ for all $\alpha \neq 0$
$[\alpha' H(\varphi^0)\alpha > 0$ for all $\alpha \neq 0]$.

Exercise 2: Write out this test for d = 1 and use that to show
$$x^2 - 1 - \log(x^2) \geq 0 \text{ for } x > 0 .$$

Example: Consider a RS of size n from the model $N(\mu, \sigma^2)$.

a) $\log L_n(\theta) =$
$$(-n/2)\log 2\pi - (n/2)\log \sigma^2 - \sum_{i=1}^{n} (x_i - \mu)^2/2\sigma^2 .$$
$$\frac{\partial L_n}{\partial \mu} = -\sum 2(x_i - \mu)(-1)/2\sigma^2 = \Sigma(x_i - \mu)/\sigma^2 .$$
$$\frac{\partial L_n}{\partial \sigma^2} = -n/2\sigma^2 + \sum (x_i - \mu)^2/2\sigma^4 .$$

The likelihood equation $\begin{bmatrix} \partial L_n/\partial\mu \\ \partial L_n/\partial\sigma^2 \end{bmatrix} = 0$ becomes the pair of equations
$$\frac{\Sigma(x_i - \hat{\mu})}{\hat{\sigma}^2} = 0 , \quad \frac{n}{\hat{\sigma}^2} = \frac{\Sigma(x_i - \hat{\mu})^2}{\hat{\sigma}^4}$$
with solution $\hat{\mu} = \Sigma x_i/n = \bar{x}$ $\quad \hat{\sigma}^2 = \Sigma(x_i - \bar{x})^2/n$.

Moreover, $\log L_n(\hat{\mu},\hat{\sigma}^2)$
$$= -(n/2)\log 2\pi - (n/2)\log \hat{\sigma}^2$$
$$- \Sigma(x_i - \hat{\mu})^2/2\Sigma(x_i - \hat{m})^2/n$$
$$= -(n/2)\log 2\pi - (n/2)\log \hat{\sigma}^2 - n/2 .$$

b) In this case, we can verify the maximality by doing some algebra. Note that for any c free of "i",

$$\Sigma(X_i - c)^2 = \Sigma(x_i - \bar{x} + \bar{x} - c)^2$$
$$= \Sigma(x_i - \bar{x})^2 + \Sigma(\bar{x} - c)^2$$

because the sum of the crossproducts is

$$2\Sigma(x_i - \bar{x})(\bar{x} - \mu) = 0 .$$

(Why?) Then, $\log L_n(\hat{\mu},\hat{\sigma}^2) - \log L_n(\mu,\sigma^2)$

$$= -(n/2)\log\frac{\hat{\sigma}^2}{\sigma^2} - (n/2) + \Sigma(x_i - \mu)^2/2\hat{\sigma}^2$$

$$= -(n/2)\log\frac{\hat{\sigma}^2}{\sigma^2} - (n/2) + \frac{\Sigma(x_i - \bar{x})^2 + n(\bar{x} - \mu)^2}{2\hat{\sigma}^2}$$

$$= -(n/2)\log\frac{\hat{\sigma}^2}{\sigma^2} - (n/2) + (n/2)\frac{\hat{\sigma}^2}{\sigma^2} + (n/2)(\bar{x} - \mu)^2/\sigma^2$$

$$= (n/2)\left[\frac{\hat{\sigma}^2}{\sigma^2} - 1 - \log\frac{\hat{\sigma}^2}{\sigma^2}\right] + (n/2)(\bar{x} - \mu)^2/\sigma^2$$

which is non–negative by exercise 2.

c) On the other hand for the Hessian, we need

$$\frac{\partial^2 L_n}{\partial\mu^2} = -n/\sigma^2 , \quad \frac{\partial^2 L_n}{\partial\sigma^2\,\partial\mu} = -\Sigma(x_i - \mu)/\sigma^4$$

$$\frac{\partial^2 L_n}{\partial(\sigma^2)^2} = n/2\sigma^4 - \Sigma(x_i - \mu)^2/\sigma^6$$

evaluated at $\hat{\mu}$, $\hat{\sigma}^2$. Then, the Hessian is

$$\begin{bmatrix} -1/\Sigma(x_i - \bar{x})^2 & 0 \\ 0 & -n^3/2\Sigma(x_i - \bar{x})^2 \end{bmatrix}$$

and this is obviously negative definite. Thus by either criterion, $\hat{\mu}$ and $\hat{\sigma}^2$ are indeed the MLEs.

Exercise 3: Find the MLEs for θ_1 and θ_2 in the multinomial density

$$\frac{n!}{x_i!\ x_2!\ x_3!}(\theta_1)^{x_1}(\theta_2)^{x_2}(1-\theta_1-\theta_2)^{n-x_1-x_2}$$

where x_1, x_2, x_3 are non–negative integers with sum n and $\theta_1, \theta_2 \in (0,1)$. Include the evaluation of the Hessian. (Ignore boundary values.)

The most used property of MLE is one of *invariance*. For this proof, we take a slightly more general definiton of "max". The function defined therein which takes γ into B_γ is sometimes called the *induced likelihood*.

Theorem: *If* $\hat{\theta} \in \Theta$ *is a MLE of* θ, *then* $\varphi(\hat{\theta})$ *is a MLE of the function with values* $\varphi(\theta)$.

Proof: The function $\varphi : \Theta \to \Gamma$ partitions Θ into orbits $\varphi^{-1}(\gamma)$ for $\gamma \in \Gamma$; hence, Θ is the union over all $\gamma \in \Gamma$ of $A_\gamma = \{\theta \in \varphi^{-1}(\gamma)\}$. For all θ_1 and θ_2 in A_γ, $\varphi(\theta_1) = \varphi(\theta_2)$. Then, $L(\hat{\theta})$, which is the "sup" of $L(\theta)$ over all $\theta \in \Theta$, can be obtained as a sup over Γ of the sup over A_γ; let

$$B_\gamma = \sup_{\theta \in A_\gamma} L(\theta) ;$$

that is, $L(\hat{\theta}) = \sup_{\theta \in \Theta} L(\theta) = \sup_{\gamma \in \Gamma} \{\sup_{\theta \in A_\gamma} L(\theta)\} = \sup_{\gamma \in \Gamma} B_\gamma$.

Let $\hat{\gamma} = \varphi(\hat{\theta})$ so that $\hat{\theta} \in \varphi^{-1}(\hat{\gamma})$. Since $\sup_{\gamma \in \Gamma} B_\gamma \geq B_\gamma$ for all $\gamma \in \Gamma$ and $\hat{\gamma} = \varphi(\hat{\theta})$ is just one point in Γ,

$$\sup_{\gamma \in \Gamma} B_\gamma \geq B_{\hat{\gamma}} = \sup_{\theta \in A_{\hat{\gamma}}} L(\theta) = L(\hat{\theta}) .$$

It follows that $L(\hat{\theta}) = B_{\hat{\gamma}}$.

Example. Let $Y_1, \cdots, Y_n$ be a RS from the log–normal distribution (Lesson 6). Here $\theta = (\mu, \sigma^2) \in \Theta = R \times (0, \infty)$. Take $\varphi : \Theta \rightarrow \Gamma = (0, \infty)^2$ to be defined by

$$\varphi(\theta) = \left[\exp\{\mu + \frac{\sigma^2}{2}\}, (e^{\sigma^2} - 1)\exp\{2\mu + \sigma^2\}\right] .$$

The RVs $X_i = \log Y_i$, $i = 1(1)n$, are a random sample from $N(\mu, \sigma^2)$ so that from a previous example, the MLE of θ is $\hat{\theta} = (\hat{\mu}, \hat{\sigma}^2)$ where

$$\hat{\mu} = \bar{x} \text{ and } \hat{\sigma}^2 = \Sigma(x_i - \bar{x})^2 / n .$$

Application of the last theorem gives the MLE of $\varphi(\theta)$ as

$$\hat{\varphi}(\theta) = \left[\exp\{\bar{x} + \tfrac{1}{2}\hat{\sigma}^2\}, (e^{\hat{\sigma}^2} - 1)\exp\{2\bar{x} + \hat{\sigma}^2\}\right] .$$

Exercise 3: Let X be Bernoulli with $\Theta = [0,1]$. From a random sample $X_1, \cdots, X_n$, find the MLE of the variance of X.

Finally, we investigate the efficiency of a MLE. We know that for $f(x;\theta) = (2\pi)^{-1/2}\exp\{-\tfrac{1}{2}(x - \theta)^2\}$, the estimator $\bar{X}$ of

θ is efficient, i.e., $\mathrm{Var}(\bar{X}) = \frac{1}{n} =$ the information lower bound (Lesson 4). From an earlier example in this lesson, we find that that $\bar{X}$ is the MLE of θ. Thus there is at least one MLE which is an efficient estimator. The following result generalizes this example in a simple case.

Theorem: *Under the regularity conditions of lesson 4, and for Θ an open subset of R, if there exists an unbiased efficient estimator T, it is given by the method of maximum likelihood.*

Proof: The information inequality is an equality when

$$\frac{\partial}{\partial\theta}\log L_n(x_1,\cdots,x_n;\theta) = k(\theta)[T - \theta] \quad \text{(with probability one)} .$$

Then, $T(x_1,\cdots,x_n)$ is a solution of the likelihood equation:

$$\frac{\partial}{\partial\theta}\log L_n = 0 .$$

The regularity conditions allow us to get

$$\frac{\partial^2}{\theta\partial^2}\log L_n = \frac{\partial k(\theta)}{\partial\theta}\cdot[T - \theta] - k(\theta) .$$

At $\theta = T$, this is $-k(\theta)$ which is negative definite (by the next exercise).

Exercise 4: Under the regularity conditions, show that $k(\theta) = n\cdot I(\theta)$ where $k(\theta)$ is the quantity considered in the proof above.

LESSON 8. CONFIDENCE INTERVAL ESTIMATION–I

Consider a statistical model

$$\mathcal{F} = \{f(x;\theta),\ x \in \mathcal{X} \subseteq R^m,\ \theta \in \Theta \subseteq R^d\},$$

and a parameter of interest $\varphi(\theta)$. In previous lessons, we have investigated a few methods of finding point estimators of $\varphi(\theta)$, i.e. a value of some statistic $T(X_1,\cdots,X_n)$ as an estimate for $\varphi(\theta)$. In practical applications, point estimation is not enough.

A very specific example is the manufacture of transistors whose resistances have to be established within certain limits rather than just "on the average". In statistical terms, this means finding $L(X)$ and $V(X)$ so that the interval $(L(X), V(X))$ contains the true parameter with high probability. More generally, we seek a set $C(X)$ which contains the true parameter. Now we are trying to build *set-estimators.*

Roughly speaking, a set–estimator of $\varphi(\theta)$ is a set depending upon the random sample $X_1, \cdots, X_n$, say $C(X_1,\cdots,X_n)$. For example, for the mean θ, $C(X_1,\cdots,X_n)$ might be $[\bar{X}_n - a, \bar{X}_n + a]$ for some $a \in R$. Such a set–estimator is usually referred to as a

confidence interval (C. I.)

for the true value θ_o; (in higher dimensions we get a *confidence region*). The "confidence" is defined to be the probability of coverage, i.e.

$$P(\theta_o \in C(X_1,\cdots,X_n)).$$

This is to be read and remembered as the probability that the set contains the parameter not the other way around.

Definition: *The random set* $C(X_1,\cdots,X_n)$ *is a confidence set for* $\varphi(\theta)$ *at confidence level* $1 - \alpha \in (0,1)$, *if for any* $\theta \in \Theta$,

$$P_\theta(\varphi(\theta) \in C(X_1,\cdots,X_n)) \geq 1 - \alpha.$$

In particular, if $\varphi : \Theta \rightarrow R$ *and* $C(X_1,\cdots,X_n)$ *is a random interval* $[T_1(X_1,\cdots,X_n), T_2(X_1,\cdots,X_n)]$ $(T_1 \le T_2$ *a.s.) such that* $P_\theta(T_1 \le \varphi(\theta) \le T_2) \ge 1 - \alpha$ *for all* $\theta \in \Theta$, *then* $[T_1, T_2]$ *is called a* $100(1 - \alpha)\%$ *confidence interval for* $\varphi(\theta)$.

Example: Consider the normal model $N(\theta,1)$, $\theta \in R$. Let the CDF be ϕ . For $\alpha \in (0,1)$, let $z_{\alpha/2}$ be such that

$$\phi(z_{\alpha/2}) = 1 - \alpha/2 .$$

Since $\bar{X}_n$ is distributed as $N(0, \frac{1}{n})$, $\sqrt{n}(\bar{X}_n - \theta)$ is $N(0,1)$. Hence,

$$P_\theta[-z_{\alpha/2} \le \sqrt{n}(\bar{X}_n - \theta) \le z_{\alpha/2}] = 1 - \alpha$$

for all $\theta \in R$. This says that $[\bar{X} - \frac{1}{\sqrt{n}} z_{\alpha/2}, \bar{X} + \frac{1}{\sqrt{n}} z_{\alpha/2}]$ is a $100(1 - \alpha)\%$ confidence interval for the mean θ .

In the example, the length of the confidence interval is

$$L(n, \alpha) = \frac{2}{n} z_{\alpha/2}$$

which is a non–increasing function of α for fixed n . That is, if we increase the confidence $(1 - \alpha)$ then the interval will become larger. Of course, for fixed α, $L(n, \alpha)$ decreases as n increases.

Exercise 1: Consider the normal model $N(\mu, \sigma^2)$.

a) If σ^2 is known, find a $100(1 - \alpha)\%$ confidence interval for μ .

b) If σ^2 is unknown, find a $100(1 - \alpha)\%$ confidence interval for μ . Hint: if $S^2 = \frac{1}{n-1}\sum_{i=1}^{n}(X_i - \bar{X})^2$,

then $\dfrac{\bar{X} - \mu}{S/\sqrt{n}}$ is distributed as Student T with $n–1$ dof and $P_\theta(T \ge t_{\alpha/2}(n-1)) = \alpha/2$.

c) Find a $100(1-\alpha)\%$ confidence interval for σ^2 (μ unknown). Hint: $\dfrac{(n-1)S^2}{\sigma^2}$ is $\chi^2(n-1)$.

d) Suppose that μ is known. Show that

$$\left[\frac{\sum_{i=1}^{n}(\bar{X}_i-\mu)^2}{\chi^2_{\alpha/2}(n)}, \frac{\sum_{i=1}^{n}(X_i-\mu)^2}{\chi^2_{1-\frac{\alpha}{2}}(n)}\right]$$

is a $100(1-\alpha)\%$ confidence interval for σ^2. Here, $P[\chi^2(n) > \chi^2_{\alpha/2}(n)] = \frac{\alpha}{2}$ and

$$P[\chi^2(n) > \chi^2_{1-\frac{\alpha}{2}}(n)] = 1-\frac{\alpha}{2}.$$

Hint: $\sum_{i=1}^{n}\left[\dfrac{X_i-\mu}{\sigma}\right]^2$ is $\chi^2(n)$.

Example: Consider the normal model $N(\mu, \sigma^2)$, $\theta = (\mu, \sigma^2)$. The sample mean $\bar{X} = \frac{1}{n}\sum_{i=1}^{n} X_i$ and the sample variance $S^2 = \frac{1}{n-1}\sum_{i=1}^{n}(X_i-\bar{X})^2$ are point estimators of μ and σ^2, respectively. Also,

$$\frac{\bar{X}-\mu}{S/\sqrt{n}} \text{ is } T(n-1) \text{ and } \frac{(n-1)S^2}{\sigma^2} \text{ is } \chi^2(n-1).$$

We seek a random rectangle in $\Theta = R \times (0, \infty)$ as a $100(1-\alpha)\%$ confidence region for θ. For any $\theta \in \Theta$,

$$P_\theta\left[\mu \in [\bar{X} - t_{\frac{\alpha}{4}}(n-1)\frac{S}{\sqrt{n}}, \bar{X} + t_{\frac{\alpha}{4}}(n-1)\frac{S}{\sqrt{n}}]\right] = 1-\frac{\alpha}{2},$$

and $P_\theta\left[\sigma^2 \in \left[\frac{(n-1)S^2}{\chi^2_{\frac{\alpha}{4}}(n-1)}, \frac{(n-1)S^2}{\chi^2_{1-\frac{\alpha}{4}}(n-1)}\right]\right] = 1 - \frac{\alpha}{2}$.

Let the set for "μ" be A and the set for "σ^2" be B so that

$$P_\theta(A) = 1 - \alpha/2 = P_\theta(B).$$

Then,

$$\begin{aligned} P_\theta(A \cap B) &= 1 - P((A \cap B)^c) \\ &= 1 - P(A^c \cup B^c) \\ &= 1 - P(A^c) - P(B^c) + P(A^c \cap B^c) \\ &\geq 1 - P(A^c) - P(B^c) = 1 - \alpha. \end{aligned}$$

Thus $A \cap B$ is a $100(1-\alpha)\%$ confidence region for the parameter $\theta = (\mu, \sigma^2)$.

Exercise 2: Let X be $N(\mu, \sigma^2)$ and Y be $N(\nu, \sigma^2)$ be independent populations. Given random samples $X_1, \cdots, X_n$ and $Y_1, \cdots, Y_m$ from these populations, find a $100(1-\alpha)\%$ confidence interval for $\mu_1 - \mu_2$.

Hint: $\bar{X}_n - \bar{Y}_m$ is $N(\mu - \nu, \sigma^2(\frac{1}{n} + \frac{1}{m}))$ and

$\frac{1}{\sigma^2}[\sum_{i=1}^{n}(X_i - \bar{X}_n)^2 + \sum_{j=1}^{m}(Y_j - \bar{Y}_m)^2]$ is $\chi^2(n+m-2)$.

Example: Let $X = (X_1, X_2)$ be a bivariate normal random vector with means $E[X_1] = \theta_1$, $E[X_2] = \theta_2$,

covariance $\sum = \begin{bmatrix} \sigma_1^2 & \rho\sigma_1\sigma_2 \\ \rho\sigma_1\sigma_2 & \sigma_2^2 \end{bmatrix}$, $Var(X_1) = \sigma_1^2$, $Var(X_2) = \sigma_2^2$

and correlation $\rho = cov(X_1, X_2)/\sigma_1\sigma_2$. Consider a random sample from X, i.e. n independent pairs

$$(X_{11},X_{21}), \cdots, (X_{1n},X_{2n}) .$$

We seek a confidence interval for $\theta_1 - \theta_2$. Note that $\theta_1 - \theta_2$ is the mean of the one–dimensional RV $D = X_1 - X_2$. This RV is normal and its variance is $\sigma_1{}^2 + \sigma_2{}^2 - 2\rho\, \sigma_1\sigma_2$. But, we also have a random sample from D , namely

$$D_1 = X_{11} - X_{21}, \cdots, D_n = X_{1n} - X_{2n} .$$

Thus if we denote by $\bar{D}$ and $S^2(D)$ the sample mean and the sample variance of this sample, we get a $100(1 - \alpha)\%$ conficence interval for $\theta_1 - \theta_2$ as:

$$[\bar{D} - t_{\alpha/2}(n-1)\frac{S(D)}{\sqrt{n}}, \bar{D} + t_{\alpha/2}(n-1)\frac{S(D)}{\sqrt{n}}] .$$

Exercise 3: Let X and Y be independent populations $N(\mu_1, \sigma_1{}^2)$ and $N(\mu_2, \sigma_2{}^2)$, respectively. Given two random samples $X_1, \cdots, X_n$ and $Y_1, \cdots, Y_m$ find a $100(1 - \alpha)\%$ confidence interval for the ratio $\sigma_1^2/\sigma_2{}^2$.

Hint: $\dfrac{S_X{}^2}{\sigma_1{}^2} / \dfrac{S_Y{}^2}{\sigma_2{}^2}$ is $F(n-1, m-1)$.

Example: Let X be Bernouilli with $\theta \in (0,1)$. A $100(1 - \alpha)\%$ confidence interval for θ can be constructed as follows. Find $a(X_1,\cdots,X_n)$ and $b(X_1,\cdots,X_n)$ such that $P(a < \theta < b) \geq 1 - \alpha$.

a) An estimator of θ is $\bar{X}_n$ with mean θ and variance $\theta(1-\theta)/n$.

b) From Chebyshev's inequality (Lesson 11, Part II),, we have first

$$P_\theta(|\bar{X}_n - \theta| < k\sqrt{(\theta(1-\theta)/n}\,) \geq 1 - 1/k^2 .$$

Substituting from $1/k^2 = a$ yields

$$P_\theta(|\bar{X}_n - \theta| < \sqrt{(\theta(1-\theta)/\alpha n)} \geq 1 - \alpha .$$

c) But the event $|\bar{X}_n - \theta| < \sqrt{\dfrac{\theta(1-\theta)}{\alpha n}}$ is realized if and only if

$$(1 + \frac{1}{\alpha n})\theta^2 - (2\bar{X}_n + \frac{1}{\alpha n})\theta + \bar{X}_n{}^2 < 0 .$$

Therefore a and b can be taken as the roots of the associated quadratic equation.

Exercise 4: Find a and b above.

LESSON 9. CONFIDENCE INTERVAL ESTIMATION–II

In the previous lesson, we were able to construct confidence intervals for $\varphi(\theta)$ by using a RV $T(X_1, \cdots, X_n; \theta)$ such that:

(i) the distribution of T was independent of θ;

(ii) the equation $t = T(X_1, \cdots, X_n; \theta)$ was solvable for θ for any t.

Recall the first example of Lesson 8.

$$T(X_1, \cdots, X_n; \theta) = \sqrt{n}(\bar{X}_n - \theta)$$

was distributed as $N(0,1)$ and for any $t \in R$,

$$\sqrt{n}(\bar{X}_n - \theta) = t \text{ implies } \theta = \bar{X}_n - t/\sqrt{n}\,.$$

For given t_1, t_2 such that $t_1 < T < t_2$, we can invert the inequality:

$$\bar{X}_n - \frac{t_2}{\sqrt{n}} < \theta < \bar{X}_n - \frac{t_1}{\sqrt{n}}\,.$$

We now describe this situation in general terms.

Definition: *A function $T(X_1, \cdots, X_n; \theta)$ is called a pivotal quantity if the distribution of T does not depend on θ.*

Example: Let $X_1, \cdots, X_n$ be a RS from a population with CDF $F(x;\theta)$ continuous in x. Set $U_j = F(X_j;\theta)$. Since the U_j's are IID, uniformly distributed over $(0,1)$, the random variable $T(X_1, \cdots, X_n; \theta) = \prod_{j=1}^{n} F(X_j;\theta)$ is a pivotal quantity.

Exercise 1: Verify the details of the example above.

Suppose now that we have such a pivotal quantity $T(X_1, \cdots, X_n; \theta)$. Can we actually pivot it? That is, if

t_1 and t_2 are such that $P(t_1 < T < t_2) = 1 - \alpha \in (0,1)$,

is the event $t_1 < T < t_2$ equivalent to

$$L(X_1,\cdots,X_n,t_1,t_2) < \varphi(\theta) < U(X_1,\cdots,t_1,t_2)\ ?$$

This will be true, like the normal case above, if $T(X_1,\cdots,X_n;\theta)$ is a continuous and a strictly monotone function of θ.

Example: In the second example of Lesson 11, we considered what is now seen to be a pivotal quantity for $\varphi(0) = \mu$:

$$T(X_1,\cdots,X_n;\theta) = \frac{\bar{X}_n - \mu}{S/\sqrt{n}}$$

is distributed as $T(n-1)$. As a function of μ, this pivotal quantity is strictly decreasing so that

$$t_1 < \frac{\bar{X} - \mu}{S/\sqrt{n}} < t_2 \text{ iff } \bar{X}_n - t_2\frac{S}{\sqrt{n}} < \mu < \bar{X}_n - t_1\frac{S}{\sqrt{n}}.$$

Even if, as above, we have a nice pivotal quantity so that

$$t_1 < T < t_2 \text{ iff } L < \varphi(\theta) < U,$$

another problem arises. There may be many points (t_1, t_2) such that $P(t_1 < T < t_2) = 1 - \alpha$; there is then generated a whole set of CIs for $\varphi(\theta)$, say $\{L(t_1, t_2), U(t_1, t_2)\}$, each with the same confidence level $1 - \alpha$. Since $[L, U]$ represents the precision of the estimation, it is reasonable to look for the narrowest interval. More precisely, we would like to choose (t_1, t_2) so that the length or at least the average of the random length $U - L$ is minimal.

Let F be the CDF of the pivotal quantity $T(X_1,\cdots,X_n;\theta)$ and let u, v be such that

$$F(v) - F(u) = P(u \le T \le v) = 1 - \alpha. \qquad (*)$$

A $100(1-\alpha)\%$ C.I. of $\varphi(\theta)$ is

$$[L(X_1,\cdots,X_n,u,v), U(X_1,\cdots,X_n,u,v)]$$

and the length of this interval is $\ell(X_1,\cdots,X_n,u,v) = U - L$. There are now two cases:

(i) if ℓ is non–random, i.e., a function of u, v only, then we want to find u, v satisfying (*) and minimizing $\ell(u, v)$;

(ii) if ℓ is random, then we want to choose u, v, satisfying (*) and minimizing the expected length

$$E_\theta\left[\ell(X_1,\cdots,X_n,u,v]\right].$$

Case (i)

Note that (*) forces v to be a function of u (or visa–versa). Take $\ell(u, v)$ as a function of u, say $\ell(u, v(u))$. Under appropriate derivative conditions, there will be a pair (u_o, v_o) giving rise to the shortest CI as a solution to the simultaneous equations

$$F'(v)\cdot dv/du - F'(u) = 0$$

$$\partial\ell/\partial u + (\partial\ell/\partial v)\cdot dv/du = 0 .$$

Example: Let X be $N(\theta,1)$. The CDF of the pivotal quantity

$$T = \sqrt{n}(\bar{X} - \theta) \text{ is } F(x) = \frac{1}{\sqrt{2\pi}}\int_{-\infty}^{x} e^{-\frac{t^2}{2}}\,dt .$$

Given $\alpha \in (0,1)$, the CI is $[\bar{X} - v/\sqrt{n}, \bar{X} - u/\sqrt{n}]$. We seek u, v in R such that

$$F(v) - F(u) = 1 - \alpha$$

and the length of $\ell(u, v) = \frac{1}{\sqrt{n}}(v - u)$ is the shortest.

For $v = v(u)$, $F'(v)\frac{dv}{du} - F'(u) = 0$

$$\frac{d\ell}{du} = \frac{dv}{du} - 1 = 0$$

leads to (u, v) as a solution of

$$F'(u) = F'(v) \text{ with } F(v) - F(u) = 1 - \alpha .$$

Since $F'(t) = \frac{1}{\sqrt{2\pi}}e^{-\frac{t^2}{2}}$ is symmetrical in $t \in R$, we have

$$v = -u = z_{\alpha/2} \quad \text{where} \quad P(N(0,1) \geq z_{\alpha/2}] = \frac{\alpha}{2}.$$

Thus the shortest C.I. for θ is (the one we already used)

$$[\bar{X} - z_{\alpha/2}\frac{1}{\sqrt{n}}, \bar{X} + z_{\alpha/2}\frac{1}{\sqrt{n}}].$$

Case (ii)

Exercise 2: Let X be $N(\mu, \sigma^2)$ with $\theta = (\mu, \sigma^2)$. Show that a $100(1-\alpha)\%$ C.I. for μ based upon the pivotal quantity

$$T(X_1,\cdots,X_n;\mu) = \sqrt{n}(\bar{X}_n - \mu)/S,$$

is given by (again an earlier result)

$$[\bar{X}_n - t_{\alpha/2}(n-1)\frac{S}{\sqrt{n}}, \bar{X} + t_{\alpha/2}(n-1)\frac{S}{\sqrt{n}}].$$

Example: In Exercise 2, the length ℓ is in fact random, namely

$$\ell = \frac{S}{\sqrt{n}}(v-u).$$

Then, $E_\theta[\ell] = E_\theta[\frac{(v-u)}{\sqrt{n}}S] = (v-u)E_\theta[\frac{S}{\sqrt{n}}]$. Since $E_\theta(\frac{S}{\sqrt{n}})$ is independent of u and v, from $\frac{d}{du}E_\theta[\ell] = 0$, we get again

$$\frac{dv}{du} = 1.$$

Let G be the central Student CDF of the pivotal quantity T; this distribution is, like the normal, symmetric about 0 so that $G'(v) = G'(u)$ leads to

$$v = -u = t_{\alpha/2}(n-1).$$

Thus the CI in exercise 2 is the (pivotal) one with the shortest expected length.

Exercise 3: Let X be $N(\mu, \sigma^2)$ with $\theta = (\mu, \sigma^2)$. Consider the pivotal quantity $T = (n-1)\frac{S^2}{\sigma^2}$ which is $\chi^2(n-1)$. Write down the system for the shortest $100(1-\alpha)\%$ C.I. for σ^2.

Exercise 4: Let X be $N(0, \sigma^2)$.

a) Show that $T = \frac{1}{\sigma^2}\sum_{i=1}^{n} X_i^2$ is a pivotal quantity.

What is its distribution?

b) Find the length and the expected length of a $100(1-\alpha)\%$ C.I. for σ^2 based upon T.

c) Find the shortest $100(1-\alpha)\%$ C.I. for σ^2.

In higher dimensions, a confidence region at level $1-\alpha$ is a *random set* $C(X_1, \cdots, X_n)$ such that for all $\theta \in \Theta$,

$$P_\theta[\theta \in C] \geq 1 - \alpha .$$

(It is possible to define a random set as a set–valued mapping thereby extending the notion of random vectors.) When C is a random set in R^d, the measure of C is defined to be the Lebesgue measure Λ_d of C in R^d. The random variable $\Lambda_d(C)$ has expectation

$$E_\theta[\Lambda_d(C(X_1, X_2, \cdots, X_n))]$$

and we are led to considering a confidence region C with smallest "volume". The following device, due to Robbins ,1944, is useful.

Let $\Theta \subset R^d$; let C be a random set in Θ. For $z \in R^d$, the one–point coverage function of C is defined as

$$\Pi_\theta(z) = P_\theta[z \in C] .$$

Then, under some measurability conditions, we have:

$$E_\theta[\Lambda_d(C(X_1, \ldots, X_n))] = \int_{R^d} \Pi_\theta(z)\, \Lambda_d(dz) .$$

See also works like Hooper, 1982.

Finally, it is possible to construct a C.I. by using appropriate statistics. Let $T(X_1, \cdots, X_n)$ be a statistic whose density $g(t, \theta)$ is known but depends on θ. Let G be the CDF of T. For $\alpha, \beta \in (0,1)$, let $u(\theta), v(\theta)$ be such that

$G(u(\theta)) = \alpha$, $F(v(\theta)) = 1 - \beta$, that is,

$$P_\theta(u(\theta) \le T \le v(\theta)) = 1 - \alpha - \beta .$$

Suppose that $u(\theta) \le T \le v(\theta)$ iff $V(T) \le \theta \le U(T)$; then $[V(T), U(T)]$ will be a $100(1 - \alpha - \beta)\%$ C.I. for θ.

For other examples and details, see lesson 20 of Part I.

LESSON 10. CONSISTENT ESTIMATORS

So far we have been concerned with properties of estimators for a fixed sample size n. In this lesson and the rest of Part V, we will investigate properties of a sequence of estimators

$$\{T_n, n \geq 1\}$$

where for each n, T_n is a statistic based on a random sample of size n, say, $X_1, \cdots, X_n$. Roughly speaking, we wish to consider properties of T_n when n is large. By abuse of language, some properties of $\{T_n\}$, are referred to as asymptotic properties of the point estimator T_n.

One property that is often desired is that as we take more observations, the value of T_n gets closer and closer to the true parameter value say $\varphi(\theta)$: "as $n \to +\infty$, T_n will tend to $\varphi(\theta)$". This desirable property of T_n is called a consistency property. But since $\{T_n, n \geq 1\}$ is a sequence of RV, there are different ways that this sequence can converge to $\varphi(\theta)$.

First, let us agree on the notation used. For a model

$$\mathcal{F} = \{f(x;\theta), x \in \mathcal{X} \subset R^m, \theta \in \Theta \subset R^d\},$$

the norm in R^d is $\|\theta\| = \left[\sum_{i=1}^{d} \theta_i^2\right]^{1/2}$ where

$$\theta = (\theta_1, \cdots, \theta_d).$$

For a random sample $X_1, \cdots, X_n$ from the model, P_θ^n stands for the probability measure with density $\prod_{i=1}^{n} f(x_i;\theta)$. The symbol E_θ denotes expectation when θ is used. P_θ^∞ refers to the product probability measure on the product space $\prod_{j=1}^{\infty} \mathcal{X}_j$, where $\mathcal{X}_j = \mathcal{X}$ for all $j \geq 1$. Note that $T_n(X_1, \cdots, X_n)$ can be regarded as a map of $\prod_{j=1}^{\infty} \mathscr{X}_j$ depending only on $(X_1,\cdots,X_n)$.

Definition: *A sequence of estimators* $\{T_n(X_1,\cdots,X_n),\ n \geq 1\}$ *of* $\varphi(\theta)$ *is:*

a) *weakly consistent if* T_n *converges in probability to* $\varphi(\theta)$*; i.e., for each* $\theta \in \Theta$ *and each* $\varepsilon > 0$*,*

$$\lim_{n \to +\infty} P_\theta^{\,n}[\|T_n - \varphi(\theta)\| > \varepsilon] = 0 .$$

b) *strongly consistent if* T_n *converges with probability one (or almost surely, a.s.) to* $\varphi(\theta)$ *; i.e.,*

for all $\theta \in \Theta$, $P_\theta^{\,\infty}[T_n \longrightarrow \varphi(\theta)] = 1$

or $T_n \longrightarrow \varphi(\theta)$, $P_\theta^{\,\infty}$*-a.s. as* $n \longrightarrow \infty$.

c) r^{th} *- mean consistent if for all* $\theta \in \Theta$*,*

$$\lim_{n \to \infty} E_\theta[\|T_n - \varphi(\theta)\|^r] = 0 .$$

In particular, if $r = 2$*, we say that* T_n *is consistent in quadratic mean (q.m.) or mean-square-error consistent.*

Since the different concepts of consistency are defined in terms of the different modes of convergence of sequences of RVs, the relations among these types are the same as those among the convergence concepts in Lesson 6 and 7, Part III.

All concepts of consistency express some form of "closeness" of the estimator T_n to the parameter $\varphi(\theta)$ when n is large. The weak consistency means that if n is large enough, T_n will be arbitrarily close to $\varphi(\theta)$ with high probability. Strong consistency implies weak consistency and states that, for n large, T_n will be in any neighborhood of $\varphi(\theta)$ with probability one. The rth mean consistency is also stronger than the weak consistency, and expresses, roughly, a form of closeness in terms of a familiar distance.

Example: Let $X_1, \cdots, X_n$ be a RS from a distribution with density $f(x;\theta)$, $x \in R$, $\theta \in \Theta \subseteq R$.

a) Let $\varphi(\theta) = E_\theta[X] = \int_R x\, f(x;\theta)dx$, which is assumed to be finite. By the strong law of large numbers (SLLN), the sample mean

$$T_n(X_1,\cdots,X_n) = \frac{1}{n}\sum_{i=1}^{n} X_i$$

is a strongly consistent estimator of $\varphi(\theta)$. (See Lesson 6, Part III).

b) Assume that

$$\varphi(\theta) = Var_\theta(X) = \int_R (x - E_\theta(X))^2 f(x;\theta)dx \text{ is finite.}$$

$$\text{Let } T_n(X_1,\cdots,X_n) = \frac{1}{n}\sum_{i=1}^{n} (X_i - \bar{X})^2$$

$$= \frac{1}{n}\sum_{i=1}^{n} X_i^2 - \bar{X}.$$

Again, by the SLLN applied to $X_1^2, \cdots, X_n^2$, we see that $\frac{1}{n}\sum_{i=1}^{n} X_i^2$ converges a.s. to $E_\theta[X^2]$; also, by a), $\bar{X} \rightarrow E_\theta[X]$, a.s. Thus by continuity, T_n is a strongly consistent estimator of

$$Var_\theta(X) = E_\theta[X^2] - (E_\theta[X])^2.$$

It also follows that $\frac{1}{n-1}\sum_{i=1}^{n} (X_i - \bar{X})^2$ is also a strongly consistent estimator of $Var_\theta(X)$.

Exercise 1: Prove the last statement about

$$S^2 = \frac{1}{n-1}\sum_{i=1}^{n} (X_i - \bar{X})^2 .$$

Exercise 2: Prove the following "continuity argument". Let h be a continuous function.

a) If $T_n \xrightarrow{P} \varphi(\theta)$, then $h(T_n) \xrightarrow{P} h[\varphi(\theta)]$.

b) If $T_n \xrightarrow{a.s.} \varphi(\theta)$, then $h(T_n) \xrightarrow{a.s.} h[\varphi(\theta)]$. (See also Exercise 9 of Lesson 8, Part II).

Exercise 3: Let X be a Bernouilli RV with $\theta \in (0,1)$. Show that $\frac{1}{n}\sum_{i=1}^{n} X_i$ is a strongly consistent estimator of θ. Find a strongly consistent estimator of $\mathrm{Var}_\theta(X)$.

Example: Let X have density $f(x;\theta)$ with finite variance. Since,

$$E_\theta[(\bar{X} - E_\theta[X])^2] = \mathrm{Var}_\theta(\bar{X}) = \frac{1}{n}\mathrm{Var}_\theta(X) \longrightarrow 0 \text{ as } n \longrightarrow \infty,$$

the sample mean is a consistent in q.m. estimator of its expectation, the population mean $E_\theta[X]$.

Exercise 4: Let T_n be an estimator of $\varphi(\theta)$.

a) Show that T_n is q.m. consistent if the bias and the variance of T_n both converge to zero.

b) Show that if T_n is unbiased, and $\mathrm{Var}_\theta(T_n)$ converges to zero, then T_n is weakly consistent.

The bias of T_n is $b(\theta,T_n) = E_\theta[T_n] - \varphi(\theta)$. If,

$$\lim_{n \to \infty} b(\theta,T_n) = 0 \text{ for each } \theta \in \Theta,$$

or equivalently, $\lim_{n \to \infty} E_\theta[T_n] = \varphi(\theta)$ for each $\theta \in \Theta$.

Then we say that the estimator T_n is *asymptotically unbiased.*

Exercise 5: Let F_θ be the CDF of X. For a random sample $X_1, \cdots, X_n$ of X, define the empirical CDF as

$$F_n(t) = \frac{1}{n}\sum_{i=1}^{n} I(X_i \le t) \text{ for each } t \in R .$$

Show that for each $t \in R$, $F_n(t)$ is a strongly consistent estimator of $F_\theta(t)$.

Next we investigate the consistency of some estimators obtained by the method of moments, simply and with the moments equation.

Consider the real RV X with density

$$f(x;\theta) , \theta \in \Theta \subseteq R^d .$$

Suppose $E_\theta[X^j]$ exists and is finite for $j = 1(1)d$. By the SLLN, the sample moments

$$m_j = \frac{1}{n}\sum_{j=1}^{n} X_i^j , j = 1(1)d,$$

are strongly consistent estimators of $E_\theta[X^j]$.

Theorem: *Consider the real RV X such that for $j = 1(1)d$, $E_\theta[X^j]$ has continuous first partial derivatives on Θ and the Jacobian of the transformation*

$h : \theta = (\theta_1,\cdots,\theta_d) \rightarrow (E_\theta[X^1],\cdots,E_\theta[X^d])$ is non-zero on Θ.

Suppose that with probability tending to one as $n \rightarrow \infty$, the moments equations

$$m_j = E_\theta[X^j] , j = 1(1)d$$

have a unique solution $T_n(X_1,\cdots,X_n)$. Then T_n is a strongly consistent estimator of θ .

Proof: Under the above assumptions, h has a local (continuous) inverse h^{-1} and so

$$T_n = h^{-1}(m_1,\cdots,m_d) \xrightarrow{a.s.} h^{-1}[h(\theta)] = \theta ,$$

as $n \rightarrow \infty$. Behind the simplicity here is the fact that

$(m_1, \cdots, m_d) \in h(\Theta)$ with probability tending to one.

Example: Let X be as in the first example. We need only to rearrange the conclusions to see that the moment estimator of the parameter $\theta = (\mu, \sigma^2)$ is

$$(\bar{X}, \frac{1}{n}\sum_{i=1}^{n}(X_i - \bar{X})^2)$$

and is strongly consistent.

Definition: *Let F be a CDF on R. For $p \in (0,1)$, the pth-quantile of F also defines*

$$F^{-1} : \xi_p = \inf\{x : F(x) \geq p\} = F^{-1}(p) .$$

The sample p^{th}-quantile based on a RS $X_1, \cdots, X_n$ is

$$\xi_{p,n} = \inf\{x : F_n(x) \geq p\} = F_n^{-1}(p)$$

where F_n denotes the empirical CDF. For $p = 1/2$, $\xi_{1/2}$ is called the median of F.

Theorem: *Let $X_1, \cdots, X_n$ be a random sample of X with a CDF F such that for $p \in (0,1)$, $p = F(\xi_p)$ has a unique solution ξ_p. Then the sample quantiles are strongly consistent estimators of the population quantiles.*

Proof: For any $\varepsilon > 0$, we have $F(\xi_p - \varepsilon) < p < F(\xi_p + \varepsilon)$. By Exercise 5 above, $F_n(x) \to F(x)$, a.s. for each x. (Since $F_n(x)$ is really a function on the probability space Ω, one should write $F_n(x)(\omega)$ but this is suppressed in the following.)

First we show that as $n \to \infty$,

$$P(F_m(\xi_p - \varepsilon) < P < F_m(\xi_p + \varepsilon), m \geq n) \to 1. \qquad (*)$$

Since $F_n(\xi_p \pm \varepsilon) \to F(\xi_p + \varepsilon)$ a.s., let

$$A_n = \{\omega : F(\xi_p - \varepsilon) - \varepsilon_1 < F_m(\xi_p - \varepsilon) < F(\xi_p - \varepsilon) + \varepsilon_1,\ m \ge n\},$$

$$B_n = \{\omega : F(\xi_p + \varepsilon) - \varepsilon_2 < F_m(\xi_p + \varepsilon) < F(\xi_p + \varepsilon) + \varepsilon_2,\ m \ge n\}.$$

Then, $P(A_n) \to 1$, $P(B_n) \longrightarrow 1$ imply $P(A_n \cap B_n) \longrightarrow 1$. (Exercise 6 below.) If $\omega \in A_n$ and ε_1 is small enough,

$$F_m(\xi_p - \varepsilon) < F(\xi_p - \varepsilon) + \varepsilon_1 < F(\xi_p) = p.$$

Also, if $\omega \in B_n$ and ε_2 is small enough,

$$F(\xi_p) < F(\xi_p + \varepsilon) - \varepsilon_2 < F_m(\xi_p + \varepsilon).$$

Thus if $\omega \in A_n \cap B_n$, then

$$\omega \in \{F_m(\xi_p - \varepsilon) < p < F_m(\xi_p + \varepsilon),\ m \ge n\}$$

and (*) follows.

Next, for any CDF G, $G^{-1}(t) = \inf\{y : G(y) \ge t\}$ and so (Exercise 7 below), $G(x) \ge t$ iff $x \ge G^{-1}(t)$. Hence,

$$F_m(\xi_p - \varepsilon) < p < F_m(\xi_p + \varepsilon) \text{ iff}$$

$$\{\xi_p - \varepsilon < F_m^{-1}(p) \le \xi_p + \varepsilon\} = \{\xi_p - \varepsilon < \hat{\xi}_{p,m} \le \xi_p + \varepsilon\}.$$

Thus, $[|\hat{\xi}_{p,m} - \xi_p| < \varepsilon,\ \text{all } m \ge n] \to 1$ as $n \longrightarrow +\infty$, i.e.

$$\hat{\xi}_{p,m} \longrightarrow \xi_p \quad \text{a.s.}$$

Exercise 6: Show that if, in general,

$$P(A_n) \longrightarrow 1 \text{ and } P(B_n) \longrightarrow 1 \text{ as } n \to \infty,$$

$$\text{then } P(A_n \cap B_n) \longrightarrow 1 \text{ as } n \longrightarrow \infty.$$

Exercise 7: Let G be a (one–dimensional) CDF. Show that

$$\text{for all } p \in (0,1),\ G(x) \ge p \text{ iff } x \ge G^{-1}(p).$$

Exercise 8: Show that $X_n \longrightarrow X$, a.s. iff for all $\varepsilon > 0$,

$$\lim_{n \to \infty} P(|X_m - X| < \varepsilon, \text{ all } m \geq n) = 1.$$

Exercise 9: Let X be distributed as $N(\mu, \sigma^2)$, $\theta = (\mu, \sigma^2)$. Let $\xi_p(\theta)$ and ρ_p denote the pth–quantiles of $N(\mu, \sigma^2)$ and $N(0,1)$, respectively.

a) Show that $\hat{\xi}_{p,n}$ is a strongly consistent estimator of $\xi_p(\theta)$.

b) Show that $\xi_p(\theta) = \rho_p \sigma + \mu$.

c) Estimate μ and σ^2 by the estimating equations

$$\hat{\xi}_{p,n} = \rho_p \sigma + \mu$$

$$\hat{\xi}_{q,n} = \rho_q \sigma + \mu \quad \text{where } p \neq q.$$

Show that the estimator of θ obtained in this way is strongly consistent.

LESSON 11. CONSISTENCY OF MAXIMUM LIKELIHOOD ESTIMATORS

As noted earlier, one of the most used methods of estimation is that of maximum likelihood; yet, for a fixed sample size n, there is no really satisfactory justification for using this method even though some plausibility can be found in discrete cases. There is a better foundation in the asymptotic theory, where, as we will show, in most practical cases, a MLE exists and is consistent.

A complete analysis of such topics is very lengthy so that for simplicity and clarity, we consider the case where $f(x;\theta)$ is a density with x, θ in R. "Identifiability" is tacitly assumed: the map $\theta \longrightarrow f(\cdot, \theta)$ is injective (one–to–one), i.e. different values of θ yield different densities. A simple counter–example is

$$\frac{1}{\sqrt{2\pi}} \exp\{-\tfrac{1}{2}(x-\theta^2)^2\} ;$$

θ is not identifiable since $\theta = 1$ and $\theta = -1$ give the same density.

In order to prove the main theorem of this lesson, we need a result known as Jensen's inequality which involves notions of convex sets and convex functions. (For details omitted in the following, see e.g. Fleming, 1977).

Definition. *Let f be a real-valued function defined on an open interval $A \subset R$. We say that f is convex if for all $x, y \in A$ and $\alpha \in [0,1]$,*

$$f(\alpha x + (1-\alpha)y) \le \alpha f(x) + (1-\alpha)f(y) .$$

f is said to be strictly convex if the above inequality is strict whenever $x \ne y$ and $\alpha \in (0,1)$. f is concave if $-f$ is convex.

Corollary: *If f is convex on A as in the definition, then f is continuous on A.*

Examples: a) $-\log(x)$ is convex on $(0, +\infty)$, in fact, strictly convex;

b) e^{tx} is convex on $(-\infty, +\infty)$;

c) for $p > 0$, $|x|^p$ is convex on $(-\infty, +\infty)$.

Theorem: *Let f be a differentiable function defined on an open* *interval A of R. Then,*

a) f *is convex on* A *if and only if* f' *is non-decreasing on* A;

b) f *is strictly convex on* A *if* f' *is increasing on* A.

Theorem: *Let* f *be a convex function on* R; *then* f *has a left and a right derivative at every* x.

The next two results are much used. For the second, we note that continuity of φ makes $\varphi(X)$ a random variable whenever X is a random variable.

Lemma: *Let* f *be a convex function on* R. *Then for each* $x_o \in R$, *there exits a number* $\Lambda(x_o)$ *such that for each* $x \in R$,

$$f(x) \geq f(x_o) + \Lambda(x_o)\cdot(x - x_o) .$$

If f is strictly convex, then this inequality is strict for $x \neq x_o$.

Proof: a) From the previous theorem, $x < x_o$ implies

$$\frac{f(x_o) - f(x)}{x_o - x} \leq D^- f(x_o)$$

whence $f(x) \geq f(x_o) + D^- f(x_o)(x - x_o)$; Similarly, $x > x_o$

implies $\dfrac{f(x) - f(x_o)}{x - x_o} \geq D^+ f(x_o)$ and

$$f(x) \geq f(x_o) + D^+ f(x_o)(x - x_o) .$$

Since $D^- f(x_o) \leq D^+ f(x_o)$, it follows that

$$f(x) \geq f(x_o) + D^+ f(x_o)(x - x_o) \text{ for all } x .$$

b) Suppose that f is strictly convex. Then for $h > 0$,

$\varphi(h) = \dfrac{f(x + h) - f(x)}{h}$ is strictly increasing.

It follows that for all $x \neq x_o$,

$$f(x) > f(x_o) + D^+f(x_o)\,(x - x_o)\,.$$

Lemma: *(Jensen's inequality) Let* φ *be a convex function on an open interval* D *. Let* X *be a RV taking values a.s. in* D *with* $E[|X|] < \infty$ *. Then,*

$$E[\varphi(X)] \geq \varphi(E[X])\,.$$

Furthermore, this inequality is strict if φ *is strictly convex and* X *is not degenerate.*

Proof: a) Take $x_o = E[X]$ in the previous Lemma applied to φ to get

$$\varphi(x) \geq \varphi(x_o) + \Lambda(x_o)\,(x - x_o) \text{ for all } x \in D\,.$$

Then, $E[\varphi(X)] \geq \varphi(x_o) + \Lambda(x_o)\cdot E[X - x_o] = \varphi(x_o) = \varphi[E(x)]$.

b) If φ is strictly convex, then

$$\varphi(x) > \varphi(x_o) + \Lambda\,(x_o)\,(x - x_o) \text{ for all } x \in D \text{ with } x \neq x_o\,.$$

Hence $E[\varphi(X)] > \varphi(x_o)$ since $P(X = x_o) < 1$.

We are now ready to prove the major

Theorem: *Let the model* $\mathscr{F} = \{f(x;\theta),\, x \in R,\, \theta \in \Theta \subseteq R\}$ *of* X *be such that :*

a) *The densities* $f(\cdot\,;\theta)$, $\theta \in \Theta$ *, have the same support;*
b) Θ *is an open interval of* R *;*
c) $f(x;\theta)$ *is continuous in* θ *for all* $x \in R$.

Then there exists a sequence of local maxima, $\hat{\theta}_n$, *of the likelihood* $L_n(\theta)$, *which converges almost surely to the true parameter* θ_o *as* $n \to \infty$.

Proof: a) Let $\varepsilon > 0$; we need to show that if n is sufficiently large, then there exists a local maximum $\hat{\theta}_n$ of $L_n(\theta)$ such that

$$\hat{\theta}_n \in (\theta_o - \varepsilon, \theta_o + \varepsilon) \text{ with probability one.}$$

For $x = (x_1, \cdots, x_n)$, let

$$L_n^*(x;\theta) = L_n^*(\theta) = \frac{1}{n}\sum_{i=1}^{n} \log\frac{f(x_i;\theta)}{f(x_i;\theta_o)}$$

$$= \frac{1}{n}\log L_n(\theta) - \frac{1}{n}\log L_n(\theta_o) .$$

It suffices to show that $L_n^*(\theta)$ has a sequence of local maxima converging a.s. to θ_o .

b) By the SLLN, for each $\theta \in \Theta$,

$$L_n^*(\theta) \to E_{\theta_o}[\log\frac{f(x;\theta)}{f(x;\theta_o)}] , P_{\theta_o}\text{-a.s.}$$

When $\theta \neq \theta_o$, Jensen's Inequality applied with the strictly convex function $-\log(x)$ yields

$$E_{\theta_o}[\log\frac{f(x;\theta)}{f(x;\theta_o)}] < \log[E_{\theta_o}(\frac{f(x;\theta)}{f(x;\theta_o)})] = 0 .$$

Let $\Gamma = \{\theta \in \Theta : \theta = \theta_0 \pm \frac{1}{k}$, for positive intergers k$\}$. Then for each $\theta \in \Gamma$, there is a set N_θ such that

$P_{\theta_o}[N_\theta^c] = 0$ and when $x \in N_\theta$,

$$L_n^*(x;\theta) \to E_{\theta_o}[\log \frac{f(x;\theta)}{f(x;\theta_0)}] < 0 . \tag{1}$$

Since Γ is countable, we have

$$P_\theta(N) = 0 \text{ where } N = \bigcup_{\theta \in \Gamma} N_\theta^c .$$

Then when $x \notin N$, (1) holds for any $\theta \in \Gamma$.

c) Let $\theta_1 = \theta_0 - \frac{1}{m}$ with $m > \varepsilon$, $\theta_2 = \theta_0 + \frac{1}{k}$ with $k > \frac{1}{\varepsilon}$. Then

$$\theta_o - \varepsilon < \theta_1 < \theta_o < \theta_2 < \theta_o + \varepsilon$$

and for $x \notin N$ but n large enough,

$$L_n^*(x;\theta_1) < 0, L_n^*(x;\theta_2) < 0, L_n^*(x,\theta_o) = 0 .$$

Since $\theta \to L_n^*(x;\theta)$ is continuous by hypothesis, this implies a $\hat{\theta}_n$ in $(\theta_1, \theta_2) \subset (\theta_o - \varepsilon, \theta_o + \varepsilon)$ which maximizes $L_n^*(x;\theta)$.

Exercise 1: Let $f(x;\theta) = (2\pi)^{-1/2}\exp\{-\frac{1}{2}(x-\theta)^2\}$, $x, \theta \in R$. Verify that this model satisfies the hypothesis of the last theorem.

Exercise 2: Let f and g be two strictly positive probability densities. Suppose that $\int_R |\log f(x)| f(x)dx < \infty$. Show that

$$\int_R [\log g(x)]f(x)dx \le \int_R [\log f(x)]f(x)dx$$

unless $g = f$ almost everywhere. Hint: for $x > 0$, $\neq 1$, $\log(x) < x - 1$.

Exercise 3: For a model satisfying the hypothesis of the last theorem, show that with probability tending to one as $n \longrightarrow \infty$, the likelihood function $L_n(X_1,\cdots,X_n;\theta)$ has a strict maximum at θ_o . Hint: show that for all $\theta \neq \theta_o$,

$$\lim_{n\to\infty} P_{\theta_o}[L_n(\theta) < L_n(\theta_o)] = 1.$$

Before imposing fancy conditions on the model $\mathcal{F}$ to obtain more precise results, let us look at the case where Θ is finite. Then we can see that if $\{T_n, n \ge 1\}$ is a sequence of estimators of θ such that $\lim_{n\to\infty} P_\theta(T_n = \theta) = 1$ for all $\theta \in \Theta$, then T_n is weakly consistent. Indeed, for all $\varepsilon > 0$,

$$P_\theta[|T_n - \theta| \le \varepsilon] \ge P_\theta[T_n = \theta] .$$

Conversely, if for all $\varepsilon > 0$, $P_\theta[|T_n - \theta| \le \varepsilon] \longrightarrow 1$ as $n \longrightarrow \infty$ for all $\theta \in \Theta$, then, since Θ is finite, by choosing ε small enough, we also have $P_\theta[T_n = \theta] \longrightarrow 1$.

Theorem: *When Θ is finite, the MLE of θ exists, and is unique with probability tending to one, and is weakly consistent.*

Proof: Let $\hat{\theta}_n$ maximize $L_n(x_1, \cdots, x_n; \theta)$ over Θ. As in Exercise 3, we have, for $\theta \neq \theta_o$,

$$P_{\theta_o}[L_n(\theta) < L_n(\theta_o)] \longrightarrow 1 \text{ as } n \longrightarrow \infty .$$

Let $A_n(\theta) = \{w : L_n(\theta) < L_n(\theta_o)]$, $\theta \neq \theta_o$. Since Θ is finite, we have, with $\Theta^* = \Theta - \{\theta_o\}$,

$$P_{\theta_o}[\cap_{\theta \in \Theta^*} A_n(\theta)] = 1 - P_{\theta_o}[\cup_{\theta \in \Theta^*} A_n^c(\theta)]$$

$$\geq 1 - \sum_{\theta \in \Theta^*} P_{\theta_o}[A_n^c(\theta)] \longrightarrow 1 \text{ as } n \longrightarrow \infty .$$

Thus $P_{\theta_o}[L_n(\theta) < L_n(\theta_o)$, for all $\theta \neq \theta_o] \longrightarrow 1$ as $n \longrightarrow \infty$; this means that $\hat{\theta}_n$ is unique and weakly consistent.

The consistency of MLE depends on the smoothness of the function $\theta \longrightarrow f(x;\theta)$ as the following example shows.

Example: Consider $f(x;\theta)$ with $\theta \in (0,1)$ and $x \in \{0,1\}$ where

$$f(x;\theta) = \begin{cases} \theta^x (1-\theta)^{1-x} & \text{if } \theta \text{ is irrational} \\ (1-\theta)^x \theta^{1-x} & \text{if } \theta \text{ is rational} \end{cases}$$

For each x in $\{0,1\}$, the function $\theta \longrightarrow f(x;\theta)$ is not continuous on $(0,1)$: given any $\theta' \in (0,1)$, $\lim_{\theta \to \theta'} f(x;\theta)$ does not exist because any neighborhood of θ' contains rational and irrational points.

For a random sample $X_1, \cdots, X_n$, the likelihood function

$$L_n(x_1,\cdots,x_n;\theta) = \begin{cases} \theta^{\Sigma x_i}(1-\theta)^{n-\Sigma x_i} & \text{if } \theta \text{ rational} \\ (1-\theta)^{\Sigma x_i}\,\theta^{n-\Sigma x_i} & \text{if } \theta \text{ irrational} \end{cases}$$

To find the MLE of θ, we proceed as follows.

For θ rational, $L_n(\theta) = \theta^{\Sigma x_i}(1-\theta)^{n-\Sigma x_i}$. Consider the function

$$\varphi(\theta) = \theta^{\Sigma x_i}(1-\theta)^{n-\Sigma x_i} \text{ defined on } (0,1).$$

Then $\varphi(\theta)$ is maximum when $\theta = \Sigma x_i/n$ (which is rational) by differentiation. Thus $\hat{\theta}_n = \Sigma x_i/n$ also maximizes $L_n(\theta)$ when θ is rational, and

$$L_n^*(\hat{\theta}_n) = \left[\frac{\Sigma x_i}{n}\right]^{\Sigma x_i} \cdot \left[1 - \frac{\Sigma x_i}{n}\right]^{n-\Sigma x_i}.$$

For θ irrational, $L_n(\theta) = (1-\theta)^{\Sigma x_i}\,\theta^{n-\Sigma x_i}$. Now consider the function

$$\psi(\theta) = (1-\theta)^{\Sigma x_i}\,\theta^{n-\Sigma x_i} \text{ for all } \theta \text{ in } (0,1).$$

By differentiation, we see that ψ is maximum at

$$\theta = 1 - \frac{\Sigma x_i}{n} \text{ (which is rational)}$$

and $\psi\left(1 - \frac{\Sigma x_i}{n}\right) = \left[\frac{\Sigma x_i}{n}\right]^{\Sigma x_i} \cdot \left[1 - \frac{\Sigma x_i}{n}\right]^{n-\Sigma x_i}$.

Thus for all θ irrational in $(0,1)$,

$$L(\theta) \le \left(\frac{\Sigma x_i}{n}\right)^{\Sigma x_i}\left(1 - \frac{\Sigma x_i}{n}\right)^{n-\Sigma x_i} = \max_{\theta \text{ rational}} L_n(\theta).$$

Therefore, $\max_{\theta \in (0,1)} L_n(\theta) = \max_{\theta \text{ rational}} L_n(\theta)$ and hence the

MLE of θ, is $\hat{\theta}_n = \frac{\Sigma x_i}{n}$ for all $\theta \in (0,1)$.

Now, by the SLLN, $\hat{\theta}_n \to \theta$ a.s. if θ is rational;

$\hat{\theta}_n \to 1 - \theta$ a.s. if θ is irrational.

This means that $\hat{\theta}_n$ is not a strongly consistent estimator of θ.

Motivated by the above "pathological" example, we now investigate the consistency of MLE when the model is smooth. Note that when $L_n(\theta)$ is differentiable in θ for almost all x, it is important to realize that a root of the log–likelihood equation $\frac{\partial}{\partial\theta} \log L_n(\theta) = 0$ need not be a MLE. However, it is a MLE in the following case:

Theorem: *Assume that:* Θ *is an open interval containing* θ_o;

the model is identifiable;

$f(\cdot;\theta)$, $\theta \in \Theta$, *all have the same support.*

If, in addition,

a) $\theta \to f(x;\theta)$ *is differentiable for all* $x \in R$;

b) the log-likelihood equation has a unique solution;

then the MLE of θ *is strongly consistent.*

Proof: By the first theorem, a sequence of local maxima $\hat{\theta}_n$ converges a.s. to θ_o. Since $\hat{\theta}_n$ is a critical point,

$$\frac{\partial}{\partial\theta} \log L_n(x;\hat{\theta}_n) = 0$$

but by b), $\hat{\theta}_n$ is the unique root; thus $\hat{\theta}_n$ is in fact a global maximum point.

Example: $f(x;\theta) = (2\pi\theta)^{-1/2}\exp\{-x^2/2\theta\}$ for $\theta > 0$. This model satisifies all the conditions of this last theorem. Indeed,

$$\frac{\partial}{\partial\theta}\log L_n(x\,;\,\theta) = 0 \text{ has a unique solution } \hat{\theta}_n = \frac{1}{n}\sum_{i=1}^{n} x_i^2 .$$

Thus $\hat{\theta}_n \to \theta$ a.s. and is a (sequence of) MLE of θ.

Exercise 4: $f(x;\theta) = e^{-\theta}\,\theta^x / x!$, $x = 0, 1, 2, \cdots$, $\theta > 0$. Show that the MLE of θ is strongly consistent.

Exercise 5: $f(x;\theta) = \theta\, x^{\theta-1} e^{-x^{\theta}}$ for x and θ positive. Show that the log–likelihood equation has a unique root which is the MLE of the θ and is strongly consistent.

INE–Exercise 6: Consider the last theorem without b). Show that with probability tending to one as $n \to \infty$, the log–likelihood equation has a root which converges in probability to θ_o .

LESSON 12. ASYMPTOTIC NORMALITY

In this lesson, we continue the discussion of asymptotic properties of estimators with details for only the case of one dimensional parameters.

Let $X, X_1, \cdots, X_n$ be IID with $E[X] = \theta$ finite. Then $\hat{\theta}_n = \bar{X}_n = \frac{1}{n}\sum_{i=1}^{n} X_i$ is a strongly consistent estimator of θ. It follows that $\hat{\theta}_n \xrightarrow{P} \theta$ and $\hat{\theta}_n \xrightarrow{D} \theta$. This is a perfectly respectable point estimator of θ but in order to extend to an interval estimate or to use $\hat{\theta}_n$ in testing hypothesis about θ, we need to know something about its distribution.

With the additional assumption that the variance σ^2 of X is finite, we know that $(\hat{\theta}_n - \theta)/(\sigma/\sqrt{n})$ has the standard normal as limiting distribution and this is about as far as we can go.

If X is $N(\theta,1)$, $\sqrt{n}(\hat{\theta}_n - \theta)$ is $N(0,1)$ but the assumption of this distribution is often questioned.

The following phraseology is primarily for convenience of expression.

Definition: *A sequence of RV T_n is asymptotically normal (AN) if there exist constants a_n, b_n such that*

$$\frac{T_n - a_n}{b_n} \xrightarrow{D} N(0,1)$$

as $n \to \infty$; this may be abbreviated as $T_n \sim AN(a_n,b_n)$. Here a_n is called the asymptotic mean and b_n^2 is called the asymptotic variance.

Note that if $E[T_n^2] < \infty$ for all $n \geq 1$, it could happen that $a_n = E[T_n]$, $b_n^2 = Var(T_n)$, but that, in general, this need not be true. When T_n is $AN(a_n, b_n)$, probabilities are approximated in terms of the standard normal CDF ϕ as

$$P(T_n \le t) = P\left(\frac{T_n - a_n}{b_n} \le \frac{t - a_n}{b_n}\right) \approx \phi\left(\frac{t - a_n}{b_n}\right).$$

The size of n needed to get a good approximation is tied in with the rate of convergence in the Central Limit Theorem and is beyond the scope of this text. Interested readers may consult Hall, 1982. A typical result is the following Berry–Esseen Theorem whose proof can be found also in texts like Chung, 1974.

Theorem: *(Berry-Esseen) Let $X_1, \cdots, X_n$ be IID with mean μ and variance σ^2. Then*

$$\sup_{x \in R} |G_n(x) - \phi(x)| \le c\, E[|X_1 - \mu|^3]/\sigma^3\sqrt{n} \qquad (*)$$

where G_n is the CDF of $\frac{1}{\sigma\sqrt{n}}[\sum_{i=1}^{n} X_i - n\mu]$, and c is a universal constant (independent of n).

When $E[|X_1 - \mu|^3] < \infty$, we read the right–hand side of (*) to say that $\sup_{x \in R} |G_n(x) - \phi(x)| \to 0$ as $n \to \infty$ at the rate of $n^{-1/2}$; then the sequence $\{\sqrt{n} \sup_{x \in R} |G_n(x) - \phi(x)|\}$ is bounded.

Exact sampling distributions of statistics are often cumbersome and difficult to find and sometimes not explicit; hence their approximations by limiting distributions are useful. We will illustrate asymptotic normality with techniques for proving it. To begin, we recall that when $X_1, \cdots, X_n$ are IID and ψ is a measurable function, then

$$Y_1 = \psi(X_1), \cdots, Y_n = \psi(X_n)$$

are also IID. The next example includes a review of some earlier results.

Example: a) If $Z_1, Z_2, Z_3, \cdots$ are IID N(0,1) and $X_i = Z_i^2$, then $X_1, X_2, X_3, \cdots$ are IID $\chi^2(1)$. For each i,

$$E[X_i] = 1 \text{ and } Var(X_i) = 2 \text{ for all } i .$$

b) If $Y_n = X_1 + X_2 + \cdots + X_n = Z_1^2 + \cdots + Z_n^2$, then Y_n is $\chi^2(n)$. $E[Y_n] = n$ and $Var(Y_n) = 2n$. $Y_n/n = \bar{X}_n$ has mean 1 and variance $2/n$.

c) By the CLT,

$$\frac{\bar{X}_n - 1}{\sqrt{(2/n)}} = \frac{Y_n - n}{\sqrt{(2n)}} = \frac{\Sigma_{i=1}^n Z_i^2 - n}{\sqrt{(2n)}}$$

is AN(0, 1).

d) The sequence $\sqrt{(2Y_n)}$ is $AN(\sqrt{(2n)}, 1)$:

$$P(\sqrt{(2Y_n)} - \sqrt{(2n)} \le t) = P(\frac{Y_n - n}{\sqrt{(2n)}} - t^2/2\sqrt{(2n)} \le t) ;$$

since $t^2/2\sqrt{(2n)} \to 0$, Slutsky's Theorem gives $\sqrt{(2Y_n)} - \sqrt{(2n)}$ the same limiting distribution as

$$(Y_n - n)/\sqrt{(2n)} \text{ , namely, } N(0,1) .$$

The following two lemmas give a general version of this technique.

Lemma: *Let $F, G_1, G_2, \cdots$ be uniformly bounded monotone functions on R such that $\lim_{n\to\infty} G_n(t) = F(t)$ for all $t \in \bar{R}$. Let F be continuous on R. Then, $G_n \to F$ uniformly.*

Proof: Without loss of generality, let the bound be 1. Given $\varepsilon > 0$, let m be an integer greater than $\frac{2}{\varepsilon}$. For $k = 0, 1, \cdots, m$, let

$$s_k = F^{-1}(\tfrac{k}{m}) = \inf\{y : F(y) \ge k/m\}$$

so that $s_0 = -\infty < s_1 < \cdots < s_{k-1} < s_k < \cdots < s_m = \infty$ and

$$F(s_k) - F(s_{k-1}) = \frac{1}{m} < \frac{\varepsilon}{2}.$$

Each $|G_n(s_k) - F(s_k)| < \frac{\varepsilon}{2}$ for n large; since there are only a finite number $s_0, \cdots, s_m$, there is one N_ε such that $n > N_\varepsilon$ implies

$$F(s_k) - \frac{\varepsilon}{2} < G_n(s_k) < F(s_k) + \frac{\varepsilon}{2} \text{ for } k = 0, 1, \cdots, m.$$

For each $t \in R$, there is a k such that $s_{k-1} \le t \le s_k$. Then

$$G_n(t) \le G_n(s_k) \text{ and } F(s_{k-1}) \le F(t).$$

Hence, $G_n(t) \le G_n(s_k) < F(s_k) + \frac{\varepsilon}{2} \le F(s_{k-1}) + \varepsilon \le F(t) + \varepsilon$.
Similarly,

$$G_n(t) \ge G_n(s_{k-1}) > F(s_{k-1}) - \frac{\varepsilon}{2} \ge F(s_k) - \varepsilon \ge F(t) - \varepsilon.$$

It follows that $n > N_\varepsilon$ implies $|G_n(t) - F(t)| < \varepsilon$ for all $t \in \overline{R}$. This is what uniform convergence means.

Exercise 1: Verify the statement involving "a finite number" in the proof of the lemma.

Lemma: *Let* $X, X_1, X_2, \cdots$ *have CDF* $F, G_1, G_2, \cdots$ *respectively. Let* $G_n(t) \to F(t)$ *for all* $t \in \bar{R}$ *; let* $t_n \to t \in \bar{R}$ *. If* F *is continuous, then* $G_n(t_n) \to F(t)$ *.*

Proof: $|G_n(t_n) - F(t)| \le |G_n(t_n) - F(t_n)| + |F(t_n) - F(t)|$. Since F is continuous, $|F(t_n) - F(t)| \to 0$ as $n \to \infty$. Now certainly

$$|G_n(t_n) - F(t_n)| \le \sup_{t \in \mathbb{R}} |G_n(t) - F(t)|.$$

But for $n > N_\varepsilon$, the first lemma yields

$$|G_n(t) - F(t)| < \varepsilon \text{ for all } t \in \overline{R}$$

so that for $n > N_\varepsilon$, $|G_n(t_n) - F(t_n)| \le \varepsilon$ whence

$$|G_n(t_n) - F(t)| < \varepsilon \text{ for } n > N_\varepsilon .$$

Exercise 2: Show that if $\sqrt{n}(\hat{\theta}_n - \theta) \xrightarrow{D} N(0,\sigma^2)$, then $\hat{\theta}_n \xrightarrow{P} \theta$. Hint: Use the lemmas above.

Exercise 3: For $n = 1(1)\infty$, let X_n be distributed as a t–distribution with n degrees of freedom and density f_n. Show that X_n is AN(0,1). Hint: use the Lebesgue dominated convergence theorem (Lesson 9, Part II) to show that

$$\lim_{n\to\infty} \int_{-\infty}^{x} f_n(t)dt = \frac{1}{\sqrt{2\pi}} \int_{-\infty}^{x} e^{-t^2/2}\, dt .$$

Exercise 4: Let $X(\lambda)$ be a Poisson RV with parameter λ. Show that $X(\lambda)$ is $AN(\lambda, \sqrt{\lambda})$ as $\lambda \to \infty$. Hint: first find the characteristic function of $(X - \lambda)/\sqrt{\lambda}$.

Example: If $X_1, \cdots, X_n$ is a RS from a distribution with finite j^{th}–moment $E[X_i^j] = \theta_1$ and finite $2j^{th}$–moment $E[X_i^{2j}] = \theta_2$, then $X_1^j, \cdots, X_n^j$ is a random sample from a distribution with mean θ_1 and variance $\theta_3 = \theta_2 - \theta_1^2$. By the CLT,

$$m_j^{(n)} = \frac{1}{n}\sum_{i=1}^{n} X_i^j \text{ is } AN(\theta_1, \sqrt{\theta_3}) .$$

Exercise 5: Assume $E[X^{2k}] < \infty$ for some k. Show that the sequence of random vectors of sample moments

$$Y_n = (m_1^{(n)}, m_2^{(n)}, \cdots, m_k^{(n)}),$$

is asymptotically normal. Hint: use the CLT for random vectors, Lesson 15, Part III.

Now we need a multidimensional version of Taylor's formula whose proof we leave to texts like Fleming, 1977.

Theorem: *Let* $f : R^m \to R$ *be a function of class* C^q *; that is, all the* q^{th} *order partial derivatives of* f *exist and are continuous. For* $x = (x_1, \cdots, x_m), x^o = (x_1{}^o, \cdots, x_m{}^o)$,

$$f(x) = f(x^\circ) + \sum_{j_1=1}^{m} \partial f/\partial x_{j_1}\rfloor_{x^o} (x_{j_1} - x_{j_1}^o)$$

$$+ \sum_{j_1=1}^{m} \sum_{j_2=1}^{m} \partial^2 f/\partial x_{j_1} \partial x_{j_2}\rfloor_{x^o} (x_{j_1} - x_{j_1}^o)(x_{j_2} - x_{j_2}^o)/2!$$

$$+ \cdots$$

$$+ \sum_{j_1=1}^{m} \cdots \sum_{j_{q-1}=1}^{m} \partial^{q-1} f/\partial x_{j_1} \cdots \partial x_{j_{q-1}}\rfloor_{x^o} \cdot$$

$$(x_{j_1} - x_{j_1}^o) \cdots (x_{j_{q-1}} - x_{j_{q-1}}^o)/(q-1)!$$

$$+ R_q(x, x^\circ)$$

where $R_q(x, x^o)$

$$= \sum_{j_1=1}^{m} \cdots \sum_{j_q=1}^{m} \partial^q f/\partial x_{j_1} \cdots \partial x_{j_q} |_{x^*} (x_{j_1} - x_{j_1}^o) \cdots (x_{j_q} - x_{j_q}^o)/q!$$

with $x^* = x^o + s(x - x^o)$ *for some* $s \in [0,1]$ *. When the* q^{th} *partial is bounded by* B, $|R_q(x, x^o)| \leq B\, m^{q/2} \cdot |h|^q/q!$ *where*

$$|h|^2 = \sum_{i=1}^{m} (x_i - x_i^o)^2$$

. When f *is defined only on some domain* $D \subseteq R^m$ *, all statements have to be taken relative to* D *.*

Theorem: *Let* $X, X_1, X_2, \cdots$ *be IID; assume* $E[|X|^4] < \infty$. *Let* $m_j^{(n)} = \frac{1}{n}\sum_{i=1}^{n} X_i^j$, *for* $j = 1, 2$. *Let* $\psi : R^2 \to R$ *be of class* C^2. *Then* $\psi(m_1^{(n)}, m_2^{(n)})$ *is asymptotically normal.*

Proof: a) By Exercise 5, $(m_1^{(n)}, m_2^{(n)})$ is asymptotically normal with "mean" $\theta = (\theta_1, \theta_2)$, $\theta_i = E[X^i]$, $i = 1, 2$, and "covariance" $\frac{1}{n}(\sigma_{ij})$,

$$\sigma_{ij} = E[X^{i+j}] - E[X^i]E[X^j]\ , i, j = 1, 2\ .$$

By the Cramer–Wold technique (Lesson 14, Part III),

$$\sqrt{n}\,[\alpha(m_1^{(n)} - \theta_1) + \beta(m_2^{(n)} - \theta_2)] \qquad (*)$$

is also asymptotically normally distributed for all real α, β. The conclusion of the theorem will follow if we show that

$$\sqrt{n}(\psi(m_1^{(n)}, m_2^{(n)}) - \psi(\theta_1, \theta_2)] \quad \text{differs from } (*)$$

by only a random variable R_n which tends to zero in probability.

b) By the Taylor formula with $|x_i - \theta_i| < \varepsilon$, $i = 1, 2$,

$$\psi(x_1, x_2) - \psi(\theta_1, \theta_2)$$

$$= \partial\varphi/\partial x_1\rfloor_{(\theta_1,\theta_2)}(x_1 - \theta_1) + \partial\varphi/\partial x_2\rfloor_{(\theta_1,\theta_2)}(x_2 - \theta_2)$$

$$+ \sum_{j_1=1}^{2}\sum_{j_2=1}^{2} \partial^2\varphi/\partial x_{j_1}\theta x_{j_2}\rfloor_{\theta^*}(x_{j_1} - \theta_{j_1})(x_{j_2} - \theta_{j_2})/2 \quad (**)$$

where θ^* has θ_1^* between x_1 and θ_1 and θ_2^*

between x_2 and θ_2. The hypothesis includes the boundedness, say by B, of the second partials so that the double sum in (**) is bouned by $4\,B\,\varepsilon^2/2$. Now Chebyshev's inequality can be applied to the sample means $m_i^{(n)}$: for each $\varepsilon > 0$,

$P(|m_i^{(n)} - \theta_i| < \varepsilon) \geq 1 - \text{Var}\, m_i^{(n)}/\varepsilon^2 = 1 - \text{Var}\, X^i/n\,\varepsilon^2$.

In particular, with $\varepsilon = n^{-1/3}$,

$P(|m_1^{(n)} - \theta_1| < n^{-1/3}, |m_2^{(n)} - \theta_2| < n^{-1/3}) \longrightarrow 1$

as $n \to \infty$. Of course, $2 \cdot B \cdot \varepsilon^2$ also tends to zero. Let

$$\alpha = \partial\varphi/\partial x_1\rfloor_{(\theta_1,\theta_2)},\ \beta = \partial\varphi/\partial x_2\rfloor_{(\theta_1,\theta_2)}.$$

Then replacing x_i by $m_i^{(n)}$, $i = 1, 2$, in (**) yields

$$\sqrt{n}[\psi(m_1^{(n)}, m_2^{(n)}) - \psi(\theta_1, \theta_2)]$$

$$= \sqrt{n}[\alpha(m_1^{(n)} - \theta_1) + \beta(m_2^{(n)} - \theta_2)] + R_n$$

where $|R_n| \leq \sqrt{n}\, 2\, B\, \varepsilon^2 \longrightarrow 0$ as $\longrightarrow \infty$.

Finally, $\psi(m_1^{(n)}, m_2^{(n)})$ is $AN(a_n, b_n)$ where

$a_n = \psi(\theta_1, \theta_2)$ and b_n^2

$$= \frac{1}{n}\sum_{j_1=1}^{2}\sum_{j_2=1}^{2} \partial\varphi/\partial x_j(\theta_{j_1+j_2} - \theta_{j_1}\theta_{j_2})\partial\varphi/\partial x_j\rfloor_{(\theta_1,\theta_2)}.$$

Of course, a similar theorem will hold (assuming appropriate moments and derivatives) for $\psi : R^k \longrightarrow R^k$ and $\sqrt{n}[\psi(m_1^{(n)}, \cdots, m_k^{(n)}) - \psi(\mu_1, \cdots, \mu_k)]$.

Exercise 6: Let ψ be a differentiable function on Θ such that $\psi'(\theta) \neq 0$ for all $\theta \in \Theta$; suppose that $\sqrt{n}(\hat{\theta}_n - \theta)$ is

AN(0, σ^2). Show $\sqrt{n}[\psi(\hat{\theta}_n) - \psi(\theta)]$ is AN(0, $\sigma^2 \cdot [\psi'(\theta)]^2$) .
Hint: use a Taylor expansion of $\psi(\hat{\theta}_n)$ on θ and Exercise 1 .

Exercise 7: Let X be a Poisson RV with parameter θ . Use Exercise 6 to verify that $\sqrt{(\overline{X}_n)}$ is AN($\sqrt{\theta}$, $\frac{1}{4n}$) where $\overline{X}_n$ is the sample mean.

LESSON 13. ASYMPTOTIC NORMALITY OF MAXIMUM LIKELIHOOD ESTIMATORS

This lesson is devoted to the proof of the most important theorem concerning the asymptotic behavior of MLE: for a class of "nice" models, the log–likelihood equation has a solution which is a weakly consistent estimator of the true parameter and which is asymptotically normal; we will see that such a solution need not be a true MLE.

We follow Kulldorff, 1957, in the following exposition for the one dimensional case. A generalization to several parameters will be stated at the end.

Definition: *A model* $\mathcal{F} = \{f(x;\theta),\ x \in R\ ,\ \theta \in \Theta \subseteq R\}$ *is regular if the following conditions hold.*

R_1: *Θ is an open interval of R ;*

R_2: *$f(\cdot\,;\theta)$ all have the same support $\mathcal{F}$;*

R_3: $\partial f(x;\theta)/\partial\theta$, $\partial^2 f(x;\theta)/\partial\theta^2$, $\partial^3 f(x;\theta)/\partial\theta^3$ *exist on Θ , for almost all x;*

R_4: $\displaystyle\int_{-\infty}^{\infty} \partial f(x;\theta)/\partial\theta\, dx = 0$ *for every* $\theta \in \Theta$;

R_5: $\displaystyle\int_{-\infty}^{\infty} \partial^2 f(x;\theta)/\partial\theta^2\, dx = 0$ *for every* $\theta \in \Theta$;

R_6: $\displaystyle -\infty < \int_{-\infty}^{\infty} [\partial^2 \log f(x;\partial)/\partial\theta^2] f(x;\theta) dx < 0,$ *for every* $\theta \in \Theta$;

R_7: *There exist a function g, positive and twice differentiable on Θ and a function H on R such that for all $\theta \in \Theta$,*

$$\int_{-\infty}^{\infty} H(x) f(x;\theta) dx < \infty ,$$

$$|\partial^2 [g(\theta)\partial \log f(x;\theta)/\partial\theta]/\partial\theta^2 | < H(x) \text{ for all } x .$$

Exercise 1: Assume R_1 and R_2 . Show the following forms for "taking the limit under the integral sign" (Lesson 11, Part II).

a) If $|\partial f/\partial\theta| < K(x)$ for all x and θ, and $E_\theta[K(X)] < \infty$, then $E_\theta[\partial \log f/\partial\theta] = 0$.

b) If in addition, $|\partial^2 f/\partial\theta^2| < L(x)$ for all x and θ, and $E_\theta[L(X)] < \infty$, then

$$E_\theta[\partial^2 \log f/\partial\theta^2] = -E_\theta[(\partial \log f/\partial\theta)^2].$$

The following comments on regularity are also appropriate.

R_1: If θ_o is the "true" parameter, we consider neighboring θ of the form $\theta = \theta_o \pm \delta$ for sufficiently small δ. The structure of Θ ensures that all these θ are in Θ.

R_2: As we will see, the requirement that the $f(\cdot;\theta)$, $\theta \in \Theta$, have the same support (equivalently that the support is independent of θ) is necessary in the proof of the theorem.

R_3: The existence of derivatives up to third order is used in an appropriate Taylor's formula.

R_4: This yields $E_\theta[\partial \log f(X;\theta)/\partial\theta]$

$$= \int[\partial \log f(x;\theta)/\partial\theta] f(x;\theta)dx = \int \partial f(x;\theta)dx/\partial\theta = 0.$$

R_5: The Fisher information amount is

$$I(\theta) = E[(\partial \log f(x;\theta)/\partial\theta)^2] = E_\theta(\partial f(x;\theta)/\partial\theta/f(x;\theta))^2].$$

Then R_5 implies that $I(\theta) = -E_\theta[\partial^2 \log f(x;\theta)/\partial\theta^2]$.

R_6: In view of previous conditions, this is equivalent to $0 < I(\theta) < \infty$.

R_7: If there is a function given as $h(x;\theta)$ and bounded by 1 such that $\partial^2 g/\partial\theta^2 = h(x;\theta)H(x)$, then $g(\theta) \equiv 1$, reduces this condition to $E_\theta[H(X)] < \infty$ and

$$|\,\partial^3 \log f(x;\theta)/\partial\theta^3\,| < H(x) \text{ for all } x \in R.$$

Exercise 2: Check that the conditions $R_1 - R_5$ are satisfied for

$$f(x;\theta) = (2\pi\theta)^{-1/2}\cdot\exp\{-x^2/2\theta\}.$$

Example: Continue with the density in Exercise 2. Let $g(\theta) = \theta^2$. Then $g(\theta) > 0$ on $\Theta = (0, \infty)$ and twice differentiable there. On the other hand, $\partial^2 [\theta^2 \partial \log f(x;\theta)/\partial\theta]/\partial\theta^2$

$$= \partial^2[\theta^2(-\frac{1}{2\theta} + \frac{x^2}{2\theta^2})]/\partial\theta^2 = \partial^2 [-\frac{\theta}{2} + \frac{x^2}{2}]/\partial\theta^2 = 0 .$$

Thus, for this model, it suffices to take $H(x) \equiv 1$, to satisfy R_7. Note that here when $g(\theta) \equiv 1$, R_7 is *not* satisfied since

$$\partial^2[g(\theta)\partial \log f(x;\theta)/\partial\theta]/\partial\theta^2 = \partial^3 \log f(x;\theta)/\partial\theta^3 = -\frac{1}{\theta^3} + \frac{3\,x^2}{\theta^4}$$

is not bounded on $(0, \infty)$.

Exercise 3: Let $X_1, \cdots, X_n$ be a RS from a regular model $\{f(x;\theta)\}$. Let $L_n(\theta) = \prod_{i=1}^{n} f(X_i;\theta)$. Use the CLT to show that $\partial\log(L_n)/\partial\theta$ is $AN(0, nI(\theta))$.

Example: Let $X_1, \cdots, X_n$ be a random sample from the model in Exercise 2. The log–likelihood equation has a unique root:

$$\hat{\theta}_n = \frac{1}{n}\sum_{i=1}^{n} X_i^2 \text{ and hence it is a MLE of } \theta .$$

By the SLLN, $\hat{\theta}_n$ is strongly consistent for θ. On the other hand, by the CLT, $\hat{\theta}_n$ is $AN(\theta, 1/nI(\theta))$ where $I(\theta) = \dfrac{1}{2\theta^2}$.

The result of this last example almost holds for all regular models. More precisely:

Theorem: *Let $\mathcal{F}$ be a regular model. Then the log-likelihood equation has a solution $\hat{\theta}_n(X_1, \cdots, X_n)$ such that*

a) $\hat{\theta}_n$ *is a weakly consistent estimator of* θ ;

b) $\hat{\theta}_n$ *is* $AN(\theta, 1/nI(\theta))$.

Proof: a) We proceed to show that, with probability tending to one, the log–likelihood equation has a consistent root. Let θ_o be the true parameter. Applying Taylor's formula to each

$$g(\theta)\partial \log f(x_i;\theta)/\partial\theta,$$

we have, in virtue of R_7,

$$\frac{1}{n} g(\theta)\theta \log L_n(X_1,\cdots,X_n;\theta)/\partial\theta$$
$$= A_o + (\theta-\theta_o)A_1 + \frac{1}{2}(\theta-\theta_o)^2 A_2,$$

$$A_o = \frac{1}{n} g(\theta_o)\sum_{i=1}^{n} [\partial \log f(X_i\,;\,\theta)]/\partial\theta\big|_{\theta=\theta_o},$$

$$A_1 = \frac{1}{n}\sum_{i=1}^{n} [\partial(g(\theta)\frac{\partial}{\partial\theta}\log f(X_i;\theta)/\partial\theta]\big|_{\theta=\theta_o},$$

$$A_2 = \frac{1}{n}\sum_{i=1}^{n} h(X_i;\tilde{\theta})\,H(X_i) \text{ with } |h(X_i;\tilde{\theta})| < 1$$

and $\tilde{\theta}$ is "between" θ and θ_o.

i) A_o converges a.s. to zero as $n \to \infty$ by the SLLN since $E_{\theta_o}[(\partial \log f(X;\theta)/\partial\theta)\big|_{\theta=\theta_o}] = 0$ by R_4.

ii) $$A_1 = g'(\theta_o)\frac{1}{n}\sum_{i=1}^{n} [\partial \log f(X_i;\theta)/\partial\theta]\big|_{\theta=\theta_o}$$
$$+ g(\theta_o)\frac{1}{n}\sum_{i=1}^{n} [\partial^2 \log f(x_i;\theta)\partial\theta^2]\big|_{\theta=\theta_o}$$

converges a.s. to $-\,g(\theta_o)I(\theta_o)$ by the SLLN and R_4, R_5.

iii) Again by the SLLN, A_2 converges a.s. to

$$E_{\theta_o}[h(X;\tilde{\theta})H(X)] = \int_{-\infty}^{\infty} h(x;\tilde{\theta})H(x)f(x;\theta)dx$$

which is finite by R_7.

iv) Consider a small neighborhood $(\theta_o - \delta, \theta_o + \delta) \subseteq \Theta$ which is an open interval by R_1. For $\theta = \theta_o \pm \delta$, we then have: $(\partial \log L_n(\theta)/\partial\theta)\big|_{\theta = \theta_o \pm \delta}$

$$= \frac{n}{g(\theta_o \pm \delta)}[A_o \pm \delta A_1 + \tfrac{1}{2}\delta^2 A_2] .$$

By R_7, g is a positive function; thus the sign of $\partial \log L_n(\theta)/\partial\theta$ at $\theta = \theta_o \pm \delta$ is that of

$$A_o \pm \delta A_1 + \tfrac{1}{2}\delta^2 A_2 .$$

The above a.s. convergences imply convergence in probability so that for all $\varepsilon > 0$ and all $\delta > 0$, and $n \geq N(\delta, \varepsilon, \theta_o)$, $P(|A_o| \geq \delta^2) < \frac{\varepsilon}{3}$,

$$P(A_1 \geq -\tfrac{1}{2}\, g(\theta_o)I(\theta_o)) < \frac{\varepsilon}{3} ,$$

$$P(|A_2| \geq 2M) < \frac{\varepsilon}{3}$$

where $0 < M = M(\theta_o) = \int_{-\infty}^{\infty} H(x)f(x;\theta_o)dx < \infty$.

v) Let S_n be the intersection of the three sets:

$$\{|A_o| < \delta^2\} , \{A_1 < -\tfrac{1}{2}g(\theta_o)I(\theta_o)\} , \{|A_2| < 2M\} .$$

Then, $P(S_n) > 1 - \varepsilon$ for $n \geq N(\delta, \varepsilon, \theta_o)$.

In S_n, $|A_o + \frac{1}{2}\delta^2 A_2| < \delta^2(1 + M)$ and

$$|\pm \delta A_1| > \tfrac{1}{2}\, g(\theta_o)I(\theta_o)\delta .$$

This will be greater than

$$\delta^2(1 + M) \text{ if } \delta < \tfrac{1}{2}\, g(\theta_o)I(\theta_o)/(1 + M) .$$

It follows that the sign of $A_o \pm \delta A_1 + \frac{1}{2}\delta^2 A_2$ is that of $\pm\delta A_1$. Therefore, in S_n,

$\partial \log L_n(\theta)/\partial\theta > 0$ if $\theta = \theta_o - \delta$ and

$\partial \log L_n(\theta)/\partial\theta < 0$ if $\theta = \theta_o + \delta$.

vi) Now $\partial \log L_n(\theta)/\partial\theta$ is a continuous function of θ since $\partial^2 f/\partial\theta^2$ exists by R_3. Hence there is a root $\hat{\theta}_n$ of the log–likelihood equation

$$\partial \log L_n(\theta)/\partial\theta = 0$$

which lies between $\theta_o - \delta, \theta_o + \delta$ when $n > N(\delta, \varepsilon, \theta_o)$. Since this is only one of the possible values $|\hat{\theta}_n - \theta_o| < \delta$, and

$$P(|\hat{\theta}_n - \theta_o| < \delta) \geq P(S_n) > 1 - \varepsilon .$$

Also, P(log–likelihood equation has a solution $\hat{\theta}_n$)

$$\geq P(S_n) > 1 - \varepsilon .$$

Now let T_n be a RV with distribution

$$P(T_n \leq t) = P(\hat{\theta}_n \leq t \text{ and } \hat{\theta}_n \text{ is a solution}) .$$

Then $P(|T_n - \theta_o| < \delta)$

$$= P(|\hat{\theta}_n - \theta_o| < \delta \text{ and } \hat{\theta}_n \text{ is a solution})$$

$$\geq P(S_n) > 1 - \varepsilon .$$

Therefore $T_n \xrightarrow{P} \theta_o$.

b) We note first that $\partial \log L_n(\hat{\theta}_n)/\partial\theta = 0$ implies

$$A_o + (\hat{\theta}_n - \theta_o)A_1 + \tfrac{1}{2}(\hat{\theta}_n - \theta_o)^2 A_2 = 0 .$$

Then, $\sqrt{n}(\hat{\theta}_n - \theta_o) = \dfrac{-\sqrt{n}\ A_o}{A_1 + A_2(\hat{\theta}_n - \theta_o)/2}$. (*)

By the CLT, $\sqrt{n}A_o = \sum_{i=1}^{n} \partial \log f(X_i;\theta)/\partial\theta\rfloor_{\theta=\theta_o} /\sqrt{n}$ converges in distribution to $N(0, I(\theta)_o)$. Hence by virtue of part ii) above,

$\dfrac{-\sqrt{n}\ A_o/g(_o)}{A_1/g(\theta_o)}$ converges in distribution to $N(0,1/I(\theta_o))$. Since

A_2 converges to a finite quantity, and $\hat{\theta}_n - \theta_o$ converges to 0 ,

$$A_2(\hat{\theta}_n - \theta_o) \text{ converges to } 0 .$$

It follows from (*) that the limiting distribution of $\sqrt{n}(\hat{\theta}_n - \theta_o)$ is $N(0, 1/I(\theta))$.

Exercise 4: Point out the different forms of Slutsky's Theorem used in the last proof.

Exercise 5: Let $f(x;\theta) = \theta e^{-\theta x} I(0 < x < \infty)$ for $\theta > 0$.

a) Verify that the model is regular.

b) Verify that the MLE of θ is $\hat{\theta}_n = 1/\overline{X}_n$.

c) Show that $\hat{\theta}_n$ is asymptotically normal.

Exercise 6: Let $\{f(x;\theta)\}$ be a regular model. Suppose that for each $\varepsilon > 0$, with probability greater than $1 - \varepsilon$ and $n \geq N(\varepsilon)$,

$$\partial \log L_n(\theta)/\partial\theta = 0 \text{ has a unique solution } \hat{\theta}_n$$

which is then a true MLE of θ . Suppose that $I(\theta)$ is differentiable. Show that:

a) $\sqrt{n}\, H_n(\theta_o) = H_n{}'(\theta^*)\sqrt{n}(\theta_o - \hat{\theta}_n)$ where

$$H_n(X_1,\cdots,X_n;\theta) = \frac{1}{n\sqrt{I(\theta)}}\, \partial \log L_n(X_1,\cdots,X_n;\theta)/\partial\theta$$

and θ^* is "between" θ_o and $\hat{\theta}_n$;

b) $H_n'(\theta^*)\sqrt{n}(\theta_o - \hat{\theta}_n)$ is AN(0,1) . Hint: apply Taylor's formula to $\sqrt{n}H_n(\theta)$ about $\hat{\theta}_n$ and use Exercise 3 .

For completeness, we state a similar result for the multidimensional case; a proof can be found in Lehmann, 1983 .

Definition: *Let* $\mathscr{F} = \{f(x;\theta), \theta \in \Theta \subseteq R^d\}$ $\theta = (\theta_1, \cdots, \theta_d)$ *. The model* $\mathscr{F}$ *is called regular if:*

1) *the mapping* $\theta \to P_\theta(dx) = f(x;\theta)dx$ *is injective;*
2) P_θ *has a common support;*
3) Θ *is an open set of* R^d*;*
4) *for almost all* x *, all third order partial derivatives*

   $\partial^3 f(x;\theta)/\partial\theta_i\partial\theta_j\partial\theta_k = T_{ijk}$ *exist for all* $\theta \in \Theta$ *;*
5) *for all* $\theta \in \Theta$ *,* $E_\theta[\partial \log f(X;\theta)/\partial\theta] = 0$ *;*
6) $I(\theta) = E_\theta[- \partial^2 \log f(X;\theta)/\partial\theta\, \partial\theta']$ *is finite and positive definite;*
7) *there exist functions* M_{ijk} *such that*

   $E_{\theta^o}[M_{ijk}(X)] < \infty$ *(where* θ^o *is the true parameter).*

Theorem: *Let* $\mathscr{F}$ *be a regular model. Then with probability tending to one, there exist solutions* $\hat{\theta}_n$ *of the likelihood equation* $\partial L_n(\theta)/\partial\theta = 0$ *such that* $\hat{\theta}_n$ *is weakly consistent and* $\sqrt{n}(\hat{\theta}_n - \theta^o)$ *is asymptotically normal with mean vector* 0 *and covariance matrix* $[I(\theta^o)]^{-1}$.

Corollary: *If* $\partial L_n(\theta)/\partial\theta = 0$ *has a unique solution* $\hat{\theta}_n$ *, then this*

$\hat{\theta}_n$ *is consistent and asymptotically normal.*

Example: $f(x;\theta) = ae^{-ax}I(0 \le x < \tau) + be^{-bx + (b-a)\tau}I(x \ge \tau)$ for $\theta = (a, b) \in \Theta = (0, \infty)^2$ and τ known. Let $X_1, \cdots, X_n$ be a random sample from this model. Set

$$T_r = \sum_{i=1}^{n} X_i I(X_i < \tau),\ r = \sum_{i=1}^{n} I(X_i < \tau),\ T_s = \sum_{i=1}^{n} X_i I(X_i \ge \tau)\ .$$

Then the MLE of $\theta = (a, b)$ is:

$$\hat{a}_n = r[T_r + (n-r)\tau]^{-1}\ ,\ \hat{b}_n = (n-r)[T_s - (n-r)\tau]^{-1}$$

(the latter undefined when $r = n$). It follows from the theorem that $\sqrt{n}(\hat{a}_n - a), \sqrt{n}(\hat{b}_n - b)$ are asymptotically normal with covariance matrix the inverse of the Fisher information matrix.

Exercise 7: Find the Fisher information matrix for this model.

Exercise 8: If τ is unknown in the above model, i.e. $\theta = (a, b, \tau) \in (0, \infty)^3$, show that this new model is not regular.

LESSON 14. ASYMPTOTIC EFFICIENCY AND LARGE SAMPLE CONFIDENCE INTERVALS

We close our study of the asymptotic theory of statistical estimation with a classical concept of *asymptotic optimality* and an application to confidence interval estimation.

Definition: *Consider the regular model* $\{f(x;\theta),\ \theta \in \Theta \subseteq R\}$. *A sequence of estimators* T_n *of* θ *is asymptotically efficient iff*

$$\sqrt{n}(T_n - \theta) \text{ is } AN(0, I^{-1}(\theta))$$

where

$$I(\theta) = E_\theta[(\partial \log f(X;\theta)/\partial\theta)^2] .$$

In the multidimensional case, $\Theta \subset R^d$, T_n *is asymptotically efficient (or best asymptotically normal, BAN) iff* T_n *is* $AN(\theta, \frac{1}{n}I^{-1}(\theta))$ *where* $I(\theta)$ *is the d by d matrix*

$$\left[E_\theta[(\partial \log f(X;\theta)/\partial\theta_i)(\partial \log f(X;\theta)/\partial\theta_j]\right] .$$

If the parameter of interest is $\varphi(\theta),\ \theta \in R$, then the asymptotic variance will be $[\varphi'(\theta)]^2 I^{-1}(\theta)$. From Lesson 13, we see that in regular models, ML solutions are asymptotically efficient; we think of them as "best" estimators among the class of AN estimators in the sense that

if $\tilde{\theta}_n$ is another estimator and $\sqrt{n}(\tilde{\theta}_n - \theta)$ is $AN(0, \sigma^2(\tilde{\theta}_n, \theta))$,

$$\text{then } \sigma^2(\tilde{\theta}_n, \theta) \geq I^{-1}(\theta) \text{ for all } \theta \in \Theta .$$

Thus one justification for MLE is that this method provides a way of finding asymptotically efficient estimators in regular models.

Example: Let X be $N(\theta,1)$ with density $f(x;\theta)$; then $I(\theta) = 1$.

a) The sample mean $\bar{X}_n$ is asymptotically efficient since $\sqrt{n}(\bar{X}_n - \theta)$ is distributed as $N(0,1)$ and so, trivially, AN(0,1) .

b) The mean θ is also the median of X so it is not unreasonable to consider the sample median as a competing estimator of θ . Let $\xi_{.5}(n)$ denote the sample median; then it can be shown (Serfling, 1980) that $\sqrt{n}(\xi_{.5}(n) - \theta)$ is $AN(0, 1/(2f^2(.5;\theta))^2$. But, since $1/4f^2(.5;\theta) > 1$, $\xi_{.5}(n)$ is not asymptotically efficient.

Exercise 1: Let X be Poisson with mean $\theta > 0$. From $X_1, \cdots, X_n$, estimate $Var_\theta(X) = \theta$ by the sample variance $\hat{\sigma}_n^2 = \frac{1}{n}\sum_{i=1}^{n}(X_i - \bar{X})^2$. Is $\hat{\sigma}_n^2$ asymptotically efficient ?

The following example (due to Hodges but published by Lecam, 1953, p. 286 – 287) shows that, even in the regular models, there exist asymptotically normal estimators with smaller asymptotic variance than that of MLE.

Example: $f(x;\theta) = (2\pi)^{-1/2}\exp\{-\frac{1}{2}(x-\theta)^2\}$ for x and $\theta \in R$.

a) The MLE of θ is $\bar{X}_n$; $\sqrt{n}(\bar{X}_n - \theta)$ is AN(0,1) for all $\theta \in R$; the asymptotic variance of $\bar{X}_n$ is $\frac{1}{n}$.

b) Now define, for each $\alpha \in (0,1)$ and $\beta \in (0, 1/2)$,

$$T_n = \bar{X}_n[\alpha I\{|\bar{X}_n| < n^{-\beta}/2\} + I\{|\bar{X}_n| \geq n^{-\beta}/2\}] .$$

c) For $\theta \neq 0$, take $\varepsilon < |\theta|/2$. Since $|\bar{X}_n| \to |\theta|$ a.s.,

$$|\ |\bar{X}_n| - |\theta|\ | < |\bar{X}_n - \theta| < \varepsilon \text{ for } n > N_\varepsilon .$$

Now take $n > \max(N_\varepsilon, |\theta|^{1/\beta})$; then

$$|\bar{X}_n| > |\theta| - \varepsilon > |\theta|/2 > n^{-\beta}/2 .$$

This means that $T_n = \bar{X}_n$ a.s. which implies that T_n is AN$(\theta, \frac{1}{n})$ for $\theta \neq 0$.

d) For $\theta = 0$, $I\{|\bar{X}_n| \geq nm^{-\beta}/2\} \xrightarrow{P} 0$ because by Chebyshev's inequality, $P(I\{|\bar{X}_n| > n^{-\beta}/2\} > \varepsilon)$

$$= P(|\bar{X}_n| \geq n^{-\beta}/2) \leq (\tfrac{1}{n})/(n^{-\beta}/2)^2$$

and the right hand side $4\, n^{2\beta-1} \to 0$ for $2\beta < 1$. It follows that $I\{|\bar{X}_n| < n^{-\beta}/2\} \xrightarrow{P} 1$ and so

$$\alpha I\{|\bar{X}_n| < n^{-\beta}/2\} + I\{|\bar{X}_n| \geq n^{-\beta}/2\} \xrightarrow{P} \alpha .$$

By Slutsky's theorem, $\sqrt{n}\, T_n$ has the same limiting distribution as $\sqrt{n}\, \alpha \bar{X}_n$, that is,

$$T_n \text{ is AN}(0, \alpha^2/n) .$$

d) Since $0 < \alpha < 1$, the asymptotic variance of T_n is *strictly less* than $\frac{1}{n}$, the asymptotic variance of the MLE $\bar{X}_n$. In other words, when $\theta = 0$, T_n is "better" than $\bar{X}_n$ asymptotically. For other topics in asymptotic efficiency see Akahira et al, 1981.

We now illustrate the application of limiting distributions of sample statistics to the construction of approximate confidence intervals for functions of the parameters.

Consider the model $\mathcal{F} = \{f(x;\theta), \theta \in \Theta \subseteq R\}$. Suppose that $T_n(X_1, \cdots, X_n)$ is a statistic which is AN$(\varphi(\theta), \sigma_n^2(\theta))$:

$$\text{for all } x \in R, \lim_{n \to \infty} P\left(\frac{T_n - \varphi(\theta)}{\sigma_n(\theta)} \leq x\right) = \phi(x)$$

where ϕ is the CDF of the standard normal random variable. Let $z_{\alpha/2}$ be the $\alpha/2^{th}$ quantile: $\phi(z_{\alpha/2}) = 1 - \frac{\alpha}{2} \in (0,1)$. Then,

$$\lim_{n \to \infty} P(-z_{\alpha/2} \le \frac{T_n - \varphi(\theta)}{\sigma_n(\theta)} \le z_{\alpha/2}) = 1 - \alpha$$

and for n sufficiently large, we can approximate

$$P(-z_{\alpha/2} \le \frac{T_n - \varphi(\theta)}{\sigma_n(\theta)} \le z_{\alpha/2}) \text{ by } 1 - \alpha .$$

From this, one may be able to pivot (solve) the above double inequality to obtain an approximate confidence interval (CI) for all $\varphi(\theta)$ like

$$L(n,\alpha,T_n) \le \varphi(\theta) \le U(n,\alpha,T_n) .$$

Example: $f(x;\theta) = \theta e^{-\theta x} I(0 < x < \infty)$ for $\theta > 0$.

$$\varphi(\theta) = E_\theta[X] = 1/\theta .$$

The sample mean $\bar{X}_n$ is $AN(1/\theta, 1/n\theta^2)$. Hence,

$$\sigma_n^2(\theta) = 1/n\theta^2 \text{ and } \theta\sqrt{n}(\bar{X}_n - \tfrac{1}{\theta}) \text{ is } N(0,1) .$$

From

$$-z_{\alpha/2} \le \theta\sqrt{n}(\bar{X}_n - \tfrac{1}{\theta}) \le z_{\alpha/2} , \qquad (*)$$

we get

$$\bar{X}_n[1 + \tfrac{1}{n} z_{\alpha/2}]^{-1} \le \tfrac{1}{\theta} \le \bar{X}_n[1 - \tfrac{1}{n} z_{\alpha/2}]^{-1} \qquad (**)$$

Exercise 2: Verify that (*) and (**) are equivalent.

Example: Let X be Poisson with $\theta > 0$ and

$$\varphi(\theta) = (\mathrm{Var}_\theta(X))^{1/2} = \sqrt{\theta} .$$

Again, $\sqrt{n}(\bar{X}_n - \theta)$ is $AN(0,\theta)$ so that $\sqrt{n/\theta}\,(\bar{X}_n - \theta)$ is $AN(0,1)$. It follows that

$$\sqrt{n}(\overline{X}_n^{1/2} - \theta^{1/2}) \text{ is } AN(0, \tfrac{1}{4}) \qquad (***)$$

The RV $2\sqrt{n}(\overline{X}_n^{1/2} - \theta^{1/2})$ is a pivotal quantity from which we get a $100(1-\alpha)\%$ approximate CI for $\sqrt{\theta}$:

$$\overline{X}_n^{1/2} - \frac{z_{\alpha/2}}{2\sqrt{n}} \le \sqrt{\theta} \le \overline{X}_n^{1/2} + \frac{z_{\alpha/2}}{2\sqrt{n}} .$$

Exercise 3: Verify (***) by noting that $[\varphi'(\theta)]^2 = \dfrac{1}{4\theta}$.

Exercise 4: a) Show that the exponential model of the above example is regular.

b) Find the corresponding MLE of θ .

c) Use the theorem on the asymptotic efficiency of MLE (lesson 12) to derive an approximate $100(1-\alpha)\%$ C.I. for θ .

Consider again the case where the statistic T_n is $AN(\theta, \frac{\sigma^2(\theta)}{n})$. Moreover, assume that a (weakly) consistent estimator of $\sigma^2(\theta)$ is available, say $S_n^{\ 2}$. Then an approximate $100(1-\alpha)\%$ CI for θ can be derived as follows. Note that

$$\sqrt{n}(T_n - \theta)/S_n = \frac{\sigma(\theta)}{S_n} \cdot \frac{\sqrt{n}(T_n - \theta)}{\sigma(\theta)} .$$

By hypothesis, $\dfrac{\alpha(\theta)}{S_n} \xrightarrow{P} 1$ and $\dfrac{\sqrt{n}(T_n - \theta)}{\sigma(\theta)} \xrightarrow{P} N(0,1)$. Thus by Slutsky's theorem, $\dfrac{\sqrt{n}(T_n - \theta)}{S_n}$ is $AN(0,1)$. Then,

$$-z_{\alpha/2} \le \frac{\sqrt{n}(T_n - \theta)}{S_n} \le z_{\alpha/2}$$

implies

$$T_n - z_{\alpha/2}\frac{S_n}{\sqrt{n}} \le \theta \le T_n + z_{\alpha/2}\frac{S_n}{\sqrt{n}} .$$

Example: Let the population be Bernoulli with $P(X = 1) = \theta$. Since θ is the population mean, we use the sample mean $\bar{X}_n$ to estimate θ. By the CLT, $\bar{X}_n$ is $AN(\theta, \theta(1-\theta)/n)$ so that $\sigma^2(\theta) = \theta(1-\theta)$. It is obvious that $S_n^{\ 2} = \bar{X}_n(1-\bar{X}_n)$ is a (strongly) consistent estimator of $\sigma^2(\theta)$. Thus,

$$- z_{\alpha/2} \le [\sqrt{n}(\bar{X}_n - \theta)][\bar{X}_n(1-\bar{X}_n)]^{-1/2} \le z_{\alpha/2}$$

has approximately the probability $1 - \alpha$ and pivots to an approximate $100(1 - \alpha)\%$ CI for θ:

$$\bar{X}_n - \frac{1}{\sqrt{n}} z_{\alpha/2}[\bar{X}_n(1-\bar{X}_n)]^{1/2} \le \theta \le \bar{X}_n + \frac{1}{\sqrt{n}} z_{\alpha/2}[X_n(1-X_n)]^{1/2}.$$

Exercise 5: Explain "It is obvious · · ·" in the example above.

Exercise 6: Derive the same approximate C.I. for θ in the Bernoulli example above by using the asymptotic efficiency of the MLE (lesson 12).

Exercise 7: If X is binomial with parameters n and θ, $(X - n\theta)[n\,\theta(1 - \theta)]^{1/2}$ is $AN(0,1)$. Solve

$$- z_{\alpha/2} \le \frac{X - n\,\theta}{\sqrt{n\theta(1-\theta)}} \le z_{\alpha/2}$$

to obtain another approximate CI for θ. Compare to the previous one.

PART VI: TESTING HYPOTHESES

Overview

A statistical hypothesis is an assumption about the distribution of a random variable; a test of a statistical hypothesis is a procedure for deciding whether or not the hypothesis is to be rejected. Although some results of similar generality are available as "decision theory", the material herein emphasizes the essentials of Neyman–Pearson theory and other likelihood ratio procedures.

Colloquially, on the basis of observations "v", one must decide which of two hypotheses

$$H_o \text{ (the null) and } H_a \text{ (the alternative)}$$

is a more "reasonable" description of the source of the observations. More formally, if $\varphi(v)$ represents the probability with which H_o is to be rejected, then among such functions with

$$E[\varphi(V) \mid H_o \text{ true}] \le \alpha \text{ (given)} \qquad (*)$$

the object is to find that one for which

$$E[\varphi(V) \mid H_a \text{ true}] \text{ is greatest .}$$

In most cases, this formulation of testing is much too general to lead to solutions so that the problem can be studied only with various modifications.

Lesson 1 begins with a review of the story on the sale of pecans (Lesson 16, Part I) which suggests testing a simple hypothesis (one which specifies the distribution completely) against another simple hypothesis. This leads to the first form of the Neyman–Pearson (N–P) Lemma:

when the underlying distribution is continuous,

a "best test" φ satisfying (*) does exist.

A detailed discussion of this case for the mean of a normal distribution is given.

Lesson 2 begins by rephrasing the acceptance sampling in Lesson 16, Part I as testing with a hypergeometric distribution. The N–P Lemma is then extended to include such discrete populations but forces us to consider a "randomized" test. This is illustrated in the Binomial and Poisson distributions.

Lesson 3 contains a broader application of the N–P Lemma in a restricted family of distributions having "exponential likelihood" :

$$c(\theta)\ h(v)\ e^{-Q(\theta)T(v)}, \text{ where } \theta, Q, T \text{ are real valued.}$$

When Q is montonic in θ, there is a U(niformly) M(ost) P(owerful) T(est) of $H_o: \theta \le \theta_o$ against $H_a: \theta > \theta_o$. Here, uniformly, means that the same test is obtained for each $\theta > \theta_o$.

Although it may have not been noticed, all the discussion so far has concerned distributions whose support does not contain parameters. Lesson 4 contains some of the easier examples when this is not the case.

In the course of this developement, it has been pointed out that some "natural" hypotheses do not have a UMPT; this is primarily because the class of possible tests is still "too large". By putting additional restrictions on the tests, the size of the class is reduced and one may be able to "optimize" as in (*). One form of this in Lesson 5 contains consideration of Unbiased Tests for which the power satisfies:

$$E[\varphi(V) \mid H_a] \ge \alpha \ge E[\varphi(V) \mid H_o].$$

This is accomplished by judicious use of sufficiency and conditioning. A very interesting discussion concerning uniformly most powerful unbiased tests has been given by Suissa and Shuster, 1984..

Lesson 6 contains some distribution theory for normal random variables which is needed in other lessons.

The L(ikelihood) R(atio) can also be looked upon as another form of the N–P Lemma which, in the discrete case, is almost like:

1) if $P(V = v|\theta_o) > P(V = v|\theta_a)$, then we are more likely to observe v when θ_o is true than when θ_a is true;

2) so having observed v , we continue as if θ_o were true.

Naturally, this theory in lesson 7 is more particular. Some earlier examples of tests are seen to be LR tests as well.

The extension of the LR test for the difference of two normal means can be extended to the comparison of more than two normal means; this becomes one–way analysis–of–variance which is the subject of lesson 8.

In Lesson 9, detailed study of LR tests for Bernoulli distributions leads to classical Pearson chi–square goodness–of–fit tests for multinomials. This can be adapted to other cases; a good, but mature discussion, is in Moore, 1978.

The study of LR tests is completed by prescription of general asymptotic distributions in Lesson 10. This includes some other results on distributions related to the normal.

Lesson 11 contains general forms of virtually all the classical tests for univariate normal populations; these techniques form the core of many statistical methods courses. This is not really old old stuff, since some relevant material appeared in Guenther, 1981.

Since so much has been written on 2 by 2 tables of classification, this topic is treated in a separate lesson (12). The three classicial interpretations are presented.

Probability tables for various distributions are given at the end of this volume.

LESSON 1. NEYMAN–PEARSON THEORY – I

In Lesson 16, Part I, we introduced testing simple hypotheses by discussion of a particular example. We manipulated the idea of a sale of a sack of pecans into a plan for deciding which of

H_o: X is hypergeometric with given parameters N,n,D_o

H_a: X is hypergeometric with given parameters N,n,D_a

was "more reasonable" when the observation is $X = x$; here $D_o < D_a$. The intuitive solution was that we reject H_o only when the x is "too big" : $x \geq r$. Fixing r also fixes the sizes of Type I and Type II errors given by:

$$\begin{aligned}\alpha &= P(\text{Type I}) = P(\text{Reject } H_o | H_o \text{ is true}) \\ &= P(X \geq r | N,n,D_o) ,\end{aligned}$$

$$\begin{aligned}\beta &= P(\text{Type II}) = P(\text{Do not reject } H_o | H_o \text{ is false}) \\ &= P(X < r | N,n,D_a).\end{aligned}$$

Generalizing this discussion, will lead to the notion of a "best test".

First off, since we usually feel better looking at "numbers", most samples are summarized in terms of random variables or vectors; we start with the *sample space* $[\mathscr{V},\mathscr{B}]$ where $\mathscr{V} = \mathbb{R}^n$ and $\mathscr{B} = \mathscr{B}_n$, $n \geq 1$. The simple *null hypothesis* H_o is that the distribution of the observable V is the completely determined probability measure P_o; the simple *alternative hypothesis* H_a is that the distribution of V is the completely determined probability measure P_a. We assume that these distributions can be represented in terms of densities (L_o and L_a) wrt a common measure m; then for each $B \in \mathscr{B}$,

$$P_o(B) = \int_B L_o(v)\, dm(v) \text{ and } P_a(B) = \int_B L_a(v)\, dm(v) .$$

In the (absolutely) continuous case, these are ordinary (Riemann) Lebesgue integrals; in the discrete case, these are sums. For brevity, we write $\int_B L_o$, $\int_B L_a$.

A *test* is a rule for choosing one of H_o and H_a. It is sensible (though technically not required) to base the test on the observable V so that we select a set $B \in \mathscr{B}$ as the *critical region*:

if the observation $v \in B$, we reject H_o;

otherwise, we do not reject H_o.

For the Type I error, "Rejecting H_o when H_o is true", the size

$$\alpha = P(V \in B | H_o) = \int_B L_o .$$

For the Type II error, "Not rejecting H_o when H_a is true", the size is

$$\beta = P(V \notin B | H_a) = \int_{B^c} L_a .$$

As in the hypergeometric distribution, a decrease in α can be accomplished only by a "decrease " in B, accompanied by an "increase" in B^c and β. Since there is *in general* no way to decrease α and β simultaneously with $\mathscr{V}$ fixed (see Lesson 17, Part I, for another approach), statisticians, in particular J. Neyman and E.S.Pearson, 1933, suggested a slight change in the problem. For a fixed size of Type I error α, find B to minimize the size of Type II error. More precisely,

find B_* in $\mathscr{C} = \{B \in \mathscr{B} : \int_B L_o \le \alpha\}$ so that

$$\int_{B_*} L_o = \alpha \text{ and } \int_{B_*^c} L_a \le \int_{B^c} L_a \text{ for all } B \in \mathscr{C}.$$

The last inequality is equivalent to $\int_{B_*} L_a \ge \int_B L_a$. If we take B_* as a critical region, then the

$$\textit{significance level } \alpha = \int_{B_*} L_o \text{ and the}$$

$$\textit{power } 1 - \beta = P(\text{Reject } H_o | H_a) = \int_{B_*} L_a$$

is "as great as can be". In this relative sense, B_* is a *best* or *most powerful* critical region (BCR).

Neyman and Pearson (N–P) derived the solution using the calculus of variations but we can see quite easily (partly here and partly in the next lesson) that it is correct.

Lemma: *Let $k > 0$ be a constant such that*

$$B_* = \{v \in \mathcal{V} : L_a(v) > k \cdot L_o(v)\} \text{ has } \int_{B_*} L_o = \alpha.$$

Then, $\displaystyle\int_{B_*} L_a \geq \int_B L_a$ *for all* $B \in \mathcal{B}$ *such that* $\displaystyle\int_B L_o \leq \alpha$.

Proof: $\displaystyle\int_{B_*} L_a - \int_B L_a$

$$= \int_{B_* \cap B} L_a + \int_{B_* \cap B^c} L_a - \int_{B_* \cap B} L_a - \int_{B_*^c \cap B} L_a$$

$$= \int_{B_* \cap B^c} L_a - \int_{B_*^c \cap B} L_a .$$

Since in B_*, $L_a > kL_o$, $\displaystyle\int_{B_* \cap B^c} L_a \geq k\int_{B_* \cap B^c} L_o$.

Since in B_*^c, $L_a \leq kL_o$, $\displaystyle\int_{B_*^c \cap B} L_a \geq -k\int_{B_*^c \cap B} L_o$.

Therefore, $\displaystyle\int_{B_*} L_a - \int_B L_a \geq k(\int_{B_* \cap B^c} L_o - \int_{B_*^c \cap B} L_o)$.

The right–hand side is equal to

$$k(\int_{B_*} L_o - \int_B L_o) \geq k(\alpha - \alpha) = 0 .$$

It follows that $\displaystyle\int_{B_*} L_a \geq \int_B L_a$ as desired.

Exercise 1: With the notation of the lemma, show that

$$\int_{B_* \cap B^c} L_o - \int_{B_*^c \cap B} L_o = \int_{B_*} L_o - \int_B L_o .$$

The lemma shows that the given B_* i s most powerful but says nothing about whether such a $k > 0$ actually exists; the hypothesis of the lemma is a sufficient condition for the integral inequalities to hold. Let us pursue this a bit more. Since

$$P_o(\{V:L_a(V) > kL_o(V), L_o(V) = 0\}) \le P_o(\{V:L_o(V) = 0\}) = 0 ,$$

$$A(k) = P_o(\{V:L_a(V) > kL_o(V)\}) = P_o(\{V:L_a(V) > kL_o(V) > 0\})$$

.

Thus,

$$A(k) = P_o(\{L_a(V)/L_o(V) > k\}) = 1 - F(k)$$

where F is the CDF of the RV $W = L_a(V)/L_o(V)$ and all the measurability conditions have been met. If F is continuous, there is a $k_\alpha > 0$ such that $1 - F(k_\alpha) = \alpha$ and the lemma applies with $k = k_\alpha$. If F is not continuous (in particular if V is discrete), it can happen that the given $1 - \alpha$ "meets a jump" in F:

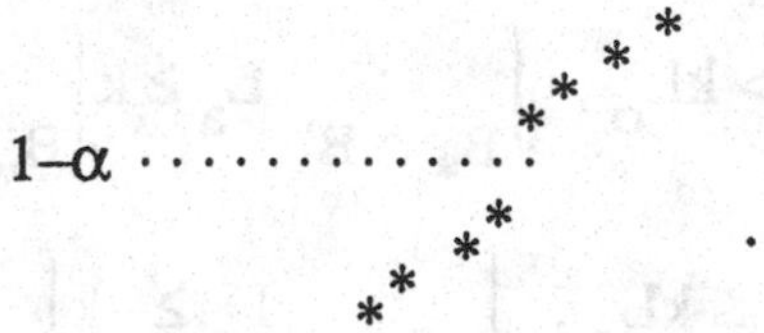

Then there is no solution $A(k) = \alpha = 1 - F(k)$. We will continue with this case in the next lesson; the rest of this lesson is devoted to a continuous case. But, first we need to say more about the notation and phraseology in applications, again by example.

Let $V = (X_1,X_2,\cdots,X_n)$. The simple null hypothesis is,

> "This is a random sample from a normal distribution with mean 1 and variance 1 ."

The simple alternative hypothesis is,

> "This is a random sample from a normal distribution with mean 2 and variance 1 ."

But methods books often use differing phraseologies such as,

"Let $V = (X_1,X_2,\cdots,X_n)$ be a RS from a normal

distribution with mean μ and variance $\sigma^2 = 1$; test $H_o : \mu = 1$ vs $H_a : \mu = 2$."

(We may also see H_o: $\mu = 1$, $n = 8$ and the like to emphasize some point.) Now rejection of H_o might just happen because one or more of the following actually obtains:

μ is not close to 1;

σ^2 is not close to 1;

the distribution is not really normal;

the sample is not really random.

A statistician's belief (hope!) is that the experiment has been performed in such a way that the last three of these conditions are "false"; then rejection of H_o is taken to imply that μ is not (close to) 1 .

Of course, this is the "application"; statistics (probability) merely gives the "odds". As with all mathematical modeling of "real life", justification for these "falsies" must come from subject matter experts and in comparing simple–simple hypotheses, these conditions are not part of the "test". A test of whether or not the distribution is normal is a test of a *composite* (*not simple*) hypothesis and such items will be taken up in later lessons. Now let's work out the N–P result.

Example: $V = (X_1, X_2, \cdots, X_n)$ is a RS from a normal distribution with mean μ and variance $\sigma^2 = 1$. Test H_o: $\mu = 1$ vs H_a: $\mu = 2$. The density referred to in the N–P lemma is the likelihood; here,

$$L_o = e^{-\Sigma(x_i - 1)^2/2}/(2\pi)^{n/2} \quad \text{and}$$

$$L_a = e^{-\Sigma(x_i - 2)^2/2}/(2\pi)^{n/2} .$$

The BCR B_* is determined by $L_a > k \cdot L_o$

$$\text{iff} \quad e^{-\Sigma(x_i - 2)^2/2} > k \cdot e^{-\Sigma(x_i - 1)^2/2} \qquad (*)$$

$$\text{iff} \quad e^{2\Sigma x_i - 2n} > k \cdot e^{\Sigma x_i - n/2}$$

$$\text{iff} \quad e^{\Sigma x_i} > k\cdot e^{3n/2}$$

$$\text{iff} \quad \Sigma x_i > \log(k) + 3n/2 = k_1$$

$$\text{iff} \quad \bar{x} > k_1/n = k_2 \qquad (**)$$

Note that once we have k_2 we can get k_1 and k . The advantage of the last inequality (**) over the first (*) is that we know the distribution of $\bar{X}$:

$$(\bar{X} - \mu)/(\sigma/\sqrt{n}) = Z \text{ is a standard normal RV.}$$

Thus, $\alpha = \int_{B_*} L_0 = P(\bar{X} > k_2 \mid \mu = 1)$

$$= P(Z > (k_2 - 1)/(1/\sqrt{n})) .$$

Now suppose that $\alpha = .05$ and $n = 16$. Then,

$$z_{.05} = 1.645 \text{ gives}$$

$$(k_2 - 1)/(1/\sqrt{16}) = 4(k_2 - 1) = 1.645$$

so that

$$k_2 \approx 1 + .411 = 1.411.$$

We reject H_0 when the observed $\bar{x} > 1.411$. Note that from

$$k_1 = 16\cdot k_2 \approx 22.576 \text{ , we get}$$

$$\log(k) = k_1 - 3(16)/2 \approx -1.424 \text{ and } k \approx .2369 .$$

But as far as a decision is concerned, we don't need the last bits of arithmetic; it is far easier to check $\bar{x} > 1.411$ than

$$e^{-\Sigma(x_i - 2)^2/2} > .2369e^{-\Sigma(x_i - 1)^2/2} .$$

Exercise 2: Continue this example. Suppose that $\alpha = .01$ and $n = 16$. Find the new k_2 and the new BCR.

You should have $k_2 \approx 1.5815$. Our lackey has just handed in a sample for which $\bar{x} = 1.5$. Now what ? We need to examine the "art" in statistics:

a) if we had chosen $\alpha = .05$ and the N–P theory, then we would reject H_o because $1.5 > 1.411$;

b) if we had chosen $\alpha = .01$ and the N–P theory, then we would not reject H_o because $1.5 < 1.5815$.

In both cases we have the "best test". But note the *subjunctive*: if $\cdots$, then $\cdots$. This is always the mood of the statistician; never do we have merely "Do this". Decision theory, in particular this N–P theory and its variations, continues to require experience, gut feelings, whatever. Here we have to choose α ; in more complicated situations, H_o and H_a must also be defined.

In recent years it has become fashionable (again) to address the subjectivity at a different point and (suprize?) with another description. We use the lemma to get the *shape* of a BCR; in the example we still want to reject H_o when $\bar{x}$ is "too big". Now suppose that we take a sample and get, say, $\bar{x} = 1.5$. Instead of using the fixed rule determined by a given α, we ask "What would α have been if the B_* were $\{\bar{X} > 1.5\}$?"

The answer is

$$P(\bar{X} > 1.5 \mid \mu = 1) = P(Z > (1.5 - 1)4)$$
$$= P(Z > 2) \approx 1 - .9773 = .0227.$$

If, in the particular application to which this sample applies, odds of 2 to 98 indicate "rare" events, then we reject H_o; now the subjectivity is in "rare". The statistician reports this *observed significance level* or *p-value* (here .0227) and lets the client make the decision.

Exercise 3: For a RS from the normal distribution with mean μ and $\sigma^2 = 1$, find the p–value for a N–P test of $H_o: \mu = 6$ vs $H_a: \mu = 8$ when

a) $n = 9, \bar{x} = 6.8.$

Hint: $P(\bar{X} > 6.8 \mid \mu = 6, n = 9) = P(Z > 2.4)$.

b) $n = 16, \bar{x} = 6.8$ c) $n = 16, \bar{x} = 7.8.$

LESSON 2. NEYMAN–PEARSON THEORY – II

We continue from Lesson 1 by picking up the N–P theme when the CDF F of $W = L_a(V)/L_o(V)$ is not continuous and $\alpha = A(k) = P(W > k) = 1 - F(k)$ does not have a solution at a jump point:

```
                                 * F
                               *
                             **
                            *
jump point: 1 − α ..........      .
                          *
                        **
                       *
                         k
                          o
```

As we did in the hypergeometric case, we could eliminate the problem by changing α to $1 - F(k_o)$; usually, this is what is done in real applications. We use the N–P lemma to get the shape of the BCR but adjust the value of α so that a solution can be obtained. Better yet, if one looks at the p–value, this problem never arises.

Example: Let $V = (X_1, X_2, \cdots, X_n)$ be a RS from a Bernoulli distribution with

$$P(X_i = 1) = \theta = 1 - P(X_i = 0) .$$

Test $H_o: \theta = \theta_o = .3$ vs $H_a: \theta = \theta_a = .2$. Here the (non–zero part of the) likelihood is

$$L_\theta(v) = (\theta)^{\Sigma x_i}(1-\theta)^{n-\Sigma x_i} .$$

Now $L_{\theta_a}(v) > k \cdot L_{\theta_o}(v)$

$$\text{iff} \qquad (.2)^{\Sigma x_i}(.8)^{n-\Sigma x_i} > k(.3)^{\Sigma x_i}(.7)^{n-\Sigma x_i}$$

$$\text{iff} \quad \left[(.2)(.7)/(.3)(.8)\right]^{\Sigma x_i} > k$$

$$\text{iff} \qquad y < k$$

where $y = \Sigma x_i$ and k is a generic constant, eliminating the subscripts $k_1, k_2, \cdots$ used in the previous lesson. Here again the final inequality has the advantage that we know the distribution of $Y = \sum_{i=1}^{n} X_i$; it is binomial with parameters n and θ . Let's take $n = 20$. Under H_o, the CDF of Y is representable by

y	$P(Y \le y \mid \theta = .3)$
0	.0008
1	.0076
2	.0354
3	.1070
. . .	.

so that these are the only attainable α . For example, if the observed value is 2 , then the p–value is

$$P(Y \le 2) = .0354 . \text{ Etc.}$$

Exercise 1: Of course, a sample of n Bernoulli trials is equivalent to a sample of one binomial trial with parameters n and θ; the non–zero part of the likelihood is now

$$\binom{n}{y} \theta^y (1-\theta)^{n-y} .$$

a) Use these ideas to derive the N–P BCR for testing $H_o: \theta = .4$ vs $H_a: \theta = .6$ with $n = 15$.

b) What is the p–value when $y = 8$?

Next we re–examine the hypergeometric experiment again. The N–P BCR is determined by

$$\binom{D_a}{x}\binom{N-D_a}{n-x} \div \binom{N}{n} > k \binom{D_o}{x}\binom{N-D_o}{n-x} \div \binom{N}{n}$$

or, $(D_o-x)!(N-D_o-n+x)!/(D_a-x)!(N-D_a-n+x)!$

$$> kD_o!(N-D_o)!/D_a!(N-D_a)!$$

say,

$$h(x) > k_1 . \qquad (*)$$

Since (see below) $h(x) > h(x-1)$, the inequality (*) is equivalent to the inequality $x > k_2$. Thus the N–P formalism leads to a critical region of the same shape as our intuition suggested in Lesson 16, Part I.

Exercise 2: Show that, (ignoring picky details at $x = 0, n$),

$$h(x) = (D_o-x)!(N-D_o-n+x)!/(D_a-x)!(N-D_a-n+x)!$$

is a strictly increasing function of the positive integer x.

To return to the general discussion, suppose that we have

$$1 - F(k_o) < \alpha < 1 - F(k_o - 0) \text{ with}$$

$$\begin{aligned} P(W = k_o) &= (1 - F(k_o - 0)) - (1 - F(k_o)) \\ &= F(k_o) - F(k_o - 0) > 0 . \end{aligned}$$

In order to give the new rule a neat form, we first evaluate these probabilities "under H_o", that is assuming L_o, and then construct a "coin" with

$$\begin{aligned} P(\text{Head}) = 1 - P(\text{Tail}) &= [\alpha - P(W > k_o)]/P(W = k_o) \\ &= [\alpha - 1 + F(k_o)]/[F(k_o) - F(k_o - 0)] = \pi_o . \end{aligned}$$

When a toss of this coin turns up head, we reject H_o; when a toss of this coin turns up tail, we do not reject H_o. In this way, we get a new process which rejects H_o with probability π_o. Now the general N–P rule is:

when $L_a(v) > k_o \cdot L_o(v)$, reject H_o;

when $L_a(v) = k_o \cdot L_o(v)$, reject H_o with probability π_o as above;

when $L_a(v) < k_o \cdot L_o(v)$, do not reject H_o.

The probability that we reject H_o when H_o is true is

$$\begin{aligned} &P_o(L_a(V) > k_o \cdot L_o(V)) + P_o(\{\text{Head}\} \cap \{L_a(V) = k_o \cdot L_o(V)\}) \\ &= P_o(W > k_o) + P(\text{Head} \mid W = k_o) \cdot P_o(W = k_o) \\ &= 1 - F(k_o) + \frac{\alpha - 1 + F(k_o)}{F(k_o) - F(k_o - 0)} \cdot [F(k_o) - F(k_o - 0)] = \alpha . \end{aligned}$$

This test involves additional randomization but we do get the exact significance level prescribed. (It is not the fault of mathematics that many statisticians' clients are not impressed by this and prefer the p–value approach!) Formally, we have

Definition: *Let* $[\mathcal{V}, \mathcal{B}, P]$ *be a probability space. The measurable function* $\tau : \mathcal{V} \to [0,1]$ *is a randomized test of size* α *for the simple hypothesis* H_o *if* $E[\tau(V)|H_o] = \alpha$. *A non-randomized* τ *takes only the values 0 and 1 and* $\tau^{-1}(\{1\})$ *is the critical (rejection) region.*

Note that for a non–randomized test, τ is the indicator function of the BCR. The proof of the following lemma is very close to that of the first lemma in lesson 1. We use more shorthand for the hypotheses.

Lemma: Let τ_* be any most powerful level α test of the simple H_o prescribed by L_o vs the simple H_a prescribed by L_a . The general N–P rule given above can be written as

$$\begin{aligned} \tau(v) &= 1 \quad \text{for } L_a(v) > k \cdot L_o(v) \\ &= \pi_o \quad \text{for } L_a(v) = k \cdot L_o(v) \\ &= 0 \quad \text{for } L_a(v) < k \cdot L_o(v) . \end{aligned}$$

Then, for the common measure m,

$$m(\{v : \tau(v) - \tau_*(v) \neq 0\} \cap \{v: L_a(v) \neq k \cdot L_o(v)\}) = 0 .$$

Proof: a) When $\tau(v) - \tau_*(v) \neq 0$ and $L_a(v) - k \cdot L_o(v) > 0$, $\tau(v) - \tau_*(v) = 1 - \tau_*(v)$ is also positive; let the set of all such "v" be A .

b) When $\tau(v) - \tau_*(v) \neq 0$ and $L_a(v) - k \cdot L_o(v) < 0$, $\tau(v) - \tau_*(v) = 0 - \tau_*(v)$ is also negative; let the set of such "v" be B .

c) If $m(A \cup B) > 0$, so is $\int (1 - \tau_*)(L_a - k \cdot L_o) dm$

$$= \int_A (1 - \tau_*)(L_a - k \cdot L_o)dm + \int_B \tau_*(k \cdot L_o - L_a)dm .$$

This implies $\int(\tau - \tau_*)L_a dm > 0$ which means that τ is more powerful than τ_* contradicting the hypothesis. Therefore, $m(A \cup B) = 0$ and this condition yields the conclusion.

As a consequence of this result, the N–P theory is saying that the most powerful test of the simple H_o given by L_o vs the simple H_a given by L_a is the τ listed in this last lemma. Further discussion can be found in the book by Lehmann, 1959.

When F is continuous, π_o is taken to be zero and, in effect, the test is to reject H_o iff $L_a > k \cdot L_o$. In a real experiment, equality might occur arithmetically and, technically, H_o is not rejected; then one might adjust α even though this violates its interpretation as a rate of error. Or, better measurements and arithmetic might remove the problem. As always, calculation of the p–value avoids that problem.

Exercise 3: Prove the following

Corollary: *Let* τ *be the N-P most powerful test as in the last Lemma with* $0 < \alpha < 1$ *. Then the power* $1 - \beta = \int \tau L_a \geq \alpha$.

Hint: what is the power of the trivial test $\tau_*(v) \equiv \alpha$?

Example: Let $V = (X_1, X_2, \cdots, X_n)$ be a RS from the Poisson distribution with parameter λ . Test

$$H_o: \lambda = 1/3 \text{ vs } H_a: \lambda = 1/2 .$$

The non–zero portion of the likelihood is $e^{-n\lambda}(\lambda)^{\Sigma x_i}/\Pi x_i!$ for $x_1, x_2, \cdots x_n$ non–negative integers. Here $L_a > k \cdot L_o$

$$\text{iff } e^{-n/2}(1/2)^{\Sigma x_i} > k \cdot e^{-n/3}(1/3)^{\Sigma x_i}$$

iff $(3/2)^{\Sigma x_i} > k$

iff $\Sigma x_i > k$.

Under H_o , $Y = \sum_{i=1}^{n} X_i$ has a Poisson distribution with parameter $n/3$.

a) Let's try $n = 6$ and $\alpha = .02$. We have:

y	$P(Y > y \mid \lambda = 2)$
0	1.0000
1	.8647
2	.5940
3	.3233
4	.1429
5	.0527
6	.0166
7	.0045
8	.0011

The significance level is

$P(Y > y_o) + [\alpha - P(Y > y_o)]/P(Y = y_o)$

$= P(Y > 6) + [.02-P(Y>6)]/[P(Y>5)-P(Y>6)] \cdot P(Y=6)$

$\approx P(Y > 6) + [.0034 / .0361] \cdot P(Y = 6)$

$= .0166 + .0526(.0361) = .02$.

If the total y is greater than 6, we reject H_o; if the total y equals 6, we toss a coin to reject H_o with probability .0526; if the total y is less than 6, we do not reject H_o. This is the most powerful test of $\lambda = 1/3$ vs $\lambda = 1/2$ at level .02 .

b) If $n = 6$ but $\alpha = .05$, the best test is:

reject H_o when $y > 6$;

toss a coin with $P(\text{Reject } H_o) = 334/361$ when $y = 6$;

do not reject H_o when $y < 6$.

Exercise 4: Find the most powerful test at level $\alpha = .05$ for

$H_o: \lambda = 1/3$ vs $H_a: \lambda = 1/4$ when the sample size $n = 6$ and the distribution is Poisson.

This is not all there is to the N–P theory which has been generalizd in several directions; we shall see two in later lessons.

Example: Again $V = (X_1, X_2, \cdots, X_n)$ is a RS from $N(\mu, \sigma)$, $\sigma = 1$; we want to test $H_o: \mu = 1$ vs $H_a: \mu = 2$. For the moment, $n = 16$, $\alpha = .05$.

a) The most powerful test has the BCR $\bar{x} > 1.411$.

b) Consider the same setup except that $H_a: \mu = 3$. Then, $L_a > k \cdot L_o$

$$\text{iff } e^{-\Sigma(x_i - 3)^2/2} > k \cdot e^{-\Sigma(x_i - 1)^2/2}$$

$$\text{iff } \quad 3\Sigma x_i > k \cdot 2\Sigma x_i$$

$$\text{iff } \quad \bar{x} > k$$

which is exactly the same shape as in a). Moreover, the value of k is determined under H_o exactly as in a) and we get the same BCR $\bar{x} > 1.411$.

Exercise 5: Continue the spirit of the last example. Show that for each $\mu_o < \mu_a$,

$$e^{-\Sigma(x_i - \mu_a)^2/2} > k \cdot e^{-\Sigma(x_i - \mu_o)^2/2} \text{ iff } \bar{x} > k \text{ (generic).}$$

Exercise 6: Let $V = (X_1, X_2, \cdots, X_n)$ be a RS from the exponential distribution with underlying density $\theta e^{-\theta x} I(x > 0)$. Consider $H_o: \theta = \theta_o$ vs $H_a: \theta = \theta_a$ and the N–P theory.

a) Show that there is only one BCR for any $\theta_a > \theta_o$.

b) Show that there is only one BCR for any $\theta_a < \theta_o$.

Example: The same phenomenon occurrs in the Bernoulli case:

$$\theta_a{}^y(1-\theta_a)^{n-y} > k\cdot\theta_o{}^y(1-\theta_o)^{n-y}$$

iff

$$[\theta_a(1-\theta_o)/\theta_o(1-\theta_a)]^y > k\,.$$

This is equivalent to $y > k_1$ when $\theta_o < \theta_a$ and to $y < k_2$ when $\theta_o > \theta_a$. Each value k_1, k_2 is determined under H_o: $\theta = \theta$ and does not depend on the exact alternative θ_a under consideration, only its relation to θ_o.

Whenever a condition obtains "for all$\cdots$", mathematicians use the adverb uniformly. For example, in exercise 5, $\bar{x} > k$ is a "uniformly" most powerful critical region for testing

$$H_o: \mu = \mu_o \text{ vs } H_a: \mu > \mu_o$$

since the same k is obtained for (any, each, every) all $\mu > \mu_o$. Of course, similar remarks apply to the other examples studied herein.

Definition: *A test* τ *is called uniformly most powerful of level* α for *testing the simple hypothesis* H_o *against the composite hypothesis* H_a*, if this test is most powerful of level* α *for each simple hypothesis in* H_a.

One sees and uses the abbreviation UMP or, emphasizing the geometry, UMPCR, with α understood.

Exercise 7: Consider one observation of S^2 where S^2/σ^2 has a chi–square distribution with n–1 degrees of freedom. Find the uniformly most powerful critical region for testing $H_o: \sigma^2 = 1$ against $H_a: \sigma^2 > 1$ with $n = 20$ and $\alpha = .05$.

LESSON 3. TESTING WITH MONOTONE LIKELIHOOD RATIOS

In addition to their having UMPCRs for certain hypotheses, the densities in the examples at the end of the previous lesson have one more thing in common. They are all "exponential families":

$$e^{(x-\mu)^2/2}/\sqrt{(2\pi)} = e^{\mu x - \mu^2/2 - x^2/2}$$

$$e^{-\lambda}\lambda^x/x! = e^{[\log(\lambda)]x - \lambda - \log(x!)}$$

$$\theta^x(1-\theta)^{1-x} = e^{[\log(\theta/(1-\theta))]x + \log(1-\theta)}$$

$$\theta e^{-\theta x} = e^{-\theta x + \log(\theta)} .$$

In this lesson, we want to get some general theorems along this line of UMPCR and to begin that process, we first realign our notions on testing.

Let the measurable space $\mathscr{V} = \mathbb{R}^n$, $\mathscr{B} = \mathscr{B}_n$ have an associated family of probability distributions $\{P_\theta\}$ indexed by a set Θ. The null hypothesis specifies that θ is in a subset and the alternative hypothesis specifies that θ is in its complement, symbolically,

$$H_o: \theta \in \Theta_o \subset \Theta \text{ vs } H_a: \theta \in \Theta_a = \Theta - \Theta_o .$$

A test function τ is a $(\mathscr{B}, \mathscr{B}_1)$ measurable map of $\mathscr{V}$ into [0,1]; its power function maps Θ into [0,1] with values

$$\mathscr{P}_\tau(\theta) = E[\tau(V) \mid \theta] .$$

The significance level α of the test is the size of type I error:

$$\alpha = \sup \{ \mathscr{P}_\tau(\theta) : \theta \in \Theta_o \} .$$

The sizes of type II error are $1 - \mathscr{P}_\tau(\theta)$ for $\theta \in \Theta_a$.

Let m be a fixed measure on $\mathscr{B}$; the $(\mathscr{B}, \mathscr{B}_1)$ measurable function L_θ on $\mathscr{V}$ is a density for P_θ wrt m iff

$$P_\theta(B) = \int_B L_\theta(v)\, dm(v) \text{ for all } B \in \mathscr{B}.$$

Then $\{P_\theta\}$ or $\{L_\theta\}$ is an exponential family if for each $\theta \in \Theta$,

$$L_\theta(v) = c(\theta)h(v)\exp \sum_{j=1}^{d} Q_j(\theta)T_j(v) .$$

Of course, all these functions are real valued and $h, T_1, \cdots, T_d$ are to be $(\mathscr{B},\mathscr{B}_1)$ measurable. The functions $1, Q_1, \cdots, Q_d$ are linearly independent on Θ and the funcitons $1, T_1, \cdots, T_d$ are linearly independent on V.

When $V = (X_1,\cdots,X_n)$ is actually a RS from an underlying population with density

$$f(x;\theta) = a(\theta)b(x)\exp\sum_{j=1}^{d} Q_j(\theta)t_j(x) ,$$

then in $L_\theta(v) = \Pi_{i=1}^{n} f(x_i;\theta)$,

$c(\theta) = (a(\theta))^n$, $h(v) = \Pi_{i=1}^{n} b(x_i)$, $T_j(v) = \Sigma_{i=1}^{n} t_j(x_i)$.

Sometimes $\{f(;\theta)\}$ is called an exponential family.

Example: The univariate normal distribution is an exponential family with $d = 2$: $e^{-(x-\mu)^2/2\sigma^2}/\sigma\sqrt{2\pi}$

$$= [e^{-\mu^2/2\sigma^2}/(2\pi\sigma^2)^{1/2}]\cdot\exp[(-1/2\sigma^2)x^2 + (\mu/\sigma^2)x]$$
$$= a(\theta)\exp[\theta_1 t_1(x) + \theta_2 t_2(x)]$$

where $\theta = (\theta_1,\theta_2)$, $Q_1(\theta) = \theta_1 = -1/2\sigma^2$, $Q_2(\theta) = \theta_2 = \mu/\sigma^2$,

$t_1(x) = x^2$, $t_2(x) = x$, $b(x) = 1$, $a(\theta) = [\exp(\theta_2^2/4\theta_1)]/(\pi/\theta_1)^{1/2}$.

For a RS $X_1,\cdots,X_n$,

$$T_1(V) = \Sigma X_i^2 , T_2(V) = \Sigma X_i .$$

Exercise 1: Find "Q,T,c,h" for the population densities with the indicated support and a RS of size n:

$\theta^x(1-\theta)^{1-x}$, $x = 0,1$

b) $\theta e^{-\theta x}$, $x > 0$

c) $e^{-\lambda}\lambda^x/x!$, $x = 0(1)\infty$

d) $(1/2\pi\sigma_1\sigma_2\sqrt{(1-\rho^2)})\exp[-W/2(1-\rho^2)]$ where x, y, μ_1, μ_2 are real, σ_1 and σ_2 are positive, $\rho^2 < 1$ and W is the quadratic form

$$(x-\mu_1)^2/\sigma_1^2 - 2\rho(x-\mu_1)(y-\mu_2)/\sigma_1\sigma_2 + (y-\mu_2)^2/\sigma_2^2 .$$

Hint: the dimension of $\mathscr{V}$ is $2n$ and $d = 5$.

Example: a) In the general exponential family, let $d = 1$ and consider testing $H_0: \theta = \theta_0$ vs $H_a: \theta = \theta_a$ via N–P theory.

$$L_{\theta_a} > k \cdot L_{\theta_0}$$

iff $\quad c(\theta_a)h(v)\exp[Q(\theta_a)T(v)] > k \cdot c(\theta_0)h(v)\exp[Q(\theta_0)T(v)]$

iff $\quad (Q(\theta_a) - Q(\theta_0))T(v) > k_1$.

When, $Q(\theta_a) - Q(\theta_0)$ is positive, this is equivalent to $T(v) > k_2$.
When $Q(\theta_a) - Q(\theta_0) < 0$, this is equivalent to $T(v) < k_2$.

b) If, more generally, we find $L_{\theta_a}/L_{\theta_0}$ as a function of the real $t = T(v)$, say $w(t)$, then $w(t) > k$ is equivalent to $t > k_2$ iff w is a non-decreasing function of t. (In all cases, we ignore v with $L_\theta(v) = 0$ as such points contribute nothing to the probabilities involved.)

Lemma: *Let* $\{L_\theta\}$ *denote a family of densities. For the simple* $H_0: \theta = \theta_0$ *consider the particular alternative*

$$H_a: \theta \in \Theta_a = \{\theta : L_\theta/L_{\theta_0} = w_\theta(t(v))\}$$

where each w_θ *is a non-decreasing function of* $t = t(v)$. *The uniformly most powerful test of* H_o *vs* H_a *is given by*

$$\tau(v) = 1\ ,\ = \pi_o\ ,\ = 0$$

according as

$$t(v) > k_o\ ,\ = k_o\ ,\ < k_o\ .$$

Proof: For each $\theta \in W_a$, the N–P best test is

$$\tau(v) = 1\ ,\ = \pi_o\ ,\ = 0$$

according as $w_\theta(t(v) > k\ ,\ = k\ ,\ < k$

equivalently $t(v) > k_o\ ,\ = k_o\ ,\ < k_o\ .$

Since π_o and k_o are determined under H_o, this same test is most powerful for each $\theta \in \Theta_a$; therefore this test is uniformly most powerful for testing H_o vs H_a.

Unfortunately, the authors are not aware of any general distributions which fit so neat a pattern. Most of the general results involving such monotonicity are for real parameters.

Definition: *The family of probability spaces* $[\mathcal{V}, \mathcal{B}, P_\theta]$ *with density* L_θ *wrt a given* m *has index set* $\Theta \subset R$. $\{P_\theta\}$ *posseses a monotone increasing likelihood ratio [MILR] iff*

1) *there is a* $(\mathcal{B}, \mathcal{B}_1)$ *measurable* $t : \mathcal{V} \to R$;
2) *for each* $\theta_o < \theta_a$ *in* Θ, *there is a non-decreasing* $(\mathcal{B}_1, \mathcal{B}_1)$ *measurable function* w_{θ_o,θ_a} *such that*

$$L_{\theta_a}/L_{\theta_o} = w_{\theta_o,\theta_a}(t)\ .$$

Written out fully, the second condition in the definition is:

$$L_{\theta_a}(v)/L_{\theta_o}(v) = w_{\theta_o,\theta_a}(t(v))$$
$$\le w_{\theta_o,\theta_a}(t(v')) = L_{\theta_a}(v')/L_{\theta_o}(v')$$

whenever $t(v) \le t(v')$ which should explain the name. If w_{θ_o,θ_a} is non–increasing, the likelihood ratio is *decreasing* [MDLR] ; note that the condition $\theta_o < \theta_a$ is not changed.

When the population density ratio $f(x;\theta_a)/f(x;\theta_o)$ is non–decreasing [non–increasing] in some $s = s(x)$, then $\{f(\ ;\theta)\}$ has, correspondingly, an MILR [MDLR] .

Example: The family of *Logistic Distributions* on R, (indexed by the real parameter θ) has density

$$f(x;\theta) = e^{-x-\theta} \div \left[1 + e^{-x-\theta}\right]^2 .$$

For $\theta_o < \theta_a$, $[\exp(-\theta_o) - \exp(-\theta_a)] > 0$; then,

$$f(x;\theta_a)/f(x;\theta_o) = \frac{\left[[\exp(-x-\theta_a)] \div [1 + \exp(-x-\theta_a)]^2\right]}{\left[[\exp(-x-\theta_o)] \div [1 + \exp(-x-\theta_o)]^2\right]}$$

$$< \frac{\left[[\exp(-x'-\theta_a)] \div [1+\exp(-x'-\theta_a)]^2\right]}{\left[[\exp(-x'-\theta_o)] \div [1 + \exp(-x'-\theta_o)]^2\right]}$$

$$= f(x';\theta_a)/f(x';\theta_o)$$

$$\text{iff} \quad [1 + \exp(-x-\theta_o)]/[1 + \exp(-x-\theta_a)]$$

$$< [1 + \exp(-x'-\theta_o)]/[1 + \exp(-x'-\theta_a)]$$

$$\text{iff} \quad e^{-x}[\exp(-\theta_o) - \exp(-\theta_a)]$$

$$< e^{-x'}[\exp(-\theta_o) - \exp(-\theta_a)]$$

$$\text{iff} \quad -x < -x' .$$

The logistic density has an MILR in $s = s(x) = -x$ or an MDLR in x . Note that this density is not that of an exponential family.

Exercise 2: Find MILR (MDLR) conditions for the densities:

a) the normal distributions with parameter $\theta = \mu$ and σ^2 known;

b) the normal distributions with parameter $\theta = \sigma$ and μ known;

c) the binomial distributions with parameter θ and n known;

d) the gamma distributions with parameter $\theta = \alpha$ and β known;

e) the gamma distributions with parameter $\theta = \beta$ and α known.

Exercise 3 : Show that for the general exponential family with $d = 1$ to have MILR or MDLR, Q must be strictly monotonic.

Example: The Cauchy family does not have an MLR in x:

$$f(x;\theta_a)/f(x;\theta_o) = [1 + (x - \theta_o)^2]/[1 + (x - \theta_a)^2]$$

has limit 1 as $x \to \infty$ and as $x \to -\infty$; therefore, this ratio cannot be monotone. Moreover,

$$f(x;\theta_a)/f(x;0) > k$$

is equivalent to $x^2(1-k) + 2\theta_a kx + (1-k-k\theta_a)^2 > 0$ so that even the shape of each BCR depends on k , θ_a and, α .

The best we can do with all this is the

Theorem: *Let* $\{\mathcal{V}, \mathcal{B}, P_\theta\}$, $\theta \in R$, *have an MILR in* $t = t(v)$; *let* $\tau(v) = 1$, $= \pi_o$, $= 0$ *according as* $t > k$, $= k$, $< k$. *Then,*

1) *the power function* $\mathcal{P}_\tau(\theta) = E[\tau(V)|\theta]$ *is non-decreasing in* θ ;

2) *any test of this form with* $\mathcal{P}_\tau(\theta_o) = E[\tau(v)|\theta_o] > 0$ *is UMP for testing* $H_o: \theta \le \theta_o$ *vs* $H_a: \theta > \theta_o \in \Theta$;

3) *for each* $0 \le \alpha \le 1$, *each* $\theta_o \in \Theta$, *there is a test of this form of level* α *and UMP for testing as in 2).*

Proof: The lemma above gives the UMP test of

$$H_{o1}: \theta = \theta_o \text{ vs } H_a: \theta > \theta_o$$

in the form $\tau(v) = 1 , = \pi_o , = 0$
according as $t(v) > k_o , = k_o , < k_o$.

a) For $\theta_b < \theta_o$, let $E[\tau(V) | \theta_b] = \alpha_b$. By the necessity part of the N–P theory, the most powerful test of $H_{o2}: \theta = \theta_b$ vs $H_{o1}: \theta = \theta_o$ at level α_b is given by $\tau_*(v) = 1 , = \pi_b , = 0$ according as $L_{\theta_o} / L_{\theta_b} > k , = k , < k$, or, again by MILR, according as $t(v) > k_b , = k_b , < k_b$. Then we have $E[\tau_*(V) | \theta_b] = \alpha_b = E[\tau(V) | \theta_b]$

or $P_{\theta_b}(t(V) > k_b) + \pi_b P(t(V) = k_b)$

$$= P_{\theta_b}(t(V) > k_o) + \pi_o P_{\theta_b}(t(V) = k_o) .$$

By properties of the distribution function (of $t(V)$), we must have $k_b = k_o$ and $\pi_b = \pi_o$ so that $\tau_* = \tau$ (almost surely).

b) For $\theta_b < \theta_o < \theta_a$, separate applications of the last corollary in the previous lesson, yield $\alpha_b \le \alpha \le \alpha_a$ or $E[\tau(V) | \theta_b] \le E[\tau(V) | \theta_o] \le E[\tau(V) | \theta_a]$

which is the montonicity of the power function.

c) Now let $\mathcal{F}$ be the set of test functions φ with

$$\sup \{E[\varphi(V) | \theta] : \theta \le \theta_o\} = \alpha ;$$

let $\mathcal{F}_o = \{\varphi : 0 \le \varphi(V) \le 1 \text{ and } E[\varphi(V) | \theta_o] = \alpha\}$. Obviously, $\tau \in \mathcal{F} \subset \mathcal{F}_o$. Since $\tau = \tau_*$ maximizes the power over $\mathcal{F}_o$, it maximizes the power over $\mathcal{F}$. Therefore, we get the same most powerful test τ for each $\theta_b \le \theta_o$ vs each $\theta_a > \theta_o$. Thus τ is UMP.

The last part of the theorem with its proof should be extracted as a

Lemma: *Suppose that $\theta \in \Theta \subset R$ is the only unknown parameter. Let τ be a UMP test of $H_o: \theta = \theta_o$ against $H_a: \theta > \theta_o$. If $\mathcal{P}_\tau(\theta) = E[\tau(V)|\theta]$ is non-decreasing for all $\theta \in \Theta$, then τ is uniformly most powerful for testing*

$$H_o: \theta \le \theta_o \text{ against } H_a: \theta > \theta_o .$$

Exercise 4: State and prove a theorem about a UMP test of $H_o: \theta \ge \theta_o$ vs $H_a: \theta < \theta_o$ under an MDLR condition.

The theorem and exercise are somewhat more general forms of results for the exponential family with d = 1; for such families, there are UMP tests of $\theta \le \theta_o$ vs $\theta > \theta_o$ iff Q is strictly monotonic. The following traditional example shows that, even then, two UMP tests for one–sided hypotheses may not combine to form a UMP test for a two–sided hypothesis.

Example: $X_1, X_2, \cdots, X_n$ is a RS from $N(\theta, \sigma^2)$, σ^2 known.

a) The UMP test for $H_o: \theta = \theta_o$ vs $H_a: \theta > \theta_o$ is to reject H_o when $\bar{x} > k_a$.

b) The UMP test of H_o vs $H_{a1}: \theta < \theta_o$ is to reject H_o when $\bar{x} < k_b$.

c) Since $P(\bar{X} > k_a | \theta = \theta_o) = \alpha = P(\bar{X} < k_b | \theta = \theta_o)$,

$k_a = \theta_o + z_\alpha \sigma/\sqrt{n}$ and $k_b = \theta_o - z_\alpha \sigma/\sqrt{n}$

where for the standard normal Z, $P(Z > z_\alpha) = \alpha$. This leads to the usual test of H_o vs $H_{a2}: \theta \ne \theta_o$: reject H_o when $\sqrt{n}|\bar{x} - \theta_o|/\sigma > z_\alpha$ with level 2α.

(For level .05 , take $\alpha = .025$, etc.) But this test cannot be UMP since it conflicts, separately, with the UMP tests of $\theta < \theta_o$ and $\theta > \theta_o$.

The expression of MLR in terms of a function

$$w_\theta(t(v))$$

is reminiscent of the expression of sufficiency:

$$L_\theta(v) = g(t(v);\theta)h(v) .$$

Then a N–P test may always be generated in terms of a sufficient statistic since

$$L_{\theta_a}(v)/L_{\theta_o}(v) \geq k \text{ iff } g(t(v);\theta_a)/g(t(v);\theta_o) \geq k .$$

The discussion in Lesson 6, Part VII, shows that, unfortunately, this does not always lead to UMP tests nor visa–versa.

L. Schmetterer, 1974, has a theorem (with a long proof) which states that if, for a real parameter, a most powerful test for a composite alternative exists for every level α , then $\{L_\theta\}$ must have an MLR. For variations on montonicity in decision problems, see S. Karlin, 1957.

LESSON 4. TESTING WHEN THE SUPPORT CONTAINS PARAMETERS

In this lesson, we consider application of the N–P theory of testing when the support of the underlying population density (with respect to Lebesgue measure) depends upon the parameters. The following are models for two very real problems.

Examples: a) "Gaps" between cars at constant speeds have min $= \theta_1$, max $= \theta_2$.

b) Crushing strengths of concrete mixes have a *triangular distribution*: the family of densities (indexed by the parameters $\theta_1 > \theta_2 > 0$) is

$$f(x;\theta) = [(x-\theta_1+\theta_2)I\{\theta_1-\theta_2 < x < \theta_1\}$$
$$+ (\theta_1+\theta_2-x)I\{\theta_1 \le x < \theta_1+\theta_2\}]/\theta_2^{\,2}.$$

Most of our discussion will concern a traditional uniform distribution: for $\theta = (\theta_1,\theta_2)$, θ_1 real and θ_2 positive,

$$f(x;\theta) = I\{\theta_1 - \theta_2 < x < \theta_1 + \theta_2\} / 2\theta_2. \qquad (*)$$

Some general results will be suggested in exercises.

If H_o limits the support to a set A, like (0,1), and H_a limits the support to a set B disjoint from A, like (1,2), we know what to do:

reject H_o iff at least one observation is in B. Theoretically,

$$\alpha = P(\text{Reject } H_o \,|\, H_o \text{ is true})$$
$$= 0 = P(\text{Do not reject } H_o \,|\, H_a \text{ is true}) = \beta$$

no matter what other properties the distribution may have. When the probability is 0 that any observation is outside some set S under H_o and under H_a ($A \cup B$ in the previous remarks), we can, and will, ignore the complement S^c in power calculations.

Exercise 1: The likelihood for a RS from the population with density (*) is

$$L_\theta(v) = I\{\theta_1 - \theta_2 < x_{(1)} \le x_{(n)} < \theta_1 + \theta_2\} / (2\theta_2)^n$$

where $x_{(1)}$ is the minimum and $x_{(n)}$ is the maximum of the observations $v = \{x_1, \cdots, x_n\}$. Show that the min and max, $X_{(1)}$ and $X_{(n)}$, are the sufficient statistics for:

a) the uniform family given above;

b) the family $f(x;\theta) = m(x)/q(\theta)$ with the same support. Note: the value $m(x)$ is "free" of θ and the value $q(\theta)$ is "free" of x.

The support for a random sample from such distributions is the n–fold Cartesian product of $(\theta_1-\theta_2, \theta_1+\theta_2)$ with itself, say W_θ. Consider testing the simple H_o: $\theta_1 = \theta_{1o}$, $\theta_2 = \theta_{2o}$ against the composite H_a: $\theta_2 = \theta_{2o}$; in effect, θ_2 is known. The rejection region will certainly include $V_o = W_{\theta o}^c$ but since $P_{\theta o}(V_o) = 0$, this contributes nothing to the size of type I error.

a) If for the alternative θ_{1a},

$$(\theta_{1a}-\theta_{2o}, \theta_{1a}+\theta_{2o}) \cap (\theta_{1o}-\theta_{2o}, \theta_{1o}+\theta_{2o}) = \phi,$$

we are finished: the BCR is V_o and $\alpha = \beta = 0$.

b) Otherwise, $V_i = W_{\theta o} \cap W_{\theta a} \neq \phi$. For our uniform distribution, $L_{\theta_a}(v)/L_{\theta_o}(v) = 1$ in V_i so that we can include in the BCR *any* part of V_i which has probability α under H_o.

Example: Let us consider a particular case:

$$H_o: \theta_1 = 0, \theta_2 = 1 \text{ vs } H_a: \theta_1 = 1, \theta_2 = 1 .$$

$W_{\theta o}$ comes from $(-1,1)$, $W_{\theta a}$ from $(0,2)$, V_i from $(0,1]$. We can draw a picture in terms of $x_{(1)}$ and $x_{(n)} > x_{(1)}$:

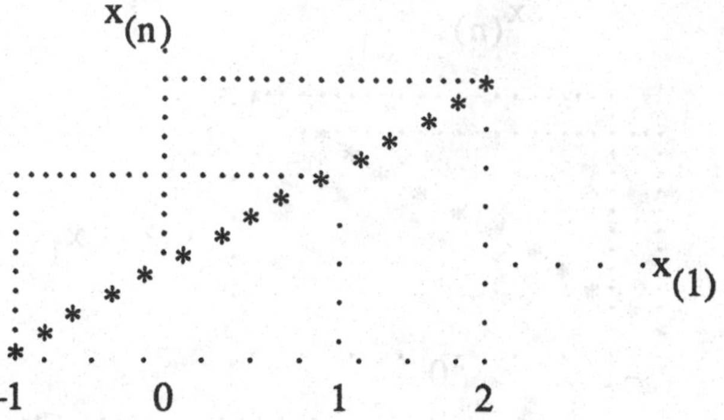

V_i is the small inner triangle. V_o is really everything outside the larger triangle over -1 to 1 but is effectively the upper trapezoid with "* on the east" ($L_{\theta o}$ and $L_{\theta a}$ are both zero outside both large triangles); their union is a set S mentioned above. Take another small triangle as the addition B_α: for

$$\int_{.5-c}^{.5+c}\int_{.5-c}^{x_n} n(n-1)(x_n-x_1)^{n-2}\,dx_1\,dx_n/2^n = (2c)^n/2^n = \alpha\,,$$

$c = \alpha^{1/n} = .5$. The power at H_a is

$$\iint_{V_o \cup B_\alpha} n(n-1)(x_n-x_1)^{n-2}\,dx_1 dx_n/2^n$$

$$= \alpha + \int_1^2\int_0^{x_n} n(n-1)(x_n-x_1)^{n-2}\,dx_1 dx_n/2^n = \alpha + 1 - 1/2^n$$

which is at most 1 for $\alpha \le 1/2^n$.

Exercise 2: Continue the specific example but use the strip between $x_{(n)} = x_{(1)}$ and $x_{(n)} = x_{(1)} + c$ as B_α . Find c .

Example: We now consider the alternative hypothesis on the other side:

$$H_o\colon \theta_1 = 0,\ \theta_2 = 1 \ \text{ vs } \ H_a\colon \theta_1 = -.25,\ \theta_2 = 1\,.$$

The picture is now:

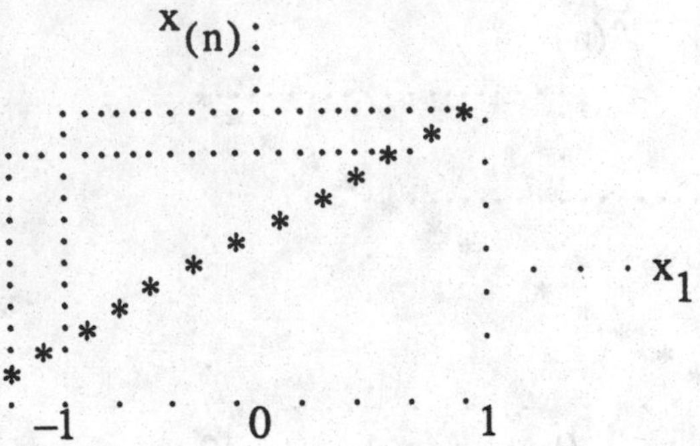

Obviously, we can select pieces of this inner triangle V_i as we did in the previous case.

For V with continuous components, a best test function τ need not be randomized but may be taken, almost surely, as the indicator function of a BCR. Now we will search for a UMPCR for testing

$$H_o: \theta_1 = \theta_{1o}, \theta_2 = \theta_{2o} \text{ vs } H_a: \theta_1 > \theta_{1o}, \theta_2 = \theta_{2o} ;$$

here we can limit rejection regions to subsets of

$$S = \{x_{(1)} > \theta_{1o} - \theta_{2o}\} .$$

In the first sketch above, imagine θ_a increasing. The corresponding "V_i" are decreasing and the "arbitrary" region selected, B_α, is forced into a similar triangular shape:

$$\int_c^{\theta_{1o}+\theta_{2o}} \int_c^{x_n} n(n-1)(x_n - x_1)^{n-2}\, dx_1\, dx_n / \theta_{2o}{}^n = \alpha$$

iff $c = \theta_{1o} + \theta_{2o} - 2\theta_{2o}\alpha^{1/n}$. That is,

$$B_\alpha = \{c < x_{(1)} \le x_{(n)} < \theta_{1o} + \theta_{2o}\}$$

and the BCR is $T \cup B_\alpha$ as in the sketch:

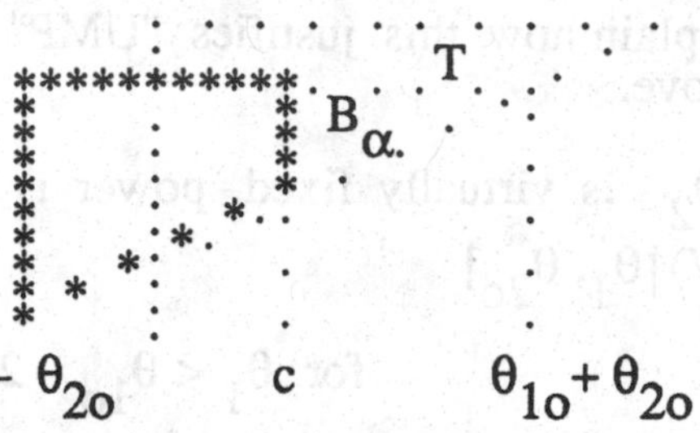

For $\theta_{1a} - \theta_{2o} < c$ (as indicated here with the "." boundary) the power is an increasing function of θ_{1a} : we integrate $n(n-1)(x_n - x_1)^{n-2}$ over regions whose top "T" is increasing. For $\theta_{1a} - \theta_{2o} \geq c$, the power is 1. Because this B_α is now independent of θ_{1a} , the UMPCR is

$$B_\alpha \cup \{x_{(n)} > \theta_{1o} + \theta_{2o}\} .$$

In other words, a uniformly most powerful test of

$$H_o : \theta_1 = \theta_{1o}, \theta_2 = \theta_{2o} \text{ against } H_a: \theta_1 > \theta_{1o}, \theta_2 = \theta_{2o}$$

is given by

$$\tau(v) = I\{\theta_{1o} + \theta_{2o} - 2\theta_{2o}\alpha^{1/n} < x_{(1)}\} + I\{\theta_{1o} + \theta_{2o} < x_{(n)}\} .$$

Intuitively, if the min and max are not "close to their limits", maybe H_o is not reasonable.

Exercise 3: Continuing with this case, let a test function be

$$\varphi(v) = 1 \text{ for } x_{(n)} > \theta_{1o} + \theta_{2o} \text{ or } (x_{(1)}, x_{(n)}) \in C \subset W_{\theta o}$$
$$= 0 \text{ otherwise}$$

a) Show that for $\theta_{1a} > \theta_{1o}$,

$E[\varphi(V) | \theta_{1a}, \theta_{2o}] > E[\tau(V) | \theta_{1a}, \theta_{2o}]$

implies $\iint_C n(n-1)(x_n - x_1)^{n-2}\, dx_1\, dx_2 > \alpha$ so that φ cannot have size α .

b) Explain how this justifies "UMP" in the paragraph above.

Since θ_2 is virtually fixed, power is a function of θ_1:

$\mathscr{P}_\tau(\theta_1) = E[\tau(V) \mid \theta_1, \theta_{2o}]$

$= 0$ for $\theta_1 < \theta_{1o} - 2\theta_{2o}\alpha^{1/n}$

$= \left[\alpha^{1/n} - \dfrac{\theta_{1o} - \theta_1}{2\theta_{2o}}\right]^n$ for $\theta_{1o} - 2\theta_{2o}\alpha^{1/n} \le \theta_1 < \theta_{1o}$

$= \alpha + 1 - \left[1 - \dfrac{\theta_1 - \theta_{1o}}{2\theta_{2o}}\right]^n$ for $\theta_{1o} \le \theta_1 < \theta_{1o} + 2\theta_2(1-\alpha^{1/n})$.

Since $\mathscr{P}_\tau(\theta_1)$ is non–decreasing in θ_1, the last lemma of lesson 3 above applies and the conclusion is that τ is a uniformly most powerful test function for

$$H_o: \theta_1 \le \theta_{1o}, \theta_2 = \theta_{2o} \text{ vs } H_a: \theta_1 > \theta_{1o}, \theta_2 = \theta_{2o}.$$

Exercise 4: Continuing with this uniform distribution, discuss testing

$$H_o: \theta_1 \ge \theta_{1o}, \theta_2 = \theta_{2o} \text{ vs } H_a: \theta_1 < \theta_{1o}, \theta_2 = \theta_{2o}.$$

So far, the discussion has been limited to tests of the location parameter θ_1; now let us consider tests of the scale parameter θ_2.

Example: a) For $H_o: \theta_1 = 0, \theta_2 = 1$ vs $H_a: \theta_1 = 0, \theta_2 = 2$, We again include the complement of $(-1,1)\times\cdots\times(-1,1)$, equivalently, $\{-1 < x_{(1)} \le x_{(n)} < 1\}^c$ in the critical region with $S = W_{\theta a}$. In this case,

$$V_i = W_{\theta o} \cap W_{\theta a} = W_{\theta o} \text{ for all } \theta_2 > \theta_{2o}$$

so that when we adjoin any subregion of size α, we get a UMPCR. Hence,

$$\tau(v) = 1 \text{ for } x_{(1)} \le -1 \text{ or } x_{(n)} \ge 1 \text{ or } v \in B_\alpha \subset V_i .$$

b) Now $W_{\theta a}$ is on two sides of $W_{\theta o}$ so, taking a cue from the work above, we choose two small triangles for B_α :

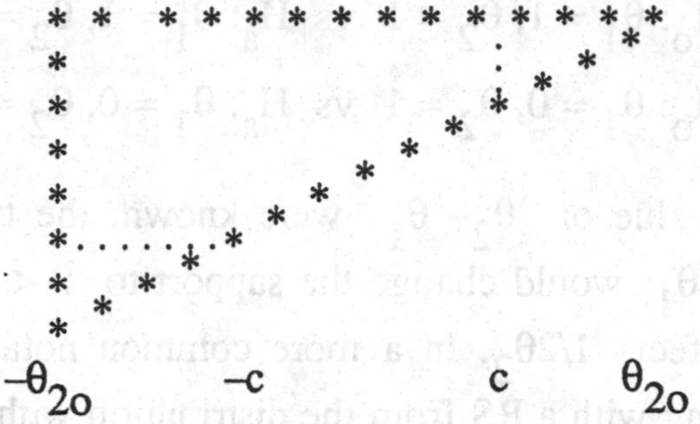

$$\int_c^{\theta_{2o}} \int_c^{x_n} n(n-1)(x_n - x_1)^{n-2}\, dx_1 dx_n/(2\theta_{2o})^n = \alpha/2$$

iff $c = \theta_{2o} - 2\theta_{2o}(\alpha/2)^{1/n}$. The power function

$$P_\tau(\theta_2) = E[\tau(V) \,|\, \theta_1 = 0, \theta_2]$$

$= 0$ for $\theta_2 < c$

$= (1 - (\theta_{2o} + 2\theta_{2o}\dfrac{(\alpha/2)^{1/n}}{\theta_2})^n/2^{n-1}$ for $c \le \theta_2 < \theta_{2o}$.

$= 1 - (\theta_{2o}/\theta_2)^n(1 - \alpha)$ for $\theta_{2o} \le \theta_2$

Since the power function is non–decreasing, the conclusion is that τ is a UMP test for

$$H_o\colon \theta_1 = 0, \theta_2 \le \theta_{2o} \text{ against } H_a\colon \theta_1 = 0, \theta_2 > \theta_{2o} .$$

c) For $H_o\colon \theta_1 = 0, \theta_2 = 2$ vs $H_a\colon \theta_1 = 0, \theta_2 = 1$,

$V_i = W_{\theta a}$ and we seem to be in trouble. Since, in general,

$$\{\theta_{1o} - \theta_2 < x_{(1)} \le x_{(n)} < \theta_{1o} + \theta_2\}$$

$$\approx \{ -\theta_2 < \theta_{1o} - x_{(n)} \le \theta_{1o} - x_{(1)} < \theta_2\} ,$$

we can get uniformly most powerful tests by considering a

sample from the distribution of $Y = \theta_{1o} - X$.

Exercise 5: Continuing with this uniform distribution, discuss BCR's for

a) $H_o: \theta_1 = 1, \theta_2 = 1$ vs $H_a: \theta_1 = 1, \theta_2 = 2$;

b) $H_o: \theta_1 = 0, \theta_2 = 1$ vs $H_a: \theta_1 = 0, \theta_2 = .25$.

If the value of $\theta_2 - \theta_1$ were known, the transformation $w = x + \theta_2 - \theta_1$ would change the support to $0 < w < 2\theta_2$ but would not affect $1/2\theta_2$. In a more common notation, we now consider testing with a RS from the distribution with density

$$f(x;\theta) = I\{0 < x < \theta\}/\theta. \qquad (**)$$

Since θ must be positive, $S = \{x_{(n)} > 0\}$ and the likelihood is

$$L_\theta(v) = I\{0 < x_{(n)} \le \theta\} / \theta^n.$$

Exercise 6: Show that the max $X_{(n)}$ is a complete sufficient statistic for a random sample from

a) the uniform distribution (**);

b) the distribution with density
$f(x;\theta) = (m(x)/q(\theta)) \cdot I\{0 < x < \theta\}$
where $m(x)$ is free of θ and $q(\theta)$ is free of x.

Now we consider $H_{o1}: \theta = \theta_o$ vs $H_{a1}: \theta = \theta_a > \theta_o$. As above, we include $V_o = \{0 < x_{(n)} < \theta_o\}^c$ and any part of its complement of size α under H_{o1}, say B_α. When $\theta_a > \theta_o$,

$$V_i = W_{\theta o} \cap W_{\theta a} = W_{\theta o}, V_o \cup B_\alpha$$

is a UMPCR for testing $H_{o1}: \theta = \theta_0$ against $H_a: \theta > \theta_o$. The test function is:

$$\tau(v) = 1 \text{ if } x_{(n)} > \theta_o, \text{ or } x_{(n)} \in B_\alpha$$
$$= 0 \text{ otherwise}$$

Exercise 7: Continuing with this last example, show that the given τ is UMP for testing

$$H_o: \theta \le \theta_o \text{ against } H_a: \theta > \theta_o .$$

In the first uniform distribution (*), it is clear that for

$$H_o: \theta_1 = \theta_{1o}, \theta_2 = \theta_{2o} \text{ vs}$$

$$H_a: \theta_1 \ne \theta_{1o}, \theta_2 = \theta_{2o} \text{ (or } \theta_1 = \theta_{1o}, \theta_2 \ne \theta_{2o}) ,$$

there can be no UMP test since, as in the normal case, the "one-sided" tests are at opposite ends of the scale. However, for this second uniform distribution (**), we are able to obtain a "two-tailed" test for the "two-sided" hypothesis.

Example: a) First we seek a test of

$$H_o: \theta = \theta_o \text{ against } H_a: \theta = \theta_a < \theta_o .$$

Here $V_i = W_{\theta_o} \cap W_{\theta_a} = W_{\theta_a}$ and $L_{\theta_a} > kL_{\theta_o}$ iff $\theta_o > k \cdot \theta_a$ imposes no other restriction on $x_{(n)}$. Intuitively, we should reject H_o when $x_{(n)}$ is "too small". Then,

$$\int_0^c nx^{n-1}\, dx/\theta_o^{\,n} = \alpha \text{ iff } c = \theta_o \alpha^{1/n} .$$

In this case, the power function

$$P_\tau(\theta) = \int_0^c I\{0 < x < \theta\} nx^{n-1}\, dx/\theta^n$$

is non-increasing. By modifying the lemma from lesson 3, we find

$$\tau(V) = I\{0 < X_{(n)} < \theta_o \alpha^{1/n}\}$$

to be a UMP test function for

$$H_o: \theta \ge \theta_o \text{ against } H_a: \theta < \theta_o .$$

b) A UMP test of $H_o: \theta = \theta_o$ against $H_a: \theta > \theta_o$ does not require conditions on $x_{(n)}$ so we may combine these two tests to get a UMP test for

$$H_o: \theta = \theta_o \text{ against } H_a: \theta \ne \theta_o ,$$

namely, $\tau_*(v) = I\{0 < x_{(n)} < \theta_o\alpha^{1/n}\} + I\{\theta_o < x_{(n)}\}$.

The power function is $E[\tau(V)\,|\,\theta]$

$$= 1 \qquad \text{for } \theta < \theta_o\alpha^{1/n}$$

$$= (\theta_o/\theta)^n \qquad \text{for } \theta_o\alpha^{1/n} \le \theta < \theta_o .$$

$$= (\theta_o/\theta)^n\alpha + 1-(\theta_o/\theta)^n \quad \text{for } \theta_o \le \theta$$

Exercise 8: Sketch the power function of the last example.

Exercise 9: State and prove a lemma like that in lesson 3 to cover the case when the power function is non–increasing.

Exercise 10; Discuss testing for θ when the density is like that in exercise 6b).

LESSON 5. UNBIASED TESTS

We first work thru some examples of the principles in lesson 3; since even under those conditions, one does not always have optimal tests, we examine one further revision of this notion to get uniformly most powerful unbiased tests; proofs of the theorems may be found in Lehmann, 1959. Numerical examples which should be expected here have been put into lesson 11 along with a summary of applications to normal distributions.

Theorem: *Suppose that the components of* $V = (X_1, \cdots, X_n)$ *constitute a RS from a population with density* $f(x;\theta)$ *having MILR in* $t(v)$. *For testing*

$$H_o: \theta \le \theta_o \text{ vs } H_a: \theta > \theta_o$$

the UMP test function is $$\varphi(v) = \begin{cases} 1 & \text{for } t(v) > k \\ \gamma & \text{for } t(v) = k \\ 0 & \text{for } t(v) < k \end{cases} \qquad (*)$$

with $E[\varphi(V) | \theta_o] = P(t(V) > k | \theta_o) + \gamma P(t(V) = k | \theta_o) = \alpha$.

If we have instead an MDLR, then the same kind of theorem is valid with the inequalities in (*) reversed. Similarly, if the inequalities in H_o and H_a are reversed, (*) gives the UMP test with an MDLR and its inequalities are reversed with an MILR. This holds in particular for exponential families and then we have in addition

Theorem: *Let the components of* $V = (X_1, \cdots, X_n)$ *be a RS from a population having density* $f(x;\theta) = c(\theta)e^{Q(\theta)t(x)}h(x)$ *with* Q *strictly monotonic for* $\theta \in \Theta \subset R$. *Let* $T(v) = \Sigma_{j=1}^{n} t(x_j)$. *Consider testing*

$$H_o: \theta \le \theta_1 < \theta_2 \text{ or } \theta \ge \theta_2 \text{ vs } H_a: \theta_1 < \theta < \theta_2 .$$

When Q *is increasing, the UMP test function is*

$$\varphi(v) = \begin{cases} 1 & \text{for } k_1 < T(v) < k_2 \\ \gamma_1\ [\gamma_2] & \text{for } T(v) = k_1\ [k_2] \\ 0 & \text{for } T(v) \le k_1 \text{ or } T(v) \ge k_2 \end{cases}$$

where for i = 1,2, $E[\varphi(V)|\theta_i]$

$$= (k_1 < T(v) < k_2|\theta_i) + \gamma_1 P(T(v) = k_1|\theta_i]$$

$$+ \gamma_2 P(T(v) = k_2|\theta_i] = \alpha.$$

When Q is decreasing, the inequalities in the domain of φ are reversed.

Example: As an application of the theorem, take $f(x;\theta)$ to be Bernoulli so that $T(v) = \Sigma_{j=1}^{n} x_j$ and $Q(\theta) = \log(\theta/(1-\theta))$ is increasing in θ.

a) $H_o: \theta \le \theta_o$; the test function $\varphi(v) = 1, = \gamma, = 0$ according as $t = \Sigma_{j=1}^{n} x_j > k, = k, < k$.

In particular, suppose that $\theta_o = .5$, $\alpha = .01$ and $n = 20$. Since the alternative is $\theta > .5$, we know that k is certainly greater than 10. We can get the null distribution from the binomial table:

t	P(T = t)	P(T > t)
11	.1601791	.2517223
12	.1201344	.1315880
13	.0739288	.0576592
14	.0369644	.0206947 .
15	.0147858	.0059090
16	.0046206	.0012884
17	.0010872	.0002012
18	.0001812	.0000203
19	.0000191	.0000010
20	.0000010	.0000000

Since $P(T > 14) > .01 > P(T > 15)$, $k = 15$. Then,

$.0059090 + \gamma(.0147858) = .01$ gives

$\gamma = (.01 - .0059090)/.0147858 = .2766844.$

(In most applications, we need to use fewer

decimals; moreover, as remarked earlier, the use of randomized test functions is usually avoided.)

b) For H_o: $\theta \le \theta_1$ or $\theta \ge \theta_2$,

$$\varphi(v) = \begin{cases} 1 & \text{if } k_1 < t < k_2 \\ \gamma_1 & \text{if } \quad t = k_1 \\ \gamma_2 & \text{if } \quad t = k_2 \end{cases} \quad \text{and 0 elsewhere .}$$

Let us take $\theta_1 = .20$, $\theta_2 = .80$, $\alpha = .05$ and $n = 20$.

Solving $P(k_1 < T < k_2 | \theta = .2)$

$+ \gamma_1 P(T = k_1 | \theta = .2) + \gamma_2 P(T = k_2 | \theta = .2) = .05$

$P(k_1 < T < k_2 | \theta = .8)$

$+ \gamma_1 P(T = k_1 | \theta = .8) + \gamma_2 P(T = k_2 | \theta = .8) = .05$

simultaneously is a trial and error process (which can be programmed). We start by looking for k_1, k_2 which will make both

$P(k_1 < T < k_2 | \ \theta_1)$ and $P(k_1 < T < k_2 | \ \theta_2)$

as close to the upper bound α (.05) as possible. Here we try pairs $20(.2) < k_1 < k_2 < 20(.8)$ and eventually find

$P(7 < T < 13 | \ .2) = P(7 < T \le 12 | \ .2)$

$= .9999848 - .9678573 = .0321275$,

$P(7 < T < 13 | \ .8) = P(7 < T \le 12 | \ .8)$

$= .0321427 - .0000152 = .0321275$.

With this, we get

$.0321275 + \gamma_1 P(T = 7 | \ .2) + \gamma_2 P(T = 13 | \ .2) = .05$

$.0321275 + \gamma_1 P(T = 7 | \ .8) + \gamma_2 P(T = 13 | \ .8) = .05$

or $\quad \gamma_1(.0545499) + \gamma_2(.0000133) = .05 - .0321275$

$\qquad \gamma_1(.0000133) + \gamma_2(.0545499) = .05 - .0321275$.

This gives $\gamma_1 = .3275559 = \gamma_2$. Note that a non–randomized test would reject for $7 < T < 13$ with

$$\alpha \approx .03 .$$

c) For $n = 20$, $\theta_1 = .2$ and $\theta_2 = .7$, the distributions are not reflections; we obtain the equations

$$.0315795 + \gamma_1(.0545499) + \gamma_2(.0004617) = .05$$
$$.0466830 + \gamma_1(.0010178) + \gamma_2(.0653696) = .05 .$$

Exercise 1: Continuing this example, take $\theta_1 = .25$, $\theta_2 = .75$, $n = 25$, $\alpha = .05$. Show that $k_1 = 10$, $k_2 = 15$ and find γ_1, γ_2.

It was seen in the normal example (lesson 3) that there is no UMP test for the mean $\theta = 0$ against the alternative $\theta \neq 0$. This carries over to other MILR families as follows:

Let $\psi(v) \equiv \alpha$; since $\mathscr{P}(\theta) = E[\varphi(V)|\theta]$ is increasing (when $\neq 1$),

$$\mathscr{P}(\theta) < \mathscr{P}(\theta_o) = \alpha \leq E[\psi(V)|\theta] \text{ for } \theta < \theta_o .$$

This means that the UMP test with $H_a: \theta > \theta_o$ is "worse" than the trivial test $\psi(v) \equiv \alpha$ when $\theta < \theta_o$; similar conclusions are reached when the inequalities in hypotheses and/or tests are reversed. Thus a "uniformity" condition must be such as to prescribe a smaller set of test functions. One of these appeared early in the history and is easy to describe.

Definition: *Let $X_1, \cdots, X_n$ be a RS from a population with density $f(x;\theta)$ for $\theta \in \Theta \subset R$; let φ be a test function based on $V = (X_1, \cdots, X_n)$. For*

$$H_o: \theta \in \Theta_o \subset \Theta \text{ vs } H_a: \theta \in \Theta_a = \Theta - \Theta_o ,$$

φ is an unbiased test of size α if

$$E[\varphi(V)|\theta] \leq \alpha \text{ for } \theta \in \Theta_o \text{ and}$$
$$E[\varphi(V)|\theta] \geq \alpha \text{ for } \theta \in \Theta_a .$$

A test is uniformly most powerful unbiased (UMPU) if it is UMP within the class of all unbiased tests.

Note that a UMP test φ is unbiased because

$$E[\varphi(V)\,|\,\theta] \geq \alpha = E[\alpha\,|\,\theta]\ .$$

Since the class of UMPU tests is a subclass of the class of UMP tests, φ is also UMPU. Again, general results can be given for the natural parameter exponential family as in

Theorem: *Let the population density be*

$$f(x;\theta) = c(\theta)e^{\theta t(x)}h(x),\ \theta \in \Theta \subset R\ ;$$

let $V = (X_1, \cdots, X_n)$ *be a RS and set* $y = \sum_{j=1}^{n} t(x_j)$. *Suppose that* θ_1, θ_2 *and* θ_o *are given.*

i) *For testing* $H_o: \theta_1 \leq \theta \leq \theta_2$ *vs* $H_a: \theta < \theta_1$ *or* θ_2, *a UMPU test is*

$$\varphi(v) = \begin{cases} 1 & \text{when } y <_1 \text{ or } y > k_2 \\ \gamma_1 & \text{when } y = k_1 \\ \gamma_2 & \text{when } y = k_2 \\ 0 & \text{otherwise.} \end{cases} \quad \text{with}$$

$$E[\varphi(V)\,|\,\theta_1] = \alpha = E[\varphi(V)\,|\,\theta_2]\ .$$

ii) *For testing* $H_o: \theta = \theta_o$ *vs* $H_a: \theta \neq \theta_o$, *a UMPU test has the same form but with*

$$E[\varphi(V)\,|\,\theta_o] = \alpha \text{ and } E[Y\varphi(V)\,|\,\theta_o] = \alpha \cdot E[Y\,|\,\theta_o]\ .$$

Example: $X_1, \cdots, X_n$ are IID $N_1(\theta, \sigma^2 = 1)$. The likelihood is

$$e^{-\Sigma(x_j - \theta)^2/2} \div (2\pi)^{n/2}$$

$$= e^{-n\theta^2/2} \cdot e^{\theta\Sigma x_j} \cdot e^{-\Sigma x_j^2/2} \cdot (2\pi)^{-n/2}$$

so that $Y = \Sigma X_j = n\bar{X}$. Consider

$$H_o: \theta = \theta_o \text{ vs } H_a: \theta \neq \theta_o .$$

The UMPU test is $\varphi(v) = 1$ if $\bar{x} < k_1$ or $> k_2$
$= 0$ otherwise

with $\alpha = E[\varphi(V)|\theta_o] = P(\bar{X} < k_1|\theta_o) + P(\bar{X} > k_2|\theta_o)$. Under H_o, $\sqrt{n}(\bar{X} - \theta_o)$ is distributed as a standard normal RV Z so that

$$\alpha = P(Z < \sqrt{n}(k_1 - \theta_o)) + P(Z > \sqrt{n}(k_2 - \theta_o)) .$$

Symmetry of Z about 0 suggests that we take "equal tails": reject H_o when $\bar{x} < \theta_o - z_{\alpha/2}/\sqrt{n}$ or $\bar{x} > \theta_o + z_{\alpha/2}/\sqrt{n}$.

When σ^2 is known but not necessarily equal 1, these inequalities are equivalent to $\sqrt{n}|\bar{x} - \theta_o|/\sigma \geq z_{\alpha/2}$.

The property of symmetry, which suggests an easy solution in the normal case, can be used to simplify matters in general as will be seen via the following lemma. Note that symmetry is required only under H_o.

Lemma: *In addition to the hypotheses for ii) of the last theorem, suppose that for each fixed b,*

$$P(Y < b - x|\theta_o) = P(Y > b + x|\theta_o) \text{ for all real } x .$$

Also take the test function φ *symmetric about b .Then the test with* $k_2 = 2b - k_1$, $\gamma_2 = \gamma_1$ *and*

$$P(Y < k_1|\theta_o) + \gamma_1 P(Y = k_1|\theta_o) = \alpha/2$$

is UMPU.

Proof: We need to verify only $E[\varphi(Y)|\theta_o] = \alpha$ and

$$E[Y\varphi(Y)|\theta_o] = \alpha E[Y|\theta_o] .$$

With the additional conditions on k_1, k_2,

$$E[\varphi(Y)\,|\,\theta_o] = P(Y < k_1\,|\,\theta_o] + \gamma_1 P(Y = k_1\,|\,\theta_o)$$
$$+ P(Y > k_2\,|\,\theta_o) + \gamma_2 P(Y = k_2\,|\,\theta_o)$$
$$= \alpha/2 + P(Y > 2b-k_1\,|\,\theta_o) + \gamma_1 P(Y = 2b-k_1\,|\,\theta_o)$$
$$= \alpha/2 + P(Y>b + (b-k_1)\,|\,\theta_o) + \gamma_1 P(Y=b + (b-k_1)\,|\,\theta_o)\,.$$

By the symmetry of the distribution of Y, this is

$$\alpha/2 + P(Y < b - (b-k_1)\,|\,\theta_o) + \gamma_1 P(Y = b - (b-k_1)\,|\,\theta_o)$$
$$= \alpha/2 + P(Y < k_1\,|\,\theta_o) + \gamma_1 P(Y = k_1\,|\,\theta_o) = \alpha\,.$$

Since φ is also symmetric about b ,

$$E[Y\varphi(Y)\,|\,\theta_o] = E[(Y-b)\varphi(Y)\,|\,\theta_o] + bE[\varphi(Y)\,|\,\theta_o] = 0 + b\alpha\,.$$

Of course, $b = E[Y\,|\,\theta_o]$.

This justifies the choice of equal tails in the normal case (lesson 3) so that that test is UMPU; for some purposes, the rejection region may be taken as

$$n(\bar{x} - \theta_o)^2/\sigma^2 > z_{\alpha/2}^2 = \chi_\alpha^2 \text{ with 1 dof}\,.$$

In many applications, while the hypotheses of interest concern a parameter θ , there are other "nuisance" parameters η in the distribution. The next theorem will isolate some tests for this case but first we want to emphasize some consequences of sufficiency.

If T is a sufficient statistic and φ is a test function, then

$$\psi(T) = E[\varphi(V)\,|\,T]$$

is also a test function with the same power since

$$E[\varphi(V)] = E[\,E[\varphi(V)\,|\,T]\,] = E[\psi(T)]\,.$$

In other words, when a sufficient statistic T exists, we can restrict attention to test functions based on T and so deal with the likelihood for T in the first place. The simplest distributions including sufficiency are the exponential families; when, in the notation below, θ is a convex set, (that is, Θ contains the line segment joining any two of its points) but not a subset in dimension less than k+1 , these exponential distributions are also "complete". We will ignore complications needed to specify particular measures; the proof of the theorem is in several parts of Lehmann, 1959.

Theorem: *Let an exponential family have* $(\theta,\eta) \in \Theta \subset R \times R^k$ *and the density* $c(\theta,\eta)e^{\theta t(x) + \Sigma_{i=1}^{k}\eta_i t_i(x)}$.
Let $X_1, \cdots, X_n$ *be a RS.*

a) *The joint distribution of the sufficient statistics*

$$v_o = \Sigma_{j=1}^{n} t(x_j),\ v_1 = \Sigma_{j=1}^{n} t_1(x_j), \cdots, v_k = \Sigma_{j=1}^{n} t_k(x_j)$$

is given by $C(\theta,\eta)\cdot exp(\theta v_o + \eta' v)$

with $\eta = (\eta_1, \cdots, \eta_k)'$, $v = (v_1, \cdots, v_k)'$.

b) *The conditional distribution of* V_o *given* v *is* $c_v(\theta)\cdot exp(\theta v_o)$.

c) *The following are UMPU tests for the hypotheses indicated.*

1) $H_o: \theta \le \theta_o$ *vs* $H_a: \theta > \theta_o$

$$\varphi_1(v_o, v) = \begin{cases} 1 & \text{for } v_o > k_1(v) \\ \gamma(v) & \text{"} \quad v_o = k_1(v) \\ 0 & \text{"} \quad v_o < k_1(v) \end{cases}$$

with $E[\varphi_1(V_o, V) \mid v, \theta_o] = \alpha$ *for all v.*

2) $H_o: \theta \le \theta_1$ *or* $\theta \ge \theta_2$ *vs* $H_a: \theta_1 < \theta < \theta_2$

$$\varphi_2(v_o, v) = \begin{cases} 1 & \text{for } k_2(v) < v_o < k_3(v) \\ \gamma_2(v) & \text{"} \quad v_o = k_2(v) \\ \gamma_3(v) & \text{"} \quad v_o = k_3(v) \\ 0 & \text{otherwise} \end{cases}$$

$$\textit{with } E[\varphi_2(V_o,V)|v,\theta_1] = E[\varphi_2(V_o,V)|v,\theta_2]$$
$$= \alpha \textit{ for all } v .$$

3) $H_o: \theta_1 \le \theta \le \theta_2$ vs $H_\alpha: \theta < \theta_1$ or $\theta > \theta_2$

$$\varphi_3(v_o,v) = \begin{cases} 1 & \text{for } v_o < k_4(v) \text{ or } v_o > k_5(v) \\ \gamma_4(v) & \text{for } v_o = k_4(v) \\ \gamma_5(v) & \text{for } v_o = k_5(v) \\ 0 & \text{otherwise} \end{cases}$$

$$\textit{with } E[\varphi_3(V_o,V)|v,\theta_1] = E[\varphi_3(V_o,V)|v,\theta_2]$$
$$= \alpha \textit{ for all } v .$$

4) $H_o: \theta = \theta_o$ vs $H_a: \theta \ne \theta_o$

$\varphi_4(v_o,v) = \varphi_3(v_o,v)$ *as above but with*

$E[\varphi_4(V_o,V)|v,\theta_o] = \alpha$

and $E[V_o\varphi_4(V_o,V)|v,\theta_o] = \alpha E[V_o|v,\theta_o]$.

Example: For two independent Poisson RVs X and Y with means μ and ρ respectively,

$$P(X = x, Y = y) = e^{-\mu-\rho}e^{x\log(\mu/\rho) + (x+y)\log(\rho)}/x!y! .$$

a) To fit this into the patterns above, take
$\theta = \log(\mu/\rho)$, $\eta = \log(\rho)$, $v_o = x$ and $v = x+y$.

The density of $V_o = X$ given $v = x+y$ is

$$P(X = x \mid V = v) = \binom{v}{x}\left[\frac{\mu}{\mu+\rho}\right]^x \left[\frac{v}{\mu+\rho}\right]^{v-x}$$

for $x = 0(1)v$; this is a binomial distribution with parameters v and $p = \mu/(\mu+\rho)$.

b) A null hypothesis $\mu \le bv$ for a given b , becomes

$$H_o: p \le b(/(b+1) .$$

We reject H_o when x is "too large". The critical point $k_1(v)$ is determined by use of binomial tables (or normal tables when v is large).

Exercise 2: Continue with the example above.

a) Show that the corresponding power function is increasing in v:

$$\mathscr{P}(\theta \mid v) = \sum_{x=k_1}^{v} \begin{bmatrix} v \\ x \end{bmatrix} p^x(1-p)^{v-x} .$$

b) Hence show that the over all power, $E[\mathscr{P}(\theta \mid V)]$, is an increasing function of θ. Hint: use the distribution of V.

Exercise 3: Let X_1, X_2 be independent binomial RVs with parameters (n_1, p_1), (n_2, p_2) respectively.

a) Show that $H_o: p_2 \le p_1$ can be expressed in terms of $\theta = \log((1-p_1)p_2/p_1(1-p_2))$.

b) Find the distribution of X_1 given $X_1 + X_2 = V = v$.

LESSON 6. QUADRATIC FORMS IN NORMAL RANDOM VARIABLES

In this lesson, we consider a few aspects of the multivariate normal distribution which are needed for distributions of some quadratic forms used in other lessons. We begin afresh with a

Definition: *Let the components of* $Z = (Z_1, Z_2, \cdots, Z_m)'$ *be IID standard normal RVs. Let* C *[* μ *] be a p by m matrix [p by 1 vector] of constants. Then* $X = CZ + \mu$ *is a p-variate normal random vector with a multivariate normal distribution.*

We sometimes shorten this to "X is multinormal" or even "X is normal". We are really considering the joint distribution of

$$X_1 = \sum_{j=1}^{m} c_{1j}Z_j + \mu_1 \,,\ X_2 = \sum_{j=1}^{m} c_{2j}Z_j + \mu_2 \,,\ \cdots,$$

$$X_p = \sum_{j=1}^{m} c_{pj}Z_j + \mu_p \,.$$

In this form, it is obvious that any subset of $\{X_i\}$, any subvector of X, is also multinormal; we are merely taking some of the rows in C and μ.

Since E is a linear operator and $E[Z_j]$ are all 0,

$$E[X_i] = \mu_i \text{ and } E[X] = \mu \,.$$

Since $E[Z_\alpha Z_\beta] = \delta_{\alpha\beta} = \begin{matrix} 1 \text{ for } \alpha = \beta \\ 0 \text{ for } \alpha \neq \beta \end{matrix}$,

$$\begin{aligned} E[(X_i - \mu_i)(X_k - \mu_k)] &= E[\sum_{\alpha=1}^{m} c_{i\alpha}Z_\alpha \cdot \sum_{\beta=1}^{m} c_{k\beta}Z_\beta] \\ &= \Sigma_\alpha \Sigma_\beta\, c_{i\alpha}c_{k\beta}\, E[Z_\alpha Z_\beta] \\ &= \Sigma_\alpha\, c_{i\alpha}c_{k\alpha} \,. \end{aligned}$$

We recognize this as the ik^{th} element of CC'. Therefore,

$$\text{Cov}(X) = CC' \,.$$

This p by p matrix is usually denoted by Σ.

We note that Σ is symmetric. Since

$$\mu'\Sigma\mu = \mu'CC'\mu \geq 0 \text{ for all } \mu ,$$

Σ is said to be *positive semi-definite*. If in fact, $\mu'\Sigma\mu > 0$ for all $\mu \neq 0$, Σ is said to be positive definite (write $\Sigma > 0$) and then Σ has an inverse. Another abbreviation is "X is $N_p(\mu,\Sigma)$". The calculation of the CF of the multinormal is mostly algebraic and follows from

Exercise 1: Show that the CF of the standard normal is $e^{-t^2/2}$.

Theorem: *Let X be a p-variate multinormal as in the definition. Let $t \in R^p$. Then the CF of X is*

$$exp[it'\mu - t'\Sigma t/2] .$$

Proof: The CF is $E[e^{it'X}] = E[e^{it'(CZ + \mu)}]$

$$= E[e^{it'\mu}e^{it'CZ}] = e^{it'\mu}E[e^{iw'Z}]$$

for $w' = t'C$. Since $iw'Z = iw_1Z_1 + \cdots + iw_mZ_m$ and the Z's are independent,

$$E[e^{iw'Z}] = E[e^{iw_1Z_1}]\cdots E[e^{iw_mZ_m}]$$

$$= e^{-w_1^2/2}\cdots e^{-w_m^2/2} = exp[-\sum_{k=1}^{m} w_k^2/2] .$$

From $w = C't$, we get $w_k = \sum_{\alpha=1}^{p} c_{\alpha k}t_\alpha$ for $k = 1(1)m$ and hence (see below) $\sum_{k=1}^{m} w_k^2 = t'CC't$. Finally,

$$E[e^{it'X}] = e^{it'\mu - t'\Sigma t/2} .$$

Exercise 2: Show that

$$t'CC't = (\sum_{\alpha=1}^{p} c_{\alpha 1}t_\alpha)^2 + \cdots + (\sum_{\alpha=1}^{p} c_{\alpha m}t_\alpha)^2 .$$

Since the CF determines the distribution, each multinormal is fixed by its parameters μ and Σ. From exercise 2, we see that

$t'\Sigma t \ge 0$ and also that the equality holds iff $\sum_{\alpha=1}^{p} c_{\alpha k} t_\alpha = 0$ for $k = 1(1)m$, that is, when $t'C = 0$ for some $t \ne 0$. If there are no such t, C must be full rank so that $p = m$ from which it follows that C and $\Sigma = CC'$ are non–singular and so $\Sigma > 0$. Then the distribution of X is also said to be *non-singular* and

Theorem: *Let X be a p-variate multinormal with $\Sigma = CC' > 0$. Let $\|\Sigma\|$ be the absolute value of the determinant of Σ. Then the density of X is*

$$((2\pi)^n \|\Sigma\|)^{-1/2} \cdot exp[-(x - \mu)'\Sigma^{-1}(x - \mu)/2] .$$

Proof: Since $X = C\cdot Z + \mu$, $Z = C^{-1}(X - \mu)$ and the Jacobian matrix of the transformation from Z to X is C^{-1}. Since $\Sigma^{-1} = C'^{-1}C^{-1}$, $\|C^{-1}\| = \|\Sigma\|^{-1/2}$. The density of Z (the joint density of its components) is

$$f(z) = (2\pi)^{-n/2} \cdot \exp[-\sum_{\alpha=1}^{m} z_\alpha^{\,2}/2] = (2\pi)^{-n/2} \cdot \exp[-z'z/2] .$$

The density of X is $f(Z(x)) \cdot \|C^{-1}\|$

$$= (2\pi)^{-n/2} \cdot \exp[-(x - \mu)'C'^{-1}C^{-1}(x - \mu)/2] \cdot \|C^{-1}\|$$

$$= ((2\pi)^n \|\Sigma\|)^{-1/2} \cdot \exp[-(x - \mu)'\Sigma^{-1}(x - \mu)/2] .$$

Example: Let $p = m = 2$. Then

$$E[X] = \begin{bmatrix} \mu_1 \\ \mu_2 \end{bmatrix} \text{ and } Cov(X) = \Sigma = \begin{bmatrix} \sigma_{11} & \sigma_{12} \\ \sigma_{21} & \sigma_{22} \end{bmatrix} .$$

More common notation is

$$\sigma_{11} = \sigma_1^2 = Var(X_1) , \sigma_{22} = \sigma_2^2 = Var(X_2) ,$$

$$\sigma_{12} = \sigma_{21} = \rho\sigma_1\sigma_2 = Cov(X_1, X_2)$$

where $\rho = \sigma_{12}/\sigma_1\sigma_2$ is the correlation of X_1 and X_2.

Then $\Sigma^{-1} = \begin{bmatrix} \sigma_{22} & -\sigma_{21} \\ -\sigma_{12} & \sigma_{11} \end{bmatrix} \div (\sigma_{11}\sigma_{22} - \sigma_{12}^{2})$

$$= \begin{bmatrix} \sigma_2^2 & -\rho\sigma_1\sigma_2 \\ -\rho\sigma_1\sigma_2 & \sigma_1^2 \end{bmatrix} \div \sigma_1^2\,\sigma_2^2(1-\rho^2)\,.$$

The quadratic form $(x-\mu)'\Sigma^{-1}(x-\mu)$ becomes

$$(x_1-\mu_1 \;\; x_2-\mu_2)\begin{bmatrix} \sigma_2^2 & -\rho\sigma_1\sigma_2 \\ -\rho\sigma_1\sigma_2 & \sigma_2^2 \end{bmatrix}\begin{bmatrix} x_1-\mu_1 \\ x_2-\mu_2 \end{bmatrix} \div \sigma_1^2\sigma_2^2(1-\rho^2)$$

$$= \left[\left[\frac{x_1-\mu_1}{\sigma_1}\right]^2 - 2\rho\left[\frac{x_1-\mu_1}{\sigma_1}\right]\left[\frac{x_2-\mu_2}{\sigma_2}\right] + \left[\frac{x_2-\mu_2}{\sigma_2}\right]^2\right] \div (1-\rho^2)\,.$$

The joint density can be written as

$$\frac{e^{-\left[\left[\frac{x_1-\mu_1}{\sigma_1}\right]^2 -2\rho\left[\frac{x_1-\mu_1}{\sigma_1}\right]\left[\frac{x_2-\mu_2}{\sigma_2}\right] + \left[\frac{x_2-\mu_2}{\sigma_2}\right]^2\right] \div 2(1-\rho^2)}}{2\pi\;\sigma_1\sigma_2\surd(1-\rho^2)}$$

Exercise 3: Verify the algebra for the inverse and quadratic form in this last example.

In this *bivariate* normal, it is easy to see that X_1 and X_2 are uncorrelated iff $\sigma_{12} = \rho\sigma_1\sigma_2 = 0$; then Σ is diagonal. Then, when $\sigma_{11}\sigma_{22} > 0$,

$$(x_1-\mu_1 \;\; x_2-\mu_2)\Sigma^{-1}\begin{bmatrix} x_1-\mu_1 \\ x_2-\mu_2 \end{bmatrix}$$

$$= (x_1-\mu_1)^2/\sigma_{11} + (x_2-\mu_2)^2/\sigma_{22}\,.$$

This makes $f(x) = e^{-(x_1-\mu_1)^2/\sigma_{11}} \cdot e^{-(x_1-\mu_2)^2/\sigma_{22}} \div 2\pi\sigma_1\sigma_2$

which means that X_1 and X_2 are independent. Thus, generalizing, but without a formal proof,

Corollary: *The joint normal RVs* $X_1, \cdots, X_p$ *are independent iff they are uncorrelated iff their covariance matrix is diagonal.*

This is not true if the variables are not jointly normal to begin with; even in general, correlation 0 does not imply independence. The joint normality cannot be omitted from the next theorem either; for additional commentary, see the article by Melnick and Tenenbein, 1982.

Theorem: *Let* X *be a p-variate multinormal with mean* μ *and covariance* Σ. *Let* A *be a* k *by* p *matrix of constants. Then* $Y = AX$ *is a k-variate multinormal with mean* $A\mu$ *and covariance* $A\Sigma A'$.

Proof: Let $X = CZ + \mu$ as in the definition. Then

$$Y = AX = A(CZ + \mu) = ACZ + A\mu$$

is of the form $C_1Z + \mu_1$ so that the definition makes Y multinormal. It is easily seen that $E[Y] = A\mu$. Now the covariance of Z is

$$E[ZZ'] = (E[Z_iZ_j]) = (\delta_{ij}) = I_m$$

and since $Y - A\mu = ACZ$, the covariance of Y is

$$E[ACZ(ACZ)'] = E[ACZZ'C'A'] = ACE[ZZ']C'A'$$

$$= ACC'A' = A\Sigma A'.$$

This theorem says that any linear combination(s) of jointly normal RVs is (are) also jointly normally distributed. In particular, if $X_1, X_2, \cdots, X_n$ are IID $N_p(\mu,\Sigma)$, then

$$\overline{X} = \sum_{j=1}^{n} X_j/n \text{ is } N_p(\mu,\Sigma/n).$$

Exercise 4: Prove the last statement about $\overline{X}$ directly. Hint: find its CF.

Exercise 5: For $j = 1(1)n$, let X_j be $N_1(\mu_j,\sigma^2)$ and independent. Let P be an orthogonal matrix ($PP'=I$) and $X = (X_1,\cdots,X_n)'$, $Y = PX$. Show that the components of Y are also independent normals with the same variance σ^2. Hint: normality follows from the theorem; find the covariance of Y .

Recall the definition of a non–central chisquare: if for $j = 1(1)m$, X_j are $N_1(\mu_j,1)$ and independent, then $\sum_{j=1}^{m} X_j^2$ is non–central chisquare with m dof and noncentrality parameter $\sum_{j=1}^{m} \mu_j^2$.

To complete our results for statistics, we need properties of the rank of a matrix A , say $\rho(A)$. In the theorem following, the first two standard results are taken without proof (Rao, 1965).

Theorem: *a) When the matrix product $A \cdot B$ is defined, its rank is at most the minimum $\{\rho(A), \rho(B)\}$.*

b) When the matrix sum $A + B$ is defined, its rank is at most the sum of the ranks of A and B .

c) When C and B are non-singular and CAB is defined, $\rho(CAB) = \rho(CA) = \rho(AB) = \rho(A)$.

Partial proof: $\rho(A) = \rho(C^{-1}CABB^{-1}) \le \rho(CAB) \le \rho(A)$ by several applications of a) . Therefore the equality holds.

Exercise 6: Complete the proof of c) in the theorem above.

In the following theorems and corollaries discovered by Fisher and Cochran, we write out the details for the sum of three C's but the results hold for an arbitrary finite number of such C's .

Theorem: *Let C_1, C_2, C_3 be p by p symmetric matrices with ranks r1, r2, r3 respectively and such that*

$$I_p = C_1 + C_2 + C_3 .$$

Then there is an orthogonal matrix P such that

$$PC_1P' = \begin{bmatrix} I_{r1} & 0 & 0 \\ 0 & 0 & 0 \\ 0 & 0 & 0 \end{bmatrix}, PC_2P' = \begin{bmatrix} 0 & 0 & 0 \\ 0 & I_{r2} & 0 \\ 0 & 0 & 0 \end{bmatrix}, PC_3P' = \begin{bmatrix} 0 & 0 & 0 \\ 0 & 0 & 0 \\ 0 & 0 & I_{r3} \end{bmatrix}$$

iff $r1 + r2 + r3 = p$.

Note: these large matrices are "partitioned" and each "0" is a matrix of appropriate dimensions with all zero elements.

Proof: *Necessity* If such a P exists,

$$I_p = PI_pP' = P(C_1 + C_2 + C_3)P' = PC_1P' + PC_2P' + PC_3P'$$

$$= \begin{bmatrix} I_{r1} & 0 & 0 \\ 0 & 0 & 0 \\ 0 & 0 & 0 \end{bmatrix} + \begin{bmatrix} 0 & 0 & 0 \\ 0 & I_{r2} & 0 \\ 0 & 0 & 0 \end{bmatrix} + \begin{bmatrix} 0 & 0 & 0 \\ 0 & 0 & 0 \\ 0 & 0 & I_{r3} \end{bmatrix} = \begin{bmatrix} I_{r1} & 0 & 0 \\ 0 & I_{r2} & 0 \\ 0 & 0 & I_{r3} \end{bmatrix}$$

which implies that $r1 + r2 + r3 = p$.

Sufficiency For $j = 1,2,3$, there exists an orthogonal matrix P_j such that $P_jC_jP_j'$ is a diagonal matix with the eigenvalues of C_j on the diagonal. We can rearrange rows and columns without changing the diagonality so that

$$P_1C_1P_1' = \begin{bmatrix} \Delta_1 & 0 & 0 \\ 0 & 0 & 0 \\ 0 & 0 & 0 \end{bmatrix}, P_2C_2P_2' = \begin{bmatrix} 0 & 0 & 0 \\ 0 & \Delta_2 & 0 \\ 0 & 0 & 0 \end{bmatrix}, P_3C_3P_3' = \begin{bmatrix} 0 & 0 & 0 \\ 0 & 0 & 0 \\ 0 & 0 & \Delta_3 \end{bmatrix} \quad (*)$$

where each Δ_j is diagonal with the rj non–zero eigenvalues of C_j. Now $P_1(C_2 + C_3)P_1' = P_1(I_p - C_1)P_1' = I_p - P_1C_1P_1'$

$$= \begin{bmatrix} I_{r1} - \Delta_1 & 0 \\ 0 & I_{r2} & 0 \\ 0 & 0 & I_{r3} \end{bmatrix}.$$

By parts b) and c) of the previous theorem, the rank of $P_1(C_2 + C_3)P_1'$ is at most $r2 + r3$; because its equivalent matrix on the right is diagonal, its rank is at least $r2 + r3$.

Together these imply that the rank is exactly $r2 + r3$ and so $I_{r1} - \Delta_1$ must be 0.

This means that the non–zero eignvalues of C_1 are all 1 and C_1 is positive semi–definite. Similarly, the $r2$ eigenvalues of C_2 and the $r3$ eigenvalues of C_3 are all 1. Then with the three "Δ" matrices above, we get

$$C_1 = P_1' \begin{bmatrix} I_{r1} & 0 & 0 \\ 0 & 0 & 0 \\ 0 & 0 & 0 \end{bmatrix} P_1,\ C_2 = P_2' \begin{bmatrix} 0 & 0 & 0 \\ 0 & I_{r2} & 0 \\ 0 & 0 & 0 \end{bmatrix} P_2,\ C_3 = P_3' \begin{bmatrix} 0 & 0 & 0 \\ 0 & 0 & 0 \\ 0 & 0 & I_{r3} \end{bmatrix} P_3.$$

Now we partition the "P" matrices as

$$P_1' = (B_1, A_2, A_3),\ P_2' = (A_4, B_2, A_5),\ P_3' = (A_6, A_7, B_3)$$

so that B_1 has r1 columns, B_2 has r2 columns, B_3 has r3 columns. Then brute force mulitplication yields

$$C_1 = B_1B_1',\ C_2 = B_2B_2',\ C_3 = B_3B_3'.$$

From
$$I_p = C_1 + C_2 + C_3 = B_1B_1' + B_2B_2' + B_3B_3' = (B_1, B_2, B_3)\begin{bmatrix} B_1' \\ B_2' \\ B_3' \end{bmatrix},$$

we see that the required orthogonal matrix is $P = (B_1, B_2, B_3)'$.

Exercise 7: With the P given in terms of the B's above, check that each PC_jP' is the appropriate diagonal matrix.

Corollary: *Let the components of* $X = (X_1, X_2, \cdots, X_p)'$ *be independent normal RVs with common variance* 1 *and means* $\mu_1, \cdots, \mu_p$ *respectively. For* $j = 1,2,3$, *let* $Q_j = X'C_jX$ *be such that* $X'X = Q_1 + Q_2 + Q_3$. *Then* Q_1, Q_2, Q_3 *are independent non-central chisquare RVs iff* $p = \rho(C_1) + \rho(C_2) + \rho(C_3)$.

Proof: The necessity is essentially obvious as the degrees of freedom of the sum of independent chisquare RVs is the sum of their dof. Since the hypothesis of this corollary implies $I = C_1 + C_2 + C_3$, application of the theorem yields an orthogonal $P = (B_1,B_2,B_3)'$ with $C_j = B_jB_j'$, $j = 1,2,3$.
Let $Y = PX$; then, with the results in the proof of the theorem, substitution turns $X'X = X'C_1X + X'C_2X + X'C_3X$ into

$$Y'Y = Y'\begin{bmatrix} I_{r1} & 0 & 0 \\ 0 & 0 & 0 \\ 0 & 0 & 0 \end{bmatrix} Y + Y'\begin{bmatrix} 0 & 0 & 0 \\ 0 & I_{r2} & 0 \\ 0 & 0 & 0 \end{bmatrix} Y + Y'\begin{bmatrix} 0 & 0 & 0 \\ 0 & 0 & 0 \\ 0 & 0 & I_{r3} \end{bmatrix} Y .$$

This makes Q_1 a sum involving the first r1 components of Y, Q_2 the next r2 components of Y and Q_3 the last r3 components of Y. Since by an early exercise, the components of Y are also independent normal RVs, the independence of Q_1,Q_2,Q_3 follows. That their distributions are non–central chisquare follows from the definition of such chi–square.

LESSON 7. LIKELIHOOD RATIO TESTS – I

In this lesson, we begin another variation of the N–P tests. First, the test of

$$H_o: \theta = \theta_o \text{ (simple) vs } H_a: \theta = \theta_a \text{ (simple)}$$

includes rejection of H_o when $L_{\theta_a}(v) > k \cdot L_{\theta_o}(v)$.

When V is discrete, this corresponds to the probability of observing v under H_a being "greater than" that of observing v under H_o; loosely speaking, we retain the hypothesis which yields the greater probability of observing what in fact we did observe. This same spirit can be followed in certain cases where H_o and H_a are composite:

$$\text{for } H_o: \theta \in \Theta_o \text{ vs } H_a: \theta \in \Theta_a = \Theta - \Theta_o,$$

reject H_o when $\sup_{\theta \in \Theta_a} L_\theta(v)$ is "greater than" $\sup_{\theta \in \Theta_o} L_\theta(v)$.

Focusing more directly on H_o, we have,

Definition: *Let $\mathcal{F} = \{L_\theta(v) : \theta \in \Theta\}$ be a family of likelihood functions for the observable random vector $V \in R^n$, $n \geq 1$; let Θ_o be a subset of Θ. Assume that $\sup_{\theta \in \Theta_o} L_\theta(v)$ and $\sup_{\theta \in \Theta} L_\theta(v)$ are both finite. A likelihood ratio test (LRT) of $H_o: \theta \in \Theta_o$ has the rejection region*

$$\lambda(v) = \sup_{\theta \in \Theta_o} L_\theta(v) / \sup_{\theta \in \Theta} L_\theta(v) \leq \lambda_o$$

with size $\alpha = \sup_{\theta \in \Theta_o} P(\lambda(V) \leq \lambda_o)$.

Of course, $\lambda(v) \leq 1$; note that $\{v: \lambda(v) \leq \lambda_o\}$ is determined only a.s. Moreover, the same λ is obtained for any

H_a which includes $\theta \notin \Theta_o$. As before, the likelihood function for a RS of a continuous or discrete RV X is the product of the density functions: $L_\theta(v) = \Pi_{i=1}^{n} f(x_i;\theta)$.

Example: If $X_1, X_2, \cdots, X_n$ is a RS from the uniform distribution on $(0,\theta)$ with $\theta \in (0,\infty)$, then

$$L_\theta(v) = I\{0 < x_{(1)} \le x_{(n)} < \theta\}/\theta^n$$

where $x_{(1)}$ is the minimum, $x_{(n)}$ is the maximum of the observed sample. Now, $L_\theta(v)$ is maximal when it is positive and its denominator is minimal:

$$\sup_{\theta>0} L_\theta(v) = 1/x_{(n)}^n ;$$

for $H_o: \theta = 1$ and $\sup_{\theta \in \Theta_o} L_\theta(v) = 1$. The LRT against any $\theta \neq 1$ is given by $\lambda(v) = x_{(n)}^n \le \lambda_o$.

For $0 < \alpha < 1$, $\alpha = P(\lambda(V) \le \lambda_o \mid \theta = 1)$

$$= P(X_{(n)} \le \lambda_o^{1/n} \mid \theta = 1) = \lambda_o .$$

The power function is

$$P(X_{(n)} \le \alpha^{1/n} \mid \theta) = 1 \text{ for } \theta < \alpha^{1/n}$$
$$= \alpha \text{ for } \theta \ge \alpha^{1/n} .$$

Exercise 1: Let $X_1, X_2, \cdots, X_n$ be a RS from the distribution with density $e^{-(x-\theta)} I\{\theta < x\}$ and θ real. Find a LRT for $H_o: \theta = 0$ vs $H_a: \theta \neq 0$.

It is often convenient to use log–likelihood

$$\mathcal{L}(\theta) = \log L_\theta(v) \text{ (base e).}$$

When $\theta \in \Theta \subset R^p$, and $\mathcal{L}$ is differentiable with respect to θ, either of

$$\partial L_\theta(v)/\partial\theta = 0 \text{ or } \partial \mathscr{L}(\theta)/\partial\theta = 0$$

may be referred to as the *likelihood equation* and a solution is denoted by $\hat{\theta}$. If this equation has more than one solution, each must be examined to find the maximum likelihood. When $\mathscr{L}$ has second order derivatives, these may be helpful in establishing a maximum.

Theorem: *For* $\Theta \subset R^p$, *let* $\mathscr{L} : \Theta \to R$ *have continuous second order partial derivatives. Then* $\mathscr{L}(\theta_o)$ *is a local maximum if at* θ_o, $\partial\mathscr{L}/\partial\theta = (\partial\mathscr{L}/\theta_1, \cdots, \partial\mathscr{L}/\theta_p)' = 0$

and the Hessian matrix $\partial^2\mathscr{L}/\partial\theta\partial\theta' = \left[\partial^2\mathscr{L}/\partial\theta_i\partial\theta_j\right] < 0$.

Proof: Taylor's formula gives

$$\mathscr{L}(\theta) = \mathscr{L}(\theta_o) + \partial\mathscr{L}/\partial\theta'\rfloor_{\theta_o}(\theta - \theta_o)$$
$$+ (\theta - \theta_o)'\partial^2\mathscr{L}/\partial\theta\partial\theta'\rfloor_\xi(\theta - \theta_o)/2$$

where ξ is "between" θ and θ_o. The (negative definite) condition on the Hessian matrix means that

$$\varphi'\partial^2\mathscr{L}/\partial\theta\partial\theta'\varphi < 0 \text{ for all } \varphi \neq 0 \text{ in } R^p$$

and all θ in a neighborhood of θ_o. Since the first derivative is 0 at θ_o, $\mathscr{L}(\theta) = \mathscr{L}(\theta_o) + 0 + (\text{a negative number})$ or $\mathscr{L}(\theta_o) > \mathscr{L}(\theta)$.

Exercise 2: Write out all details, including more common notation, of the proof above when p = 1, 2 .

Example: Let $X_1, X_2, \cdots, X_n$ be a RS from a normal distribution with mean θ in R and positive known variance (which we may as well take as 1). Then,

$$\mathscr{L}(\theta) = -\Sigma(x_i - \theta)^2/2 - (n/2)\log 2\pi .$$

Consider testing $H_o: \theta = 0$ vs $H_a: \theta \neq 0$.

a) First, $\sup_{\theta \in \Theta_o} \mathscr{L}(\theta) = -\Sigma x_i^2/2 - (n/2)\log 2\pi$.

Since $\partial\mathscr{L}/\partial\theta = -\Sigma(x_i - \theta)$,

$$\sup_{\theta \in \Theta} \mathscr{L}(\theta) = -\Sigma(x_i - \bar{x})^2/2 - (n/2)\log 2\pi .$$

Then $\log \lambda(v) = \sup_{\theta \in \Theta_o} \mathscr{L}(\theta) - \sup_{\theta \in \Theta} \mathscr{L}(\theta)$

$= -\Sigma x_i^2/2 + \Sigma(x_i - \bar{x})^2/2 < \log \lambda_o$

iff $-n\bar{x}^2/2 < \log \lambda_o$ iff $\bar{x}^2 > k$ (generic).

This should agree with your intuition: if the sample mean is "large", then the population mean is probably not 0.

b) If the significance level (size) α is prescribed, we have $\alpha = P(\bar{X}^2 > k \mid \theta = 0)$

$$= P(Z^2 > nk) = P(|Z| > \surd(nk)) \qquad (*)$$

where Z is a standard normal RV. Recall that Z^2 has a chisquare distribution with 1 degree of freedom. Then (*) means that,

$$nk = \chi^2_\alpha \text{ or } \surd(nk) = z_{\alpha/2} .$$

For example, with $\alpha = .05$, $\chi^2_{.05} = 3.84$ and

$$z_{.025} = 1.96 .$$

Exercise 3: Refer to this last example.

a) Explain "which we may as well take as 1". Hint: X/σ is observable when σ is known.

b) Check that $\partial^2\mathscr{L}/\partial\theta^2 < 0$ at $\hat{\theta} = \bar{x}$.

Example: Let $X_1, X_2, \cdots, X_n$ be a RS from a normal distribution with mean θ_1 and variance θ_2.

$$\mathcal{L}(\theta) = -\Sigma(x_i - \theta_1)^2/2\theta_2 - (n/2)\log \theta_2 .$$

Consider $H_o: \theta_1 = 0, \theta_2 > 0$ vs $H_a: \theta_1 \neq 0, \theta_2 > 0$.

a) Under H_o, $\hat{\theta}_{1o} = 0$ and

$$\partial \mathcal{L}/\partial\theta_2 = -(\Sigma x_i^2/2)(-1/\theta_2^2) - n/2\theta_2$$

yields $\hat{\theta}_{2o} = \Sigma x_i^2/n > 0$. At $\hat{\theta}_{2o}$,

$$\partial^2 \mathcal{L}/\partial^2\theta_2 = -\Sigma x_i^3/\theta_2^3 + n/2\theta_2^2$$

$$= -n\hat{\theta}_{2o}/\hat{\theta}_{2o}^3 + n/2\hat{\theta}_{2o}^2 = -n/\hat{\theta}_{2o}^2$$

which is negative so that the max is attained:

$$\mathcal{L}(\hat{\theta}) = -n/2 - (n/2)\log \Sigma x_i^2/n .$$

b) Under H_a, $\partial \mathcal{L}/\partial\theta_1 = \Sigma(x_i - \theta_1)/\theta_2$,

$$\partial^2 \mathcal{L}/\partial\theta_1^2 = -1/\theta_2 ,$$

$$\partial^2 \mathcal{L}/\partial\theta_2\partial\theta_1 = \Sigma(x_i - \theta_1)/\theta_2^2 ,$$

$$\partial \mathcal{L}/\partial\theta_2 = \Sigma(x_i - \theta_1)^2/2\theta_2^2 - n/2\theta_2 ,$$

$$\partial^2 \mathcal{L}/\partial\theta_2^2 = -\Sigma(x_i - \theta_1)^2/\theta_2^3 + n/2\theta_2^2 .$$

The likelihood equation yields

$$\hat{\theta}_{1a} = \Sigma x_i/n , \quad \hat{\theta}_{2a} = \Sigma(x_i - \bar{x})^2/n .$$

The value of the Hessian matrix at $\hat{\theta}_{1a}, \hat{\theta}_{2a}$ is

$$\begin{bmatrix} -n/\Sigma(x_i - \bar{x})^2 & 0 \\ 0 & -n/2\Sigma\,(x_i - \bar{x})^2 \end{bmatrix}$$

which is obviously negative definite.

$\mathscr{L}(\hat{\theta}_a) = -\,n/2 - (n/2)\log \Sigma(x_i - \bar{x})^2/n$.

c) Since $\log \lambda = \log \mathscr{L}(\hat{\theta}_o) - \log \mathscr{L}(\hat{\theta}_a)$ is

$-n/2 - (n/2)\log \Sigma x_i^2/n + n/2 + (n/2)\log \Sigma(x_i - \bar{x})^2/n$,

$\lambda \le \lambda_o$

iff $\log \lambda \le \log \lambda_o$

iff $\log (\Sigma(x_i - \bar{x})^2/\Sigma x_i^2 \le k$

iff $\Sigma x_i^2 / \Sigma(x_i - \bar{x})^2 \ge k$

$$\text{iff} \quad \frac{\Sigma(x_i - \bar{x})^2 + n\bar{x}^2}{\Sigma(x_i - \bar{x})^2} \ge k \qquad (*)$$

iff $n\bar{x}^2/\Sigma(x_i - \bar{x})^2 \ge k$.

Under H_o , $n\bar{X}^2 \div \Sigma(X_i - \bar{X})^2/(n-1)$ is distributed as the square of a central Student–T RV with n–1 degrees of freedom (dof). This leads to:

reject H_o when $\sqrt{n}|\bar{x}|/s > t_{\alpha/2}$

where $s^2 = \Sigma(x_i - \bar{x})^2/(n-1)$, $P(|T| > t_{\alpha/2}) = \alpha$.

d) Under H_a , $\sqrt{n}\,\bar{x}/s$ has a non–central T distribution but the power function depends on the parameters only thru $\delta = \theta_1/\theta_2$.

Exercise 4: Check the equivalence of the inequalities at (*) .

Exercise 5: Let Z be a normal RV with mean δ and variance 1 ; let V be a chisquare RV with f dof. If Z and V are independent, the density of $T = Z/\sqrt{(V/f)}$ is given by

$$\frac{\int_0^\infty y^{(f-1)/2} e^{-y/2} e^{-(t\sqrt{(y/f)}-\delta)^2/2} dy}{2^{(f+1)/2}\Gamma(f/2)\cdot\sqrt{(\pi f)}}$$

where Γ is the gamma function. Prove this by first finding the joint density of T and V from that of Z and V .

This example with the normal distribution extends immediately to the classical (Student–Fisher) procedure: let $X_1, X_2, \cdots, X_n$ be a RS from a normal distribution with mean μ and finite non–zero variance σ^2. Reject $H_o: \mu = \mu_o$ in favor of $H_a: \mu \neq \mu_o$ when the observed

$$|t| = |\bar{x} - \mu_o|/(s/\sqrt{n}) > t_{\alpha/2} \cdot$$

If $\sqrt{n}|\bar{x} - \mu_o|/s = b$, the p–value is $P(|T| > b)$ where this central Student–T has n–1 dof.

Exercise 6: Find the p–value for a test of the hypothesis that the mean is 3.025 when the following sample is from a normal distribution.

–2.052 3.092 2.728 2.488 2.508 5.094 4.023 1.828
3.958 2.866 2.458 5.230 3.806 5.812 4.468 1.768
1.656 .200 2.268 1.326

When extended to certain other normal populations, this procedure turns out to be an analysis of variance (ANOVA) even though the principal hypotheses concern means; perhaps it should be analysis by variance. We illustrate the simplest case termed "one–way" in the next lesson.

LESSON 8. ONE–WAY ANALYSIS OF VARIANCE

The lesson title is that of the simplest case of ANOVA, comparison of the means of several independent normal populations. Although our model is Fisherian, our discussion is not quite the classical one; we use results from both Lessons 6 and 7 .

The problem is to compare the effect of several treatments (paints on wooden fences or fertilizers on wheat fields or oven temperatures on pottery, or textbooks on students, or $\cdots$). It is somewhat obvious that the *experimental units* (fences, fields, pots, students,$\cdots$) to which the treatments are to be applied must be "independent and homogeneous" at least with respect to characteristics which might affect or be affected by the treatment. In physical laboratories, one can "control" components so that in a macroscopic sense this requirement can be met; in the "field", including groups of people, this attainment is difficult. Since so many of these difficulties are closely intertwined with subject matter, we will not pursue the construction of a "good design".

Instead, we suppose that treatment 1 is applied to n_1 experimental units, treatment 2 to n_2 experimental units, etc., and we assume that the observables can be represented by an *additive model*: $X_{ij} = \mu + \alpha_i + \varepsilon_{ij}$, $j = 1(1)n_i$, $i = 1(1)r$, where μ is a general mean, $\alpha_1,\cdots,\alpha_r$ are (fixed or non–random) treatment effects, and ε_{ij} are the random errors ("noise" in some contexts). To get a LRT, we assume that these error terms are IID normal RVs with mean 0 and finite positive variance σ^2 ; the parameter $\theta = (\mu, \alpha_1, \alpha_2,\cdots, \alpha_r, \sigma^2)$ belongs to the set $\Theta = R^{r+1}\times(0,\infty)$. Let $n_. = \Sigma_{i=1}^{r} n_i$. The log likelihood function

$$\mathscr{L}(\theta) = -\sum_{i=1}^{r}\sum_{j=1}^{n_i}(x_{ij}-\mu-\alpha_i)^2/2\sigma^2 - n_.(\log 2\pi\,\sigma^2)/2 \;.(0)$$

The partial derivatives are:

$$\partial\mathscr{L}/\partial\mu = \Sigma_i\,\Sigma_j\,(x_{ij} - \mu - \alpha_i)/\sigma^2 \;; \qquad (1)$$

for $i = 1(1)r$, $\partial \mathscr{L}/\partial\alpha_i = \Sigma_j (x_{ij} - \mu - \alpha_i)/\sigma^2$; (2)

$$\partial \mathscr{L}/\partial\sigma^2 = \Sigma_i \Sigma_j (x_{ij} - \mu - \alpha_i)^2/2\sigma^4 - n_. /2\sigma^2 . \qquad (3)$$

Exercise 1: Verify the derivatives above.

Since $\sum_{i=1}^{r} \partial \mathscr{L}/\alpha_i = \partial \mathscr{L}/\partial\mu$, there are only $r+1$ linearly independent equations in (1,2,3) while there are $r+2$ components in θ . However, we don't have to determine the values of all the parameters to determine the max. From each equation in (2), we get

$$\sum_j^{n_i}(x_{ij} - (\hat{\mu} + \hat{\alpha}_i)) = 0 ,$$

and so

$$\hat{\mu} + \hat{\alpha}_i = \sum_{j=1}^{n_i} x_{ij}/n_i = \bar{x}_{i.} .$$

Putting this in (3) yields

$$\Sigma_i \Sigma_j (x_{ij} - \hat{\mu} - \hat{\alpha}_i)^2 / \hat{\sigma}^4 = n_. /\hat{\sigma}^2$$

whence $\hat{\sigma}^2 = \Sigma_i \Sigma_j (x_{ij} - \bar{x}_{i.})^2/n_.$.

In this paragraph, we show that it is simpler to verify such maximal conditions by using a little bit of algebra than by using second order partial derivatives. For $c_1, c_2, \cdots, c_r$, independent of

$$j , \sum_{j=1}^{n_i} (x_{ij} - c_i)^2 = \sum_{j=1}^{n_i} (x_{ij} - \bar{x}_{i.} + \bar{x}_{i.} - c_i)^2$$

$$= \sum_{j=1}^{n_i} (x_{ij} - \bar{x}_{i.})^2 + \sum_{j=1}^{n_i} (\bar{x}_{i.} - c_i)^2 \qquad (4)$$

since

$$\sum_{j=1}^{n_i} (x_{ij} - \bar{x}_{i.})(\bar{x}_{i.} - c_i) = (\bar{x}_{i.} - c_i)\sum_{j=1}^{n_i} (x_{ij} - \bar{x}_{i.}) = 0 .$$

Thus, in general, $\sum_{j=1}^{n_i} (x_{ij} - c_i)^2$ is minimal when $c_i = \bar{x}_{i.}$.

Note that algebraically, the only condition on the x's is that they be finite real numbers.

In particular, $\Sigma_i \Sigma_j (x_{ij} - (\mu + \alpha_i))^2$ is minimized when $\mu + \alpha_i = c_i = \bar{x}_{i.}$ for i = 1(1)r . Since $\mu, \alpha_1, \alpha_2, \cdots, \alpha_r$ are functionally independent of σ^2, we get the same $\hat{\sigma}^2$ as above. It follos that

$$\sup_{\theta \in \Theta} \mathcal{L}(\theta) = \max_{\theta \in \Theta} \mathcal{L}(\theta) = -n_{.}/2 - n_{.}(\log 2\pi\, \hat{\sigma}^2)/2 .$$

For this, we need the condition that the x's are not all equal, which is of course true, almost surely.

Exercise 2: Give details to verify the "little bit of algebra" above.

The most common null hypothesis in these situations is that there is no difference in the (means of the) treatments:

$$H_o : \mu + \alpha_1 = \mu + \alpha_2 = \cdots = \mu + \alpha_r = \mu_o \text{ (unknown).}$$

In an equivalent more formal expression, $\theta \in \Theta_o \subset \Theta$,

$$\Theta_o = \left[(\mu_o, \mu_o - \mu, \mu_o - \mu, \cdots, \mu_o - \mu, \sigma^2) \text{ for } \mu\text{'s} \in R \text{ and } \sigma^2 \in (0, \infty) \right]$$

Exercise 3: Show that

$$\max_{\theta \in \Theta_o} \mathcal{L}(\theta) = -n_{.}/2 - n_{.}(\log 2\pi\, \hat{\sigma}_o^2)/2$$

where $\bar{x}_{..} = \Sigma_i \Sigma_j x_{ij}/n_{.}$ and $\hat{\sigma}_o^2 = \Sigma_i \Sigma_j (x_{ij} - \bar{x}_{..})^2/n_{.}$.

Continuing with the above, the LRT is

$$\log \lambda(v) \le k$$

iff $\sup_{\theta \in \Theta_o} \mathcal{L}(\theta) - \sup_{\theta \in \Theta} \mathcal{L}(\theta)$

$$= -n_{.}/2 - n_{.}(\log 2\pi\, \hat{\sigma}^2)/2 + n_{.}/2 + n_{.}(\log 2\pi\, \hat{\sigma}_o^2)/2 \le k$$

$$\text{iff } \hat{\sigma}^2/\hat{\sigma}_o^2 = \frac{\Sigma_i \Sigma_j (x_{ij} - \bar{x}_{..})^2}{\Sigma_i \Sigma_j (x_{ij} - \bar{x}_{i.}^2)^2} \geq k .$$

Since $\bar{x}_{..}$ is independent of j, we may apply the principle in (4) to get

$$\Sigma_i \Sigma_j (x_{ij} - \bar{x}_{..})^2 = \Sigma_i \Sigma_j (x_{ij} - \bar{x}_{i.})^2 + \Sigma_i \Sigma_j (\bar{x}_{i.} - \bar{x}_{..})^2 .$$

Then we see that the LRT is equivalent to

$$\frac{\Sigma_i \Sigma_j (\bar{x}_{i.} - \bar{x}_{..})^2}{\Sigma_i \Sigma_j (x_{ij} - \bar{x}_{i.})^2} \geq k \tag{5}$$

Under H_o, $X_{ij} = \mu_o + \varepsilon_{ij}$ so that

$$\Sigma_i \Sigma_j (\bar{X}_{i.} - \bar{X}_{..})^2 = \Sigma_i \Sigma_j (\bar{\varepsilon}_{i.} - \bar{\varepsilon}_{..})^2 \tag{6}$$

and

$$\Sigma_i \Sigma_j (X_{ij} - \bar{X}_{i.})^2 = \Sigma_i \Sigma_j (\varepsilon_{ij} - \bar{\varepsilon}_{i.})^2 .$$

Thus the null distribution of the ratio in (5) does not depend upon the parameters μ , μ_o .

Exercise 4: Verify the algebra for the two equations at (6).

As will be shown below, $\Sigma_i \Sigma_j (\bar{\varepsilon}_{i.} - \bar{\varepsilon}_{..})^2/\sigma^2$ is distributed as chisquare with $r - 1$ dof independently of $\Sigma_i \Sigma_j (\varepsilon_{ij} - \bar{\varepsilon}_{i.})^2/\sigma^2$ which is distributed as chisquare with $\Sigma_i(n_i - 1) = n_. - r$ dof. Consequently, under H_o,

$$\Sigma_i \Sigma_j (\bar{\varepsilon}_{i.} - \bar{\varepsilon}_{..})^2/(r-1) \div \Sigma_i \Sigma_j (\varepsilon_{ij} - \bar{\varepsilon}_{i.})^2/(n_. - r) \tag{7}$$

is distributed as a central F random variable with r–1 and

$n_{\cdot} - r$ dof. The LRT becomes:
reject H_o(no differences in the treatment means) when

$$\frac{\Sigma_i \Sigma_j (\bar{x}_{i\cdot} - \bar{x}_{\cdot\cdot})^2/(r-1)}{\Sigma_i \Sigma_j (x_{ij} - \bar{x}_{i\cdot})^2/(n_{\cdot}-r)} > F_\alpha$$

where

$$P(F > F_\alpha) = \alpha .$$

When H_o is false, this ratio has a non-central F distribution.

The last ratio above is that of the "variance between treatments" to the "variance within treatments". There are many texts which treat ANOVA in detail; two of the best are Scheffé, 1959, and Hocking, 1985. The rest of this lesson completes the distribution theory for this test.

Theorem: *If the components of* $X = (X_1, X_2, \cdots, X_p)'$ *are IID* $N_1(0,1)$, *then* $Q = X'CX$ *is a chisquare RV with* r *dof iff* $C = C' = C^2$ *with rank* r.

Proof: *Sufficiency* When $C = C' = C^2$, $I-C$ is also symmetric idempotent. For such matrices, rank = trace so that

$$\rho(C) + \rho(I-C) = tr(C) + tr(I-C) = tr(I) = p .$$

As $X'X = X'CX + X'(I-C)X$, application of the last corollary of Lesson 6 yields:

$X'CX$, $X'(I-C)X$ are independent chisquare RVs with r, $p-r$ dof respectively.

Necessity When $C = C'$, there is an orthogonal matrix P such that for $Y = PX$,

$$X'CX = Y'PCP'Y = \lambda_1 Y_1^2 + \lambda_2 Y_2^2 + \cdots + \lambda_m Y_m^2 \quad (8)$$

where $\lambda_1, \cdots, \lambda_m$ are the non-zero eigenvalues of C. Since the components of Y are IID $N_1(0,1)$, $Y_1^2, Y_2^2, \cdots, Y_m^2$ are independent chisquare RVs each with 1 dof. The CF of the right-hand side of (8) is

$$\sum_{j=1}^{m} \lambda_j Y_j^2 = \Pi_{j=1}^{m} (1 - 2it\lambda_j)^{-1/2} \text{ for t real .}$$

Since $X'CX$ is to have a chisquare distribution with r dof, its CF must be

$$(1 - 2it)^{-r/2} \text{ for t real .}$$

The two polynomials in t

$$\Pi_{j=1}^{m} (1 - 2itr_j) , (1 - 2rt)^r$$

can be identical iff all λ_j equal 1 and $r = m$. Thus the rank of C is r and the diagonal matrix $PCP' = \Lambda$ contains only zeroes and ones. It follows that $\Lambda \cdot \Lambda = \Lambda$, equivalently,

$$PCP'PCP' = PCP' \text{ whence } C \cdot C = C .$$

Exercise 5: Let $A = A' = A^2$ and $B = B' = B^2$ be p by p matrices. Show that $(A+B)^2 = A+B$ iff $A \cdot B = 0 = B \cdot A$. Hint: for necessity, first show that

$$x'ABx = -x'BAx = -x'ABx \text{ for all vectors x .}$$

Theorem: *Let the components of* $Y = (Y_1, \cdots, Y_p)'$ *be IID* $N_1(0,1)$. *Let* C_1 *and* C_2 *be constant symmetric matrices. If* $Y'C_1'Y_1$ *and* $Y'C_2Y$ *are chisquare RVs, then they are independent iff* $C_1C_2 = 0$.

Proof: a) By the last theorem, the chisquare hypothesis implies that C_1 and C_2 are symmetric idempotent. When $C_1C_2 = 0$, $C_2C_1 = 0$ so that $I–C_1–C_2$ is also symmetric idempotent. Hence

$$\rho(C_1) + \rho(C_2) + \rho(I–C_1–C_2) = tr(C_1) + tr(C_2) + tr(I–C_1–C_2)$$
$$= tr(I) = p .$$

The distribution is established by the last corollary of Lesson 6.

b) When these quadratic forms are independent chisquare Rvs, their sum $Y'(C_1 + C_2)Y$ is also chisquare. Then $C_1 + C_2$ must be symmetric idempotent and the last exercise can be applied.

Exercise 6: Complete the details for part a) of the proof above.

It is convenient to have the following notation: for positive integers t , I_t is a t by t identity matrix;

1_t is a column vector of all ones of length t ;

$J_t = 1_t 1_t'$ is a t by t matrix of all ones ;

$K_t = J_t/t$; $Y = (y_1, y_2, \cdots, y_n)'$;

$$\sum_{j=1}^{t} y_j = Y'1_t = 1_t'Y \,;\ \sum_{j=1}^{t} y_j^2 = Y'Y \,.$$

For any matrices A,B,C, $\langle A, B, C\rangle = \begin{bmatrix} A & 0 & 0 \\ 0 & B & 0 \\ 0 & 0 & C \end{bmatrix}$ is a block diagonal matrix filled in with appropriately dimensioned zero matrices.

Exercise 7: Consider the matrices

$$I_7 - \langle K_3, K_4\rangle \,,\ \langle K_3, K_4\rangle - K_7 \,,\ K_7 \,.$$

Show that

a) each of these matrices is symmetric idempotent;

b) these matrices are pairwise orthogonal (ie, they have products 0).

For i = 1(1)r, let $Y_i' = (\varepsilon_{i1}, \varepsilon_{i2}, \cdots, \varepsilon_{in_i})/\sigma^2$ and then

$$Y' = (Y_1', Y_2', \cdots, Y_r') \,.$$

Note that the $\{Y_{ij} = \varepsilon_{ij}/\sigma^2\}$ are IID and $N_1(0,1)$. The quadratic forms in (7) can be written as

$$\Sigma_i \Sigma_j (\varepsilon_{ij} - \bar{\varepsilon}_{i.})^2/\sigma^2 = \Sigma_i(\Sigma_j \varepsilon_{ij}^2 - n_i\bar{\varepsilon}_{i.}^2)/\sigma^2$$

$$= \Sigma_i \Sigma_j \varepsilon_{ij}^2/\sigma^2 - \Sigma_i (\Sigma_j \varepsilon_{ij}/\sigma)^2/n_i = Y'Y - \Sigma_i(Y_i'1_{n_i}1_{n_i}'Y_i)/n_i$$

$$= Y'Y - (Y_1', Y_2', \cdots, Y_r')\langle K_{n_1}, K_{n_2}, \cdots, K_{n_r}\rangle \begin{bmatrix} Y_1 \\ \vdots \\ Y_r \end{bmatrix}$$

$= Y'(I_{n_.} - \langle K_{n_1}, K_{n_2}, \cdots, K_{n_r}\rangle)Y;$

$\Sigma_i \Sigma_j (\bar{\varepsilon}_{i.} - \bar{\varepsilon}_{..})^2/\sigma^2 = \Sigma_i n_i(\bar{\varepsilon}_{i.}{}^2 - 2\bar{\varepsilon}_{i.}\bar{\varepsilon}_{..} + \bar{\varepsilon}_{..}{}^2)/\sigma^2$

$= \Sigma_i (\Sigma_j \varepsilon_{ij}/\sigma)^2/n_i - (\Sigma_i \Sigma_j \varepsilon_{ij}/\sigma)^2/n_.$

$= Y'\langle K_{n_1}, K_{n_2}, \cdots, K_{n_r}\rangle Y - Y'1_{n_.}1_{n_.}'Y/n_.$

$= Y'(\langle K_{n_1}, K_{n_2}, \cdots, K_{n_r}\rangle - K_{n_.})Y.$

Exercise 8: Show that $Y'Y = Y'(K_{n_.})Y$

$+ Y'(\langle K_{n_1}, \cdots, K_{n_r}\rangle - K_{n_.})Y + Y'(I_{n_.} - \langle K_{n_1}, \cdots, K_{n_r}\rangle)Y$

and find the ranks of the corresponding matrices.

This last exercise shows that the ranks of the matrices satisfy $n = 1 + r + (n - r)$ so that the last corollary in Lesson 6 can be applied to complete the distribution of the ratio in (7).

LESSON 9. LIKELIHOOD RATIO TESTS – II

We continue the discussion of LRT with another classical example although its original developement by Karl Pearson, 1900, was based on more intuitive ideas. We begin with a generalized Bernoulli distribution where the classification of data is into more than the two categories success/failure.

Definition: *The family of k-cell Bernoulli distributions (indexed by the parameter*

$$\theta = (\theta_1, \cdots, \theta_k)' \text{ , } 0 \le \theta_j \le 1, j = 1(1)k, \sum_{j=1}^{k} \theta_j = 1)$$

has density $\theta_1^{x_1} \cdot \theta_2^{x_2} \cdots \theta_k^{x_k}$ *,* $x_j \in \{0,1\}$ *and* $\sum_{j=1}^{k} x_j = 1$ *. This may be abbreviated as* $f(x;\theta) = \Pi_{j=1}^{k} \theta_j^{x_j}$.

We will need some moments of this distribution. Since $\sum_{j=1}^{k} x_j = 1$, there are only k–1 (linearly) independent variables in the distribution. Formally, we may account for this by taking

$$f(x;\theta) = \left[\prod_{j=1}^{k-1} \theta_j^{x_j}\right] \cdot \left[1 - \sum_{j=1}^{k-1} \theta_j\right]^{1-\Sigma_{j=1}^{k-1} x_j} .$$

Theorem: *a) For* $t = (t_1, \cdots, t_{k-1})' \in R^{k-1}$*, a MGF of the k-cell Bernoulli distribution is*

$$M(t) = \sum_{j=1}^{k-1} \theta_j exp(t_j) + (1 - \Sigma_{j=1}^{k-1} \theta_j) .$$

b) For $j \ne m = 1(1)k$ *,*

$$E[X_j] = \theta_j,\ Var(X_j) = \theta_j(1-\theta_j),\ Cov(X_j,X_m) = -\theta_j\theta_m .$$

Proof: $M(t) = \sum{}' (\exp(\sum_{j=1}^{k-1} t_j x_j)) \cdot \Pi_{j=1}^{k} \theta_j^{x_j}$ where the primed sigma indicates that the sum is for those x's in {0,1} with $x_1 + \cdots + x_k = 1$. For $\theta_k = 1-\Sigma_{j=1}^{k-1}\theta_j$,

$$M(t) = \theta_1 \exp(t_1) + \theta_2 \exp(t_2) + \cdots + \theta_{k-1}\exp(t_{k-1}) + \theta_k .$$

For $j \neq m = 1(1)(k-1)$,

$$\partial M/\partial t_j = \theta_j \exp(t_j)\ ,\ \partial^2 M/\partial t_j^{\,2} = \theta_j \exp(t_j)\ ,\ \partial^2 M/\partial t_j \partial t_m = 0 .$$

Evaluating these at $t = 0$ yields

$$E[X_j] = \theta_j\ ,\ E[X_j^{\,2}] = \theta_j\ ,\ E[X_j X_m] = 0 .$$

Hence for positive integers $j \neq m \leq k-1$,

$$\mathrm{Var}(X_j) = \theta_j - \theta_j^{\,2}\ ,\ \mathrm{Cov}(X_j, X_m) = -\theta_j \theta_m .$$

Since the choice of "k as the leftover cell" is arbitrary, similar operations for $X_2, \cdots, X_k$ yield the final moments.

Exercise 1: Let $(X_{11}, \cdots, X_{1k}), \cdots, (X_{n1}, \cdots, X_{nk})$ denote a RS from the k–cell Bernoulli distribution. Let

$$W_n' = \left[\sum_{i=1}^{n} X_{i1}, \cdots, \sum_{i=1}^{n} X_{ik}\right] .$$

Find means , variances–covariances of the components of W_n .

For example, $E[\sum_i X_{i1} \cdot \sum_j X_{j2}] = \sum_i \sum_j E[X_{i1} \cdot X_{j2}]$

$$= \sum_i E[X_{i1} X_{i2}] + \sum_{i \neq j} E[X_{i1} \cdot X_{j2}] = \sum_i (0) + \sum_{i \neq j} \theta_1 \theta_2$$

$$= n(n-1)\theta_1 \theta_2 .$$

Exercise 2: Herein, the prime indicates that the summation is for non–negative integers y1, y2, y3 such that y1 + y2 + y3 = n , fixed. Prove by induction:

$$(a + b + c)^n = \sum{}' n!\, a^{y1} \cdot b^{y2} \cdot c^{y3} / y1! \cdot y2! \cdot y3! .$$

For us, the following theorem also defines the *multinomial* distribution.

Theorem: *For a RS from the k-cell Bernoulli distribution, the distribution of* W_n *(from exercise 1) is the k-cell multinomial with density* $n!\cdot\Pi_{j=1}^{k}\ \theta_j^{w_j}\ /\ w_j!$ *where* $w_1,\cdots,w_k$ *are non-negative integers such that* $w_1 + \cdots + w_k = n$ *.*

Proof: By independence, the MGF of W_n is

$$\Pi_{i=1}^{n} M(t) = (\theta_1\exp(t_1) + \cdots + \theta_{k-1}\exp(t_{k-1}) + \theta_k)^n .$$

By the (mathematical) multinomial theorem generalizing exercise 2, this becomes the sum

$$\sum{}' n!\left[\Pi_{j=1}^{k-1} (\theta_j\exp(t_j))^{w_j}\right]\theta_k^{w_k} / \Pi_{j=1}^{k} w_j! .$$

Rewriting this as

$$\sum{}'\left[\Pi_{j=1}^{k-1} \exp(t_j w_j)\right] n!\cdot\Pi_{j=1}^{k}\theta_j^{w_j} / \Pi_{j=1}^{k} w_j!$$

shows the form of the density (by uniqueness of the MGF).

The following example is a classical one discussed in, among other places, Stern and Sherwood, 1966. After we derive a LRT for the genral case, you will be asked to calculate "chi–square" using this real numerical example.

Example: Mendel had, among others, a theory on cross–breeding certain plants which stated that the results should be in the ratio of 9:3:3:1 for the classes of smooth–yellow, smooth–green, wrinkled–yellow, wrinkled– green peas. One sample report listed the corresponding frequencies 315 108 101 32 . Do the data support the theory?

Example: For a sample from the k–cell Bernoulli distribution, consider a simple null hypothesis $H_o: \theta_1 = \theta_{1o}, \cdots, \theta_k = \theta_{ko}$ vs a composite alternative H_a: at least two of these equalities are false. Here the likelihood of a sample is equivalent to the density of the multinomial. Under H_o, this is

$$L_{\theta_o}(v) = n!\ \Pi_{j=1}^{k}\left[\theta_{jo}\right]^{w_j}/w_j!\ .$$

Under H_a, a log–likelihood is

$$\mathscr{L}(\theta) = \log(n!/\Pi w_j!) + \sum_{j=1}^{k} w_j \log(\theta_j)\ .$$

Remember that we are deriving a LRT. In differentiating the log–likelihood to find critical points, we use the fact that $\Sigma_j\ \theta_j = 1$ by taking $\theta_k = 1-\theta_1 - \cdots - \theta_{k-1}$. Then for $j \le k-1$, $\partial\mathscr{L}/\partial\theta_j = w_j/\theta_j + w_k/\theta_k(-1)$ yields

$$\hat{\theta}_1\ w_k = \hat{\theta}_k\ w_1$$
$$\vdots$$
$$\hat{\theta}_{k-1}\ w_k = \hat{\theta}_k\ w_{k-1}$$

Obviously,

$$\hat{\theta}_k\ w_k = \hat{\theta}_k\ w_k\ .$$

By addition of these equations, we get

$$1\ w_k = \hat{\theta}_k\ \Sigma w_j \text{ so that } \hat{\theta}_k = w_k/n\ .$$

We can manipulate with $\theta_{k-1} = 1-\theta_1 - \cdots - \theta_{k-2} - \theta_k$ etc. to get $\hat{\theta}_j = w_j/n$ for $j = 1(1)k$.

Exercise 3: In the last example, check the non–negativity of the matrix of second partial derivatives of $\mathscr{L}(\theta)$ with respect to $\theta_1, \cdots, \theta_{k-1}$.

Exercise 4: Show that for $a > 0$, a Taylor's formula is

$$\log(a) = (a-1) - (a-1)^2/2 + (a-1)^3/3\xi^3$$

for some ξ between a and 1.

In general, $\sup_{\theta\in\Theta} \mathscr{L}(\theta) = \log(n!/\Pi\ w_j!) + \Sigma_j\ w_j \cdot \log(w_j/n)$.

Under H_o, $\sup_{\theta \in \Theta_o} \mathscr{L}(\theta) = \mathscr{L}(\theta_o) = \log(n!/\Pi\, w_j!) + \Sigma_j\, w.\log(\theta)$.

The critical region for the LRT can be taken as

$$\Sigma_j\, w_j \cdot \log(w_j/n\theta_{jo}) > c\,. \qquad (*)$$

Applying the result of exercise 4 to each term of the sum on the left of (*), we get

$$\sum_j w_j \cdot \log(w_j/n\theta_{jo}) = \sum_j w_j \cdot [(w_j/n\theta_{jo}-1) - (w_j/n\theta_{jo}-1)^2/2$$
$$+ (w_j/n\theta_{jo}-1)^3/3\xi_j^{\,3}]\,.$$

Now we replace the multiplier w_j by an equivalent

$$n\theta_{jo} + (w_j - n\theta_{jo})$$

and expand the sums to get

$$\sum_j (w_j - n\theta_{jo}) - \sum_j (w_j - n\theta_{jo})^2/2n\theta_{jo}$$
$$+ \sum_j (w_j - n\theta_{jo})^3/3(n\theta_{jo})^2\xi_j^{\,3} + \sum_j (w_j - n\theta_{jo})^2/n\theta_{jo}$$
$$- \sum_j (w_j - n\theta_{jo})^3/2(n\theta_{jo})^2 + \sum_j (w_j - n\theta_{jo})^4/3(n\theta_{jo})^3\xi_j^{\,3}\ .$$

Note that the first sum is identically 0 ; if we consider only second–order terms, a critical region is

$$\chi^2 = \sum_{j=1}^{k} (w_j - n\theta_{jo})^2/n\theta_{jo} > c\,. \qquad (**)$$

Exercise 5: In Mendel's example given above,

$$\theta_o = (9/16,3/16,3/16,1/16)\,,$$

$$w = (315,108,101,32)\ ,\ n = 556\,.$$

Compute the corresponding value of chisquare (**).

Exercise 6: Show that in (**),

$$\sum_{j=1}^{k}(w_j - n\theta_{jo})^2/n\theta_{jo} = \sum_{j=1}^{k} w_j^2/n\theta_{jo} - n .$$

To find the critical value c, we need the distribution of χ^2. Some exact values can be obtained easily when k is 2 for then the multinomial is just a binomial. For larger values of k, the arithmetic is very involved so that we usually rely on the limiting distribution which comes from the

Theorem: *Let $X = (X_1, \cdots, X_m)$ be a normal random vector with mean zero and covariance matrix $\Sigma > 0$. Let $Y = X'AX$ for a constant symmetric matrix A. Then*

$$Y \text{ is distributed as } \sum_{j=1}^{m} \lambda_j \cdot Z_j^2$$

where $Z_1^2, \cdots, Z_m^2$ are independent chi- square RVs each with 1 dof and $\lambda_1, \cdots \lambda_m$ are the eigenvalues of $A^{1/2}\Sigma A^{1/2}$.

Proof: See Rao, 1965.

From exercise 1, we get the covariance matrix of the sum

$$W_n \text{ to be } \begin{bmatrix} n\theta_1(1-\theta_1) & -n\theta_1\theta_2 & \cdots & -n\theta_1\theta_k \\ -n\theta_2\theta_1 & n\theta_2(1-\theta_2) & \cdots & -n\theta_2\theta_k \\ \cdots & \cdots & \cdots & \cdots \\ -n\theta_k\theta_1 & -n\theta_k\theta_2 & \cdots & n\theta_k(1-\theta_k) \end{bmatrix}$$

$$\text{which can be written as } \left[\begin{bmatrix} \theta_1 & & \\ & \ddots & \\ & & \theta_k \end{bmatrix} - \begin{bmatrix} \theta_1 \\ \cdots \\ \theta_k \end{bmatrix} \begin{bmatrix} \theta_1 & \cdots & \theta_k \end{bmatrix}\right]$$

or, with an obvious abbreviation, $n(D_\theta - \theta\theta')$. Now W_n/n is a sample mean so that the multivariate form of the CLT can be applied to get $\sqrt{n}(W_n/n - \theta)$ AN$(0,\Sigma)$ with $\Sigma = D_\theta - \theta\theta'$. Let's write this as

$$\sqrt{n}(W_n/n - \theta) \xrightarrow{D} U .$$

Since $g(y) = y'D_\theta^{-1}y$ is a continuous function of y,

$g(\sqrt{n}(W_n/n - \theta))$ converges in distribution to $U'D_\theta^{-1}U$ (Lesson 7, Part III). To apply the theorem quoted above, we need eigenvalues of $D_\theta^{-1/2}\cdot(D_\theta - \theta\theta')D_\theta^{-1/2}$.

Exercise 7: Let $d = D_\theta^{-1/2}\theta$. Show that $dd'dd' = dd'$.

With the notation of exercise 7, we can write

$$D_\theta^{-1/2}(D_\theta - \theta\theta')D_\theta^{-1/2} = I_k - dd'.$$

Then $(I_k - dd')\cdot(I_k - dd') = I_k - dd' - dd' + dd'dd'$

$$= I_k - dd'$$

shows that this matrix is idempotent; it is obviously symmetric. By a theorem in Lesson 6, we know that the eigenvalues are 0 and 1 and that the number of ones is the rank which is also its trace. Let J' denote a row vector of all ones. Then by properties of the trace, $tr(I_k - dd')$

$$= k - tr(D_\theta^{-1/2}\theta\theta'D_\theta^{-1/2})$$

$$= k - tr(\theta\theta'D^{-1}) = k - tr(\theta J')$$

$$= k - tr(J'\theta) = k - tr(1) = k - 1.$$

Hence, $\Sigma_j(w_i - n\theta_{jo})^2/n\theta_{jo}$ is asymptotically distributed as

$$\Sigma_j \lambda_j Z_j^2 = Z_1^2 + \cdots + Z_{k-1}^2$$

which is a chisquare RV with $k-1$ dof.

If we put all this together and take significance level α, our LRT reduces to:

reject the null hypothesis that $\theta_j = \theta_{jo}$

when Pearson's chisquare $\Sigma(w_j - n\theta_{jo})^2/n\theta_{jo} > \chi^2_\alpha$

$$\text{where } P(\chi^2 > \chi^2_\alpha) = \alpha .$$

Of course, the power function will involve the non–central chisquare distribution.

Example: For Mendel's case above, you should have obtained the observed chisquare value .47 which is certainly non–significant at any reasonable level. In fact, virtually all of Mendel's results lead to similar small values indicating good fits between theory and practice; further discussion can be found in the Stern–Sherwood reference.

Exercise 8: Travelers entering Texas from Mexico must also check thru liquor control; "other things being equal", is there a preference for any of the four "liquor–gates" at this port–of–entry when the observed frequencies are 888, 856, 829, 783 ? Hint: all $\theta_{jo} = 1/4$.

There is one point which we have ignored so far; namely, since the distribution is only approximately chisquare, one should expect some conditions on its use. There is no absolute rule but generally cells with small expectation should be grouped with adjacent cells; the dof is reduced appropriately. A good discussion of this can be found in Snedecor and Cochran, 1980; we will use the criterion $n\theta_{jo} < 2$.

Example: Could the following observations be from a Poisson distribution with mean $\theta = 5$?

# of "raisins":	0	1	2	3	4	5	6	7	8	≥9
# of "cookies":	0	4	9	14	18	19	14	10	6	6

Of course, we assume "independent cookies" etc; note that the last cell contains all possibilities above 8 since that is the range of this RV. Here the sample size is

$$n = 0 + 4 + 9 + 14 + 18 + 19 + 14 + 10 + 6 + 6 = 100 .$$

First we need to obtain the probabilities $n\theta_{jo}$ then the corresponding expectations $n\theta_{jo}$. For $\theta = 5$, the Poisson distribution yields

w_j	p_{jo}	np_{jo}
0	.0067	.67
1	.0337	3.37
2	.0842	8.42
3	.1404	14.04
4	.1755	17.55
5	.1755	17.55 .
6	.1462	14.62
7	.1044	10.44
8	.0653	6.53
9	.0681	6.81

Since the first cell has an expectation $.67 < 2$, we combine it with the second cell to get

observed	expected
5	4.04
10	8.42
15	14.04
19	17.55
21	17.55 .
17	14.62
10	10.44
2	6.53
1	6.81

The observed chisquare value $\Sigma(o_i - e_i)^2/e_i$

$$= (5-4.04)^2/4.04 + \cdots + (1-6.81)^2/6.81 = 9.894 ;$$

the dof is 8 . The tabular value $\chi^2_{.10}(8) = 13.362$ is greater then the observed 9.984 so we do not reject H_o: this could be a sample from a Poisson distribution with mean 5 .

Exercise 9: Adapt the technique of this last example for testing the same H_o with the data

w_j	0	1	2	3	4	5	6	7	8	9	10	11	12	≥13 .
freq	3	6	12	16	16	16	15	10	5	3	4	2	1	1

Hint: the dof will be 9 .

LESSON 10. LRT – ASYMPTOTIC DISTRIBUTIONS

In this lesson, we give additional examples of the LRT, some of which suggested the search for a general result to apply to any such test; we will explain that result but its proof is beyond the scope of our text. It will be convenient to have Young's form of the Taylor formula but first we review "little oh".

Example: a) Since $(\sin nx)/n \to 0$ as $n \to \infty$,

$$\frac{\sin nx}{n} = o(1) \text{ and } \sin nx = o(n) .$$

b) Let f be a real valued function on some open subset of R, differentiable at the interior point c. Then,

$$\frac{f(x) - f(c)}{x - c} \to f'(c) \text{ as } x \to c$$

$$\text{or, } \left| \frac{f(x) - f(c) - f'(c)(x - c)}{x - c} \right| \to 0 \text{ as } x \to c$$

$$\text{or, } f(x) - f(c) - f'(c)(x - c) = o(|x - c|) .$$

c) In general, $o(\alpha)$ is a function of α for which

$\lim_{\alpha \to w} o(\alpha)/\alpha = 0$ where the limit "w" may be ∞ or a finite number including 0 .

Exercise 1: Prove the following by mathematical induction. Let f be a real valued function on an interval [a,b] whose nth derivative, $f^{(n)}$, exists at some point $a < c < b$. Then,

$$f(x) = f(c) + f'(c)(x-c) + f''(c)(x-c)^2/2 + \cdots$$
$$+ f^{(n)}(c)(x-c)^n/n! + o(|x-c|^n) .$$

Example: Let $X_1, \cdots, X_n$ be RS from a normal distribution with mean μ and variance σ^2. Consider testing

$$H_o: \sigma = \sigma_o \text{ vs } H_a: \sigma \neq \sigma_o ;$$

note that there are no restrictions on μ so that both null and altrnative hypotheses are composite. The log–likelihood for

$\theta = (\mu, \sigma^2) \in \Theta = R \times (0,\infty)$ is

$$\mathscr{L}(\theta) = \log L_\theta(v) = -\Sigma_{i=1}^n (x_i - \mu)^2/2\sigma^2 - (n/2)\log 2\pi\sigma^2 .$$

In Θ , the unrestricted case, $\partial\mathscr{L}/\partial\mu = -\Sigma_i (x_i - \mu)(-2)/2\sigma^2$

so that one MLE is $\hat{\mu} = \bar{x}$; as

$$\partial\mathscr{L}/\partial\sigma^2 = -\Sigma_i (x_i - \mu)^2/2\sigma^4 - n/2\sigma^2 ,$$

the MLE $\hat{\sigma}^2 = \Sigma_i (x_i - \bar{x})^2/n$. Then,

$$\sup_{\theta\in\Theta} L_\theta(v) = e^{-n/2}/(2\pi\hat{\sigma}^2) .$$

Under H_o , $\Theta_o = R \times \{\sigma_o^2\}$ leads to MLEs $\hat{\mu}_o = \bar{x}$, $\hat{\sigma}_o^2 = \sigma_o^2$ and then

$$\sup_{\theta\in\Theta_o} L_\theta(v) = e^{-n\hat{\sigma}^2/2\sigma_o^2}/(2\pi\sigma_o^2)^{n/2} .$$

The LRT is to reject H_o when

$$\lambda = L_{\hat{\theta}_o}(v)/L_{\hat{\theta}}(v) \le \lambda_o ,$$

equivalently, $\lambda = e^{-(n/2)(\hat{\sigma}^2/\sigma_o^2 - 1)}(\hat{\sigma}^2/\sigma_o^2)^{n/2} \le \lambda_o$.

For $w_n = n\hat{\sigma}^2/\sigma_o^2$, this is equivalent to

$$\lambda^* = (w_n/n)\exp(-w_n/n) \le k .$$

From the graph of λ^* as a function of w_n , we see that this last inequality is equivalent to

$$w_n \le k_1 \text{ or } w_n \ge k_2 .$$

Under H_o , $W_n = \Sigma (X_i - \bar{X})^2/\sigma_o^2$ has a chisquare distribution with $n-1$ dof so that (theoretically) we can find k_1, k_2 to satisfy the significance level:

$$P(W_n \le k_1) + P(W_n \ge k_2) = \alpha .$$

Exercise 2: The probability statement just made has the representation

$$\int_0^{k_1} w^{(n-1)/2 - 1} \cdot e^{-w/2} dw/2^{(n-1)/2}\Gamma((n-2)/2)$$

$$+ \int_{k_2}^{\infty} w^{(n-1)/2 - 1} \cdot e^{-w/2} dw/2^{(n-1)/2}\Gamma((n-1)/2) = \alpha .$$

Show that to minimize the length $k_2 - k_1$ (for α fixed), we also need the restriction

$$(k_1/k_2)^{(n-1)/2 - 1} = e^{(k_1 - k_2)/2} .$$

Although we shall be using the results of the following lemmas and exercises in further discussion of the LRT, we believe that they should be of interest in themselves.

Lemma: *For each positive integer* $n \ge 2$, *let* W_n *have a chisquare distribution with* n-1 *dof. Then*

$$(W_n - n + 1)/\surd(2n-2) \text{ is } AN(0,1) .$$

Proof: The MGF of $(W_n - n + 1)/\surd(2n-2)$ is

$$M(t) = [\exp(-t(n-1)/\surd(2n-2)] + (1 - 2t/\surd(2n-2))^{(n-1)/2}$$

$$= \left[\frac{e^{-2t/\surd(2n-2)}}{1 - 2t/\surd(2n-2)}\right]^{(n-1)/2} .$$

By exercise 1, for each $t > 0$ and $n > 1 + 2t^2$,

$$\log(1 - 2t/\surd(2n-2)) = -2t/\surd(2n-2) - (2t/\surd(2n-2))^2/2$$

$$+ o((2t/\surd(2n-2))^2) ,$$

so that

$$\log M(t) = ((n-1)/2)[-2t/\sqrt{(2n-2)} - \log(1 - 2t/\sqrt{(2n-2)})]$$
$$= ((n-1)/2)[t^2/(n-1) - o(2t^2/(n-1))]$$
$$= t^2/2 + o(1) \text{ as } n \to \infty .$$

In other words, $\lim_{n \to \infty} M(t) = e^{t^2/2}$ which is the MGF of the standard normal. The proof is completed by application of a continuity theorem (in Lesson 14, Part III).

Exercise 3: Prove the following generalization. Let $a_1, a_2, \cdots$ and $b_1, b_2, \cdots$ be sequences of real numbers such that

$n-1-a_n \to 0$ and $(n-1)/b_n^2 \to 1/2$ as $n \to \infty$. Let $\{W_n\}$ be a sequence of chisquare RVs as in the lemma above. Then $(W_n - a_n)/b_n$ is AN(0,1). Hint: use Slutsky's technique (Lesson 6, Part III).

Lemma: *If the sequence of RVs* $\{T_n\}$ *converges in distribution to the standard normal RV* Z *, then the sequence* $\{T_n^2\}$ *converges in distribution to the chisquare RV with 1 dof.*

Proof: For $w < 0$, the CDF $G_n(w) = P(T_n^2 \le w) = 0$. Otherwise,

$$G_n(w) = P(-\sqrt{w} \le T_n \le \sqrt{w}) = P(T_n \le \sqrt{w}) - P(T_n < -\sqrt{w}) .$$

Since the limiting distribution is continuous, the limit of this last difference is

$$G(w) = P(Z \le \sqrt{w}) - P(Z \le -\sqrt{w}) = \int_{-\sqrt{w}}^{\sqrt{w}} \exp(-z^2/2)\, dz/\sqrt{2\pi} .$$

The corresponding density is

$$G'(w) = (1/2\sqrt{w})e^{-w/2}/\sqrt{2\pi} - (-1/2\sqrt{w})e^{-w/2}/\sqrt{2\pi}$$
$$= w^{-1/2}e^{-w/2}/2^{1/2}\Gamma(1/2)$$

as desIred.

Definition: *For* $\{a_n\}$ *constants,* $o_p(a_n)$ *represents a sequence of RVs, say* $\{H(n)\}$, *such that* $H(n)/a_n$ *converges in probability to* 0 *as* $n \to \infty$ *; symbolically,* $o_p(a_n)/a_n \xrightarrow{P} 0$.

Exercise 4: Let $\{W_n\}$ be a sequence of chisquare RVs as in the first lemma above. Prove that as $n \to \infty$,

a) $(W_n - n)/\sqrt{2n}$ is AN(0,1);

b) $(W_n - n)/n = o_p(1)$; c) $(W_n - n)^3/3n^2 = o_p(1)$.

Example: Continuing with the LRT technique for testing the normal σ, consider

$$-2\log \lambda^* = n(w_n/n - 1) - n\log(w_n/n)$$

$$= (w_n - n) - n\log(1 + (w_n - n)/n) . \qquad (1)$$

Another form of a Taylor expansion says that for some ξ between 0 and x ,

$$\log(1+x) = x - x^2/2 + x^3/3(1+\xi)^3 .$$

Hence, for each $n = 2(1)\infty$, there is some ξ_n between 0 and $(w_n - n)/n$, such that,

$$\log(1 + (w_n - n)/n) = (w_n - n)/n - (w_n - n)^2/2n^2$$

$$+ (w_n - n)^3/3n^3(1 + \xi_n)^3 .$$

Substituting in (1) makes $-2\log \Lambda^*$

$$= (W_n - n) - n[(W_n - n)/n - (W_n - n)^2 2n^2$$

$$+ (W_n - n)^3/3n^3(1 + \xi_n)^3]$$

$$= (W_n - n)^2/2n + (W_n - n)^3/3n^2(1 + \xi_n)^3$$

$$= (W_n - n)^2/2n + o_p(1) .$$

Since $(W_n - n)/\sqrt{(2n)} \overset{D}{\to} Z$, $(W_n - n)^2/2n \overset{D}{\to} \chi^2$. It follows that $-2\log \lambda^*$ is asymptotically distributed as chisquare.

If in the proof we used more terms of the Taylor expansion, the representation of the asymptotic distribution would be more accurate. But the magic of it all is that statisticians were able to generalize these examples to obtain results for the "regular cases" with the same basic hypotheses as for the asymptotic theory of MLE (Lesson 13, Part V). The paper of Wald, 1943, contains a rigorous developement which we summarize.

Direct form: The parameter θ has $p+q$ components, say $\theta = (\theta_p, \theta_q) \in R^{p+q}$. We denote the likelihood

$$\Pi_{i=1}^{n} f(x_i; \theta) = L(\theta_p, \theta_q)$$

and we want to test

$$H_o: \theta_p = \theta_{po} \text{ (known) against } H_a: \theta_p \neq \theta_{po}.$$

Under H_o, the MLE of θ_p is θ_{po} but there are no restrictions on θ_q so that $\hat{\theta}_{qo}$ is determined to maximize $L(\theta_{po}, \theta_q)$. Under H_a, we must determine $\hat{\theta}_p$ and $\hat{\theta}_q$ to maximize $L(\theta_p, \theta_q)$. The LRT is to reject H_o when

$$\lambda = L(\theta_{po}, \hat{\theta}_{qo})/L(\hat{\theta}_p, \hat{\theta}_q) \leq \lambda_o .$$

The general result is that the asymptotic distribution of $-2\log(\lambda)$ is chisquare with $(p+q) - (q) = p$ dof. The distribution is central under H_o but non-central otherwise; the non-centrality parameter is $(\theta_p - \theta_{po})'V^{-1}(\theta_p - \theta_{po})$ where V is the variance-covariance matrix of $\hat{\theta}_p$ under Θ.

Indirect form: H_o specifies linear relations on Θ so that $\theta \in \Theta_o \subset \Theta \subset R^d$ where Θ_o is a set in a smaller dimensional subspace, say $\dim q$. (As an example, take

$$\Theta_o = \{(\theta_1,\theta_2) : \theta_1 = \theta_2\} \subset \Theta \subset R^2 ;$$

if θ_1 and θ_2 are means, Θ_o could be a one–dimensional subspace but if these are variances, Θ_o is at most a subset of R^1.) H_a is merely $\theta \in \Theta - \Theta_o$. Now

$$\lambda = \sup_{\theta\in\Theta_o} L(\theta) / \sup_{\theta\in\Theta} L(\theta)$$

and $-2\log(\lambda)$ is asymptotically chisquare with $d-q = p$ dof. Under H_o, the distribution is of course central; to get the non–centrality parameter for H_a, one may need to transform the components of θ to fit the direct form.

Example: Consider random samples from k independent normal populations with means $\mu_1,\cdots,\mu_k$ and variances $v_1,\cdots,v_k$; for $i = 1(1)k$, the sample sizes are $n_i \geq 2$; also, $n_. = \Sigma_{i=1}^{k} n_i$. Here

$$\theta = (\mu_1,\cdots,\mu_k,v_1,\cdots,v_k) \in \Theta = R^k \times (0,\infty)^k \text{ and } d = 2k .$$

The log–likelihood function is

$$-\sum_{i=1}^{k}\sum_{j=1}^{n_i} (x_{ij} - \mu_i)^2/2v_i - (n_./2)\log(2\pi) - \sum_{i=1}^{k} (n_i/2)\log(v_i) .$$

Consider the null hypothesis that the variances are equal. For Θ_o or Θ, the MLEs of the means are

$$\hat{\mu}_i = \bar{x}_{i.} = \sum_{j=1}^{n_i} x_{ij}/n_i , i = 1(1)k .$$

For Θ, the MLE

$$\hat{v}_i = \sum_{j=1}^{n_i} (x_{ij} - \bar{x}_{i.})^2/n_i , i = 1(1)k ,$$

are the traditional "*within*" sample variances usually denoted by s_i^2. Now H_o: $v_1 = \cdots = v_k = v$ (unknown) and one way to write this "linearly" is:

$$v_1 - v_2 = 0, v_1 - v_3 = 0, \cdots, v_1 - v_k = 0 .$$

Here $q = k + 1$ (the means are still unknown) and the reduced log–likelihood is

$$-\Sigma_i \Sigma_j (x_{ij} - \bar{x}_{i.})^2/2v - (n_./2)\log(2\pi) - (n_./2)\log(v) ;$$

from this we get the "*pooled*" sample variance estimate

$$\hat{v} = s^2 = \Sigma_i \Sigma_j (x_{ij} - \bar{x}_{i.})^2/n_. = \Sigma_i n_i s_i^2/n_. .$$

Finally,
$$-2\log(\lambda) = -2\, \mathcal{L}(\hat{\theta}_o) + 2\, \mathcal{L}(\hat{\theta})$$
$$= n_. \log(s^2) - \Sigma_i n_i \log(s_i^2) .$$

Its asymptotic distribution is chisquare with
$(2k) - (k+1) = k-1$ dof.

For the special case we treat below, we need a consequence of:

Exercise 5: Let b and c be positive constants. Sketch the graph of $(1 + a)^{-b}(1 + 1/a)^{-c}$ for $a > 0$.

Example: Let us continue the last example with $k = 2$.

a) The LRT becomes

$$\lambda = \left[s^2\right]^{-n_./2} \div \left[s_1^2\right]^{-n_1/2} \left[s_2^2\right]^{-n_2/2} \le \lambda_o$$

equivalently,

$$(1+n_2 s_2^2/n_1 s_1^2)^{-n_1/2}\, (1+n_1 s_1^2/n_2 s_2^2)^{-n_2/2} \le \lambda \quad .$$

Since the LHS of this inequality is like the function in exercise 5, the critical region can be taken as

$$n_1 s_1^2/n_2 s_2^2 \le \lambda_1 \text{ or } n_1 s_2^2/n_2 s_2^2 \ge \lambda_2 .$$

b) Since the samples are independent, $n_1 S_1^2/\sigma_1^2$ and $n_2 S_2^2/\sigma_2^2$ are independent chisquare RVs and their ratio is distributed as a multiple of an F RV. In particular, we may find λ_1 and λ_2 such that for

$$\alpha_1 + \alpha_2 = \alpha\ ,\ P(n_1S_1^2/n_2S_2^2 \le \lambda_1 \mid \sigma_1 = \sigma_2) = \alpha_1\ ,$$
$$P(n_1S_1^2/n_2S_2^2 \ge \lambda_2 \mid \sigma_1 = \sigma_2) = \alpha_2$$

by using $P(F \le \lambda_1(n_2-1)/(n_1-1)) = \alpha_1$,

$$P(F \ge \lambda_2(n_2-1)/(n_1-1)) = \alpha_2$$

where F is an F RV with n_1-1 and n_2-1 dof. It is customary, but not necessary, to take $\alpha_1 = \alpha_2 = \alpha/2$.

c) For such λ_1 and λ_2 , it is also true that

$$P(\lambda_1 \le n_1S_1^2/\sigma_1^2 \div n_2S_2^2/\sigma_2^2 \le \lambda_2) = 1 - \alpha\ .$$

This is equivalent to

$$P(S_1^2/\lambda_2S_2^2 \le \sigma_1^2/\sigma_2^2 \le S_1^2/\lambda_1S_2^2) = 1 - \alpha$$

from which we get the $100(1-\alpha)\%$ CI:

$$(\ s_1^2/\lambda_2s_2^2\ ,\ s_1^2/\lambda_1s_2^2)\ .$$

When $k \ge 3$, probabilities for the LRT statistic λ can be obtained only numerically. (For some early details on this point, see Rogers, 1964).

LESSON 11. SUMMARY OF TESTS FOR NORMAL POPULATIONS

In this lesson, we illustrate applications of previous theorems specifically for the testing of hypotheses in univariate normal populations; a few of these have appeared in earlier examples and exercises but you may like the idea of having them all together in more traditional notation. Since we are dealing with a continuous distribution, all the tests are non–random; we will give only the non–zero part of each test function φ.

We start with $V = (X_1,\cdots,X_n)'$ whose components are IID $N_1(\mu,\sigma^2)$; here $0 < \sigma < \infty$ and parameters with subscripts represent known (given) values. As usual, the sample mean and variance are $\bar{X} = \sum_{j=1}^{n} X_j/n$, $S^2 = \sum_{j=1}^{n} (X_j - \bar{X})^2/(n-1)$; the observed values of RVs are written in lower case. We also adopt the following notation:

$\mathscr{Z}$ is the standard normal RV; $P(\mathscr{Z} > z_\alpha) = \alpha$.

$\chi^2(k)$ is a central chisquare RV with k dof;

$$P(\chi^2(k) > \chi^2(\alpha,k)) = \alpha .$$

$\mathscr{T}(m)$ is a central Student RV with m dof;

$$P(\mathscr{T}(m) > t(\alpha,m)) = \alpha .$$

$\mathscr{F}(n,d)$ is a central F RV with n and d dof;

$$P(\mathscr{F}(n,d) > f(\alpha,n,d)) = \alpha .$$

1) The original simple–simple case for the mean.

$$H_o: \mu = \mu_o , \sigma^2 = \sigma_o^2 \text{ vs } H_a: \mu = \mu_1 > \mu_o , \sigma^2 = \sigma_o^2 .$$

The MP test is $\varphi(v) = 1$ for $\bar{x} > k$;

$$\alpha = P(\bar{X} > k \mid \mu_o,\sigma_o) = P(\mathscr{Z} > \sqrt{n}(k - \mu_o)/\sigma_o)$$

implies $k = \mu_o + z_\alpha\sigma_o/\sqrt{n}$. The power function $\mathscr{P}(\mu)$

$$= P(\overline{X} > \mu_o + z_\alpha \sigma_o/\sqrt{n} \mid \mu) = P(Z > \sqrt{n}(\mu_o - \mu)/\sigma_o + z_\alpha)) .$$

Example: a) Take $\alpha = .025$, $\mu_o = 3$, $\mu_1 = 5$, $\sigma_o^2 = 9$, $n = 25$.

If $\bar{x} > 3 + 1.96(3/5)$, reject H_o . $\mathscr{P}(5) \approx P(\mathscr{Z} > -1.37) \approx .91$.

2) Sample size in the simple–simple case. The principles of the MP test lead to

$$\alpha = \mathscr{P}(\mu_o) = P(\overline{X} > k \mid \mu_o, \sigma_o) = P(\mathscr{Z} > \sqrt{n}(k - \mu_o)/\sigma_o)$$

$$\beta = 1-P(\mu_1) = P(\overline{X} \le k \mid \mu_1, \sigma_o) = P(Z \le \sqrt{n}(k - \mu_1)/\sigma_o)) ;$$

these imply $\sqrt{n}(k - \mu_o)/\sigma_o = z_\alpha$ and

$$\sqrt{n}(k - \mu_1)/\sigma_o = -z_\beta . \qquad (*)$$

If we now prescribe "producer's risk" to be α and the "consumer's risk" to be β , we get the sampling plan "n,k" by solving the equations (*):

$$k = (\mu_o z_\beta + \mu_1 z_\alpha)/(z_\beta + z_\alpha) ,$$

$$\sqrt{n} = (z_\beta + z_\alpha)\sigma_o/(\mu_1 - \mu_o) .$$

(A non–integral n is rounded up.)

Exercise 1: Continuing with 2), take

$$H_o: \mu_o = 3, \sigma_o = 3 \text{ vs } H_a: \mu_1 = 5, \sigma_o = 3 .$$

Find the sampling plan (k and n) for:

a) $\alpha = .025, \beta = .05$; b) $\alpha = .05, \beta = .01$.

3) The simple–simple case for the variance.

$$H_o: \sigma = \sigma_o , \mu = \mu_o \text{ vs } H_a: \sigma = \sigma_1 > \sigma_o , \mu = \mu_o .$$

The MP test is $\varphi(v) = 1$ for $\sum_{j=1}^{n} (x_j - \mu_o)^2 > k$.

$$\alpha = P(\textstyle\sum_{j=1}^{n}(X_j - \mu_o)^2/\sigma_o^2 > k/\sigma_o^2)$$

implies $k = \chi^2(\alpha,n)\sigma_o^2$ and power

$$\mathscr{P}(\sigma^2) = P(\textstyle\sum_{j=1}^{n}(X_j - \mu_o)^2/\sigma^2 > k/\sigma^2)$$
$$= P(\chi^2(n) > \chi^2(\alpha,n)\sigma_o^2/\sigma^2) .$$

Example: Take $\mu_o = 0$, $\sigma_o^2 = 9$, $\sigma_1^2 = 16$, $n = 25$, $\alpha = .025$.

If $\Sigma_j x_j^2 > 40.646(9)$, reject H_o.

$\mathscr{P}(16) = P(\chi^2(25) > 40.646(9/16)) \approx .55$.

4) The simple–composite case with σ^2 known and

$$H_o: \mu = \mu_o \text{ vs } H_a: \mu > \mu_o$$

is contained in the case with

$$H_o: \mu \le \mu_o \text{ vs } H_a: \mu > \mu_o .$$

The UMP test is $\varphi(v) = 1$ for $\bar{x} > k$. The remaining discussion follows as for 1).

5) A test for "large/small" values of the mean with σ^2 known as σ_o^2 has $H_o: \mu \le \mu_1$ or $\mu \ge \mu_2$ vs $H_a: \mu_1 < \mu < \mu_2$. The UMP test is $\varphi(v) = 1$ for $k_1 < \bar{x} < k_2$. For

$$\alpha = P(k_1 < \bar{X} < k_2 | \mu_1,\sigma_o) = P(k_1 < \bar{X} < k_2 | \mu_2,\sigma_o) ,$$

the power function is

$$\mathscr{P}(\mu) = P(\sqrt{n}(k_1 - \mu)/\sigma_o < Z < \sqrt{n}(k_2 - \mu)/\sigma_o) .$$

Because of the symmetry of the normal distribution,

$$k_1 = \mu_1 + \mu_2 - k_2$$

and this means that we need to find only k_2 such that

$$P(\mathscr{Z} \le \sqrt{n}(k_2-\mu_1)/\sigma_o) - P(\mathscr{Z} \le \sqrt{n}(\mu_2-k_2)/\sigma_o) = \alpha .$$

Example: For $n = 25$, $\sigma = 2$, $\theta_1 = -1$, $\theta_2 = 1$, some heavy arithmetic yields:

with	$\alpha = .05$	.025	.01
$k_2 = -k_1$	$= .344$	.224	.108 .

Exercise 2: Verify that k_2 , k_1 yield the given α in the example just above.

6) Tests on σ^2 with μ known.

a) For $H_o: \sigma \le \sigma_o$ vs $H_a: \sigma > \sigma_o$, the UMP test is

$$\varphi(v) = 1 \text{ for } \Sigma(x_j - \mu)^2 > k .$$

$$\alpha = P(\Sigma(X_j - \mu)^2 > k \,|\, H_o) = P(\chi^2(n) > k/\sigma_o^2)$$

yields the power function

$$\mathscr{P}(\sigma^2) = P(\chi^2(n) > \chi^2(\alpha,n)\sigma_o^2/\sigma^2) \text{ (free of } \mu) .$$

b) For $H_o: \sigma \le \sigma_1$ or $\sigma \ge \sigma_2$ vs $H_a: \sigma_1 < \sigma < \sigma_2$, the UMP test is $\varphi(v) = 1$ for $k_1 < \Sigma(x_j - \mu)^2 < k_2$.

$$\mathscr{P}(\sigma^2) = P(\chi^2(n) < k_2/\sigma^2) - P(\chi^2(n) < k_1/\sigma^2)$$

with k_2, k_1 determined by trial and error at $\alpha = \mathscr{P}(\sigma_o^2)$.

Exercise 3: Sketch $\mathscr{P}(\sigma^2)$ in 6a) when $\alpha = .05$,$n = 20$, $\sigma_o = .1$.

7) Two–sided test for μ when σ is known:

$$H_o: \mu = \mu_o \text{ vs } H_a: \mu \ne \mu_o .$$

The UMPU test is $\varphi(v) = 1$ for $\bar{x} < k_1$ or $\bar{x} > k_2$.

Because of symmetry under H_o, this is equivalent to

$$\varphi(v) = 1 \text{ for } n(\bar{x} - \mu_o)^2/\sigma^2 > k_3 \text{ , } = 0 \text{ otherwise.}$$

$$\alpha = P(n(\bar{X} - \mu_o)^2/\sigma^2 > \chi^2(\alpha,1))$$

but it is easier to get the power function from the normal distribution: $\mathscr{P}(\mu)$

$$= 1 - P(-z_{\alpha/2} + \sqrt{n}(\mu_o - \mu)/\sigma < \mathscr{Z} < z_{\alpha/2} + \sqrt{n}(\mu_o - \mu)/\sigma) .$$

Exercise 4: Verify the power function in 7).

8) One–sided tests of σ with μ unknown.

a) For $H_o: \sigma \le \sigma_o$ vs $H_a: \sigma > \sigma_o$, the UMP test is

$$\varphi(v) = 1 \text{ for } \Sigma(x_j - \bar{x})^2 > k .$$

The power function is

$$\mathscr{P}(\sigma^2) = P(\chi^2(n-1) > \chi^2(\alpha,n-1)\sigma_o^2/\sigma^2) .$$

b) For $H_o: \sigma \ge \sigma_o$ vs $H_a: \sigma < \sigma_o$, the UMPU test is

$$\varphi(v) = 1 \text{ for } \Sigma(x_j - \bar{x})^2 > k .$$

The power function is

$$\mathscr{P}(\sigma^2) = P(\chi^2(n-1) < \chi^2(\alpha,n-1)\sigma_o^2/\sigma^2) .$$

Example: For effective vaccinations, the variance of the amount injected by a "gun" must be less than .10 units. A random sample of size 10 yields $\bar{x} = 1.01$ and $s^2 = .12$. Is the "gun" performing adequately?

a) For such problems, it is customary (and almost necessary) to assume that the underlying population is normal; then the test in 8a) may be applied. To "play it safe", we take $\alpha = .20$ so that $\chi^2(.20,9) =$ 12.2 . Since $s^2 = .12 < k = 12.2(.10)/9 \approx .136$, we do not reject H_o . At $\sigma^2 = .15$, the power is

$$P(\chi^2(9) > .136(9)/.15) \approx .52 .$$

b) If we are reasonably sure of μ, we can also use the procedure in 6a). Say $\mu = 1$; then

$$\Sigma(x_j - \mu)^2 = \Sigma(x_j - \bar{x})^2 + n(\bar{x} - \mu)^2$$

yields $\Sigma(x_j - 1)^2 = 9(.12) + 10(.01)^2 = 1.081$

so that $\Sigma(x_j - 1)^2/.10 = 10.81 < \chi^2(.20,10) = 13.4$.

At $\sigma^2 = .15$, the power is

$$P(\chi^2(10) > 13.4(.10)/.15) \approx .55 .$$

9) The other two–sided tests on σ are also only UMPU. For $H_o: \sigma \le \sigma_1$ or $\sigma \ge \sigma_2$ vs $H_a: \sigma_2 < \sigma < \sigma_2$,

$$\varphi(v) = 1 \text{ for } k_1 < \Sigma(x_j - \bar{x})^2 < k_2 .$$

For $H_o: \sigma_1 \le \sigma \le \sigma_2$ vs $H_a: \sigma > \sigma_1$ or $\sigma > \sigma_2$,

$$\varphi(v) = 1 \text{ for } \Sigma(x_j - \bar{x})^2 < k_1 \text{ or } \Sigma(x_j - \bar{x})^2 > k_2 .$$

For $H_o: \sigma = \sigma_o$ vs $H_a: \sigma \ne \sigma_o$,

$$\varphi(v) = 1 \text{ for } \Sigma(x_j - \bar{x})^2 < k_1 \text{ or } \Sigma(x_j - \bar{x})^2 > k_2 .$$

Now k_1 and k_2 are determined (by trial and error) from

$$\int_{k_1/\sigma_o^2}^{k_2/\sigma_o^2} g_{n-1}(t)\, dt = 1 - \alpha \ ,$$

$$\int_{k_1/\sigma_o^2}^{k_2/\sigma_o^2} t g_{n-1}(t)\, dt/(n - 1) = 1 - \alpha \qquad (**)$$

where $g_m(t)$ is the value of the chisquare density with m dof.

Exercise 5: Show that (**) is equivalent to

$$\int_a^b g_{n-1}(t)\,dt = 1 - \alpha = \int_a^b g_{n+1}(t)\,dt .$$

Example: Note that the popular equal–tailed test is not exactly UMPU since for n = 20,

$$\int_{8.907}^{32.852} g_{19}(t)dt \approx .94999 > .94204 \approx \int_{8.907}^{32.852} g_{21}(t)dt .$$

10) One–sided tests for the mean with unknown σ .

a) For $H_o: \mu \le \mu_o$ vs $H_a: \mu > \mu_o$, the UMPU test is

$$\varphi(v) = 1 \text{ for } t = \sqrt{n}(\bar{x} - \mu_o)/s > k .$$

Under H_o , T has a central Student distribution with n–1 dof and

$P(\mathcal{T}(n-1) > t(\alpha,n-1)) = \alpha$ gives $k = t(\alpha,n-1)$.

The power function

$$\mathcal{P}(\mu) = P(\sqrt{n}(\bar{X} - \mu_o)/S > t(\alpha,n-1) \mid \mu)$$

uses the noncentral Student distribution.

b) For $H_o: \mu \ge \mu_o$ vs $H_a: \mu < \mu_o$, the UMPU test is

$$\varphi(v) = 1 \text{ when } t < -t(\alpha,n-1) .$$

The power function is based again on non–central t .

Exercise 6: Show that the likelihood ratio test for the hypotheses in 10a) leads to the very same φ .

11) The Student T test for $H_o: \mu = \mu_o$ vs $H_a: \mu \ne \mu_o$ is also UMPU with

$$\varphi(v) = 1 \text{ for } t < -\,t(\alpha/2,n-1) \text{ or } t > t(\alpha/2,n-1) .$$

We note in passing that the region of "acceptance" is the 100(1–α)% confidence interval

$$\bar{x} - t(\alpha/2,n-1)s/\sqrt{n} \,,\ \bar{x} + t(\alpha/2,n-1)s/\sqrt{n} .$$

Example: A soft–drink dispensing machine is supposed to deliver 8 ounces per cup. "Under filling" leads to customer complaints and "over filling" leads to loss of income. Does the following sample indicate that this particular machine is performing reasonably well? 7.8 8.1 8.1 7.9 8.2 7.7
7.7 7.8 7.9 8.0 8.1 7.6

We first need justification for assuming a normal population; perhaps this was tested last week. With that in mind, we find

$$\bar{x} = -.1 \,,\; s^2 = .51/10 \,,\; t \approx -.1/\sqrt{.004636} \approx -1.47 \,.$$

For $\alpha = .10$, $t(.05,10) = 1.815$ and we do not reject $H_o: \mu = 8$.

We switch now to consideration of the parameters of two independent normal populations working with independent random samples $V = (X_1, \cdots, X_n)$ from $N_1(\mu_1, \sigma_1^2)$,

$$W = (Y_1, \cdots, Y_m) \text{ from } N_1(\mu_2, \sigma_2^2) \,.$$

Note: now known values will be designated μ_{1o}, etc.

12) Tests of the ratio $\tau = \sigma_2^2/\sigma_1^{\,2}$.

For $H_o: \tau \le \tau_o$ vs $H_a: \tau > \tau_o$, the UMPU test is

$$\varphi(v,w) = 1 \text{ for } \Sigma(y_j - \bar{y})^2/\Sigma(x_i - \bar{x})^2 > k \,.$$

$$\begin{aligned} \alpha &= P(\Sigma(Y_j - \bar{Y})^2/\Sigma(X_i - \bar{X})^2 > k \,|\, \tau_o) \\ &= P(\mathcal{F}(m-1,n-1) > k(n-1)/\tau_o(m-1)) \end{aligned}$$

leads to $k = f(\alpha,m-1,n-1)\tau_o(m-1)/(n-1)$. The power function involves the noncentral F distribution.

For $H_o: \tau \ge \tau_o$ vs $H_a: \tau < \tau_o$, the UMPU test is

$$\varphi(v,w) = 1 \text{ for } \Sigma(y_j - \bar{y})^2/\Sigma(x_i - \bar{x})^2 < k_1$$

where $k_1 = \tau_o(m-1)f(1-\alpha,m-1,n-1)/(n-1)$.

Example: Let $n = 14$, $m = 26$, $\alpha = .05$.

a) For $H_o: \tau = 1$ vs $H_a: \tau > 1$, we solve

$$P(\mathscr{F}(25,13) > f(.05,25,13)) = .05$$

to get $f(.05,25,13) = 2.41$. Then $k = 2.41(25)/13$.

b) For $H_o: \tau = 2$ vs $H_a: \tau < 2$, we solve

$$P(\mathscr{F}(25,13) < a) = .05$$

to get $a = .468$. Then $k_1 = 2(25)(.468)/13$. Note: some authors give other tables by using the fact $1/\mathscr{F}(m-1,n-1)$ is distributed as $\mathscr{F}(n-1,m-1)$; then

$$k_1 = \tau_o(m-1)/(n-1)F(1-\alpha,n-1,m-1).$$

Exercise 7: Verify the value of k_1 in the last sentence of this last example.

13) The following contains the test for equality of variances. For $H_o: \tau = \tau_o$ vs $H_a: \tau \neq \tau_o$, the UMPU test is

$$\varphi(v,w) = 1 \text{ for } \Sigma(y_j-\bar{y})^2/\tau_o/(\Sigma(y_j-\bar{y})^2/\tau_o + \Sigma(x_i-\bar{x})^2) > k_1 \text{ or } < k_2.$$

One needs tables of the incomplete beta function to determine k_1, k_2. When $\tau_o = 1$, the ratio in φ is equivalent to $\Sigma(y_j - \bar{y})^2/\Sigma(x_i - \bar{x})^2$.

14) Tests on the difference of the means when the variances are known. For $H_o: \mu_1 - \mu_2 \leq \delta_o$ vs $H_a: \mu_1 - \mu_2 > \delta_o$, the UMP test is

$$\varphi(v,w) = 1 \text{ for } (\bar{x} - \bar{y} - \delta_o)/\sqrt{(\sigma_{1o}^2/n + \sigma_2^2/m)} > z_\alpha.$$

This follows from 4) by considering $D = \bar{X} - \bar{Y}$ to be a random sample of size 1 from a normal distribution with mean $\delta = \mu_2 - \mu_1$ and known variance $\sigma_{1o}^2/n + \sigma_{2o}^2/m$. A similar discussion applies to

$$H_o: \delta = \delta_o \text{ vs } H_a: \delta \neq \delta_o .$$

In either case, the most common δ_o is 0.

15) Differences of the means when the variances are unknown but equal.

a) For H_o: $\mu_1-\mu_2 \le \delta_o$ vs H_a: $\mu_1-\mu_2 > \delta_o$, the UMPU test is

$$\varphi(v,w) = 1 \text{ for } t = \frac{\bar{x} - \bar{y} - \delta_0}{\sqrt{\Sigma(x_i-\bar{x})^2+\Sigma(y_j-\bar{y})^2}} > k$$

For $\alpha = P(\mathcal{T}(n+m-2) > t(\alpha,n+m-2))$,

$$k = t(\alpha,n+m-2)\sqrt{[(n+m)/nm(n+m-2)]} .$$

The power function involves the noncentral t distribution.

b) For H_o: $\mu_1 - \mu_2 = \delta_o$ vs H_a: $\mu_1 - \mu_2 \neq \delta_o$, the UMPU test is $\varphi(v,w) = 1$ for $|t| > k$. From

$$\alpha = P(\mathcal{T}(n+m-2) > t(\alpha/2,n+m-2)) ,$$

we can find this k. The power function is again non-central. Note that this is the popular equal-tailed Student T test.

Exercise 7: Consider the testing problem in 15a) under the assumption that $\sigma_2^2 = \rho\sigma_1^2$ where the multiplier ρ is known. Find the corresponding LRT.

If $\sigma_1^2 \neq \sigma_2^2$, the tests in 15) are not UMPU. Testing the difference of means is then known as the Behrens–Fisher problem for which there is no definitive solution; some approximate solutions have been proposed (see Lehmann, 1986).

LESSON 12. TESTS FOR TWO–BY–TWO TABLES

Herein, we discuss the special case of categorical data in the title of this lesson. There are three kinds of experiments which have such a summary; we consider exact and approximate probabilities associated with tests of "independence". In all cases, the observations can be displayed in a table:

$$\begin{array}{cc|l} n_{11} & n_{12} & n_{1\cdot} \\ n_{21} & n_{22} & n_{2\cdot} \\ \hline n_{\cdot 1} & n_{\cdot 2} & n_{\cdot\cdot} = n \end{array} \quad . \qquad (*)$$

Case I Two dichotomous characteristics are observed on n independent experimental units.

Example: One thousand people are asked whether they voted Democratic or Republican; each is also observed to be either male or female. (This is an exit poll of an election and "other" responses are passed over. To enhance independence between observations, close relatives and/or friends are also passed by.) The results of a sample of voters might be:

	Dem	Rep .
Male	459	131
Female	284	126

This model assumes independence of the voters included (the experimental units) in a four cell multinomial with the total n observations fixed; the cell parameters are:

$$\theta_{11} = P(\text{Male} \cap \text{Dem}) \qquad \theta_{12} = P(\text{Male} \cap \text{Rep})$$
$$\theta_{21} = P(\text{Female} \cap \text{Dem}) \qquad \theta_{22} = P(\text{Female} \cap \text{Rep}) .$$

We also use:

$$P(\text{Male}) = \theta_{11} + \theta_{12} = \theta_{1\cdot} \quad P(\text{Dem}) = \theta_{11} + \theta_{21} = \theta_{\cdot 1}$$
$$P(\text{Female}) = \theta_{21} + \theta_{22} = \theta_{2\cdot} \quad P(\text{Rep}) = \theta_{12} + \theta_{22} = \theta_{\cdot 2}$$
$$\theta_{1\cdot} + \theta_{2\cdot} = 1 = \theta_{\cdot 1} + \theta_{\cdot 2} .$$

In this case, only $n = n_{..}$ is fixed; all the other "n" are random.

The traditional null hypothesis is that of independence of sex and party. This translates to

$$H_o: \theta_{ij} = \theta_{i.}\theta_{.j} \text{ for } i,j = 1,2 \text{ vs } H_a: \text{otherwise .}$$

By the results of the following example (and exercise), we see that H_o is equivalent to $\theta_{11}\theta_{22} = \theta_{12}\theta_{21}$. There is a "positive association" when θ_{11} and θ_{22} are substantially larger than θ_{12} and θ_{21}. This is often expressed as

$$\text{lods} = \text{log odds} = \log(\theta_{11}\theta_{22}/\theta_{12}\theta_{21}) > 0 .$$

Example: Now $\theta_{11} = \theta_{1.}\theta_{.1}$, implies

$$\begin{aligned}\theta_{11} &= (\theta_{11} + \theta_{12})(\theta_{11} + \theta_{21}) \\ &= \theta_{11}(\theta_{11} + \theta_{12} + \theta_{21}) + \theta_{12}\theta_{21} \\ &= \theta_{11}(1 - \theta_{22}) + \theta_{12}\theta_{21}\end{aligned}$$

whence $\theta_{11}\theta_{22} = \theta_{12}\theta_{21}$.

Exercise 1: Show that $\theta_{12} = \theta_{1.}\theta_{.2}$, $\theta_{21} = \theta_{2.}\theta_{.1}$, $\theta_{22} = \theta_{2.}\theta_{.2}$ are each equivalent to $\theta_{11}\theta_{22} = \theta_{12}\theta_{21}$.

Now we use the likelihood ratio principle to build a test. Without restrictions, the likelihood

$$L(\theta) = (n!/n_{11}!\, n_{12}!\, n_{21}!\, n_{22}!)\, \theta_{11}^{n_{11}} \theta_{12}^{n_{12}} \theta_{21}^{n_{21}} \theta_{22}^{n_{22}}$$

contains 3 free parameters. In Lesson 9 above, we found the MLEs under H_a : $\hat{\theta}_{ij} = n_{ij}/n$. Under H_o , the parametric portion of the likelihood is

$$(\theta_{1.}\theta_{.1})^{n_{11}}(\theta_{1.}\theta_{.2})^{n_{12}}(\theta_{2.}\theta_{.1})^{n_{21}}(\theta_{2.}\theta_{.2})^{n_{22}}$$

$$= \theta_{1\cdot}^{n_{1\cdot}} \theta_{\cdot 1}^{n_{\cdot 1}} \theta_{\cdot 2}^{n_{\cdot 2}} \theta_{2\cdot}^{n_{2\cdot}}$$

$$= \theta_{1\cdot}^{n_{1\cdot}} (1 - \theta_{2\cdot})^{n_{2\cdot}} \theta_{\cdot 1}^{n_{\cdot 1}} (1 - \theta_{\cdot 1})^{n_{\cdot 2}}$$

so that $L(\theta)$ contains only 2 free parameters and the MLEs are

$$\hat{\theta}_{1\cdot} = n_{1\cdot}/n, \ \hat{\theta}_{2\cdot} = n_{2\cdot}/n .$$

The likelihood ratio criterion is based on

$$\lambda = L(\hat{\theta})/L(\hat{\Theta})$$

$$= n_{\cdot 1}^{n_{\cdot 1}} n_{\cdot 2}^{n_{\cdot 2}} n_{1\cdot}^{n_{1\cdot}} n_{2\cdot}^{n_{2\cdot}} \div n_{11}^{n_{11}} n_{12}^{n_{12}} n_{21}^{n_{21}} n_{22}^{n_{22}} . \qquad (**)$$

Since under H_o, we still have two unknown parameters, we cannot find exact probabilities; however, as in Lesson 10, the asymptotic distribution of $-2\ln(\lambda)$ is chisquare with

$3 - 2 = 1$ dof and so we reject H_o when $-2\ln(\lambda) > \chi^2_\alpha(1)$.

Exercise 2: In Lesson 11, we also saw that $-2\ln(\lambda)$ was asymptotic to $n(\Sigma_i \Sigma_j n_{ij}^2/n_{i\cdot} n_{\cdot j} - 1)$. Show that in the 2 by 2 case, the double sum in these parentheses is equal to

$(n_{11}n_{22} - n_{12}n_{21})^2/n_{\cdot 1} n_{\cdot 2} n_{1\cdot} n_{2\cdot}$ (a form used by Fisher).

Exercise 3: Complete the test of independence for the election poll in the first example; take $\alpha = .05$.

Case II One dichotomous characteristic is observed on two independent samples of experimental units.

Example: Each screw in a box of 50 "Made in USA" is classified as defective/non-defective. Each of a box of 50 screws "Foreign made" is classified by a similar process. The data is

	Def	Non
USA	1	49 .
FOR	5	45

Here we take the model to be that of independent Bernoulli experiments with $\theta_U = \theta_{11}$ and $\theta_F = \theta_{21}$ as the corresponding proportions of defectives. The likelihood $L(\theta)$ is

$$\begin{bmatrix} n_{1\cdot} \\ n_{11} \end{bmatrix} \theta_{11}^{n_{11}} (1-\theta_{11})^{n_{1\cdot}-n_{11}} \begin{bmatrix} n_{2\cdot} \\ n_{21} \end{bmatrix} \theta_{21}^{n_{21}} (1-\theta_{21})^{n_{2\cdot}-n_{21}} .$$

The data table is the same (*) but in this case, $n_{1\cdot}$ and $n_{2\cdot}$ are fixed (at 50 in the example) while $n_{\cdot 1}$ and $n_{\cdot 2}$ are random. The MLEs for this case are

$$\hat{\theta}_{11} = n_{11}/n_{1\cdot} \text{ and } \hat{\theta}_{21} = n_{21}/n_{2\cdot}$$

and these are independent sample proportions (because the samples are independent).

The hypotheses which are analogous to those in Case I are

$$H_o: \theta_{11} = \theta_{21} = \theta_c \text{ (unknown) vs } H_a: \theta_{11} < \theta_{21} .$$

Under H_o, the likelihood is

$$L(\theta) = \begin{bmatrix} n_{1\cdot} \\ n_{11} \end{bmatrix} \theta_c^{n_{11}} (1-\theta_c)^{n_{1\cdot}-n_{11}} \begin{bmatrix} n_{2\cdot} \\ n_{21} \end{bmatrix} \theta_c^{n_{21}} (1-\theta_c)^{n_{2\cdot}-n_{21}}$$

and the MLE is $\hat{\theta}_c = n_{\cdot 1}/n$. The likelihood ratio criterion is exactly the same form (**) but, of course, $n_{1\cdot}$ and $n_{2\cdot}$ are not random. Under H_o, we still have 1 unknown parameter and so the asymptotic distribution of $-2\ln(\lambda)$ is chisquare with $2 - 1 = 1$ dof.

Exercise 4: a) Verify the reduction of the likelihood ratio criterion.

b) Show that under H_o,

$$\mathrm{Var}(\hat{\theta}_{11} - \hat{\theta}_{21}) = \theta_c(1 - \theta_c)(1/n_{1\cdot} + 1/n_{2\cdot}) .$$

c) Show that under H_o,

$$E[\hat{\theta}_c(1 - \hat{\theta}_c)] = (n-1)\theta_c(1 - \theta_c)/n .$$

Now we can make use of the fact that under H_o, the asymptotic distribution of

$$t = (\hat{\theta}_{11} - \hat{\theta}_{21})/\sqrt{[(\hat{\theta}_c(1 - \hat{\theta}_c)(1/n_1 + 1/n_2)]}$$

is the standard normal Z (via another form of the CLT which we have not spelled out). This ratio t reduces to

$$(n_{11}\, n_{22} - n_{21}\, n_{12})(n_{1\cdot}\, n_{\cdot 1}\, n_{2\cdot}\, n_{\cdot 2})^{1/2} ;$$

we can take the one–tailed rejection region to be

$$t < -z_\alpha .$$

Exercise 5: Complete the LRT and the one–tailed test for the data on screws. Take $\alpha = .05$.

Case III Two independent observers use a dichotomous characteristic to separate experimental units into two sets of known sizes.

This is one model for "taste tests" , a popular example suggested by Fisher, 1935.

Example: A lady is presented 8 cups, four of which had the tea poured in first, then the milk; in the other four, the pattern was reversed. She knows that there are four of each kind and she is to identify them by taste. One outcome is

	Lady says	Tea first	Milk first .
Maker says	Tea first	1	3
	Milk first	3	1

In this case, both sets of marginal totals are fixed and $n_{1\cdot} = n_{\cdot 1}$, $n_{2\cdot} = n_{\cdot 2}$. It is obvious that there is only one cell observation which can be "random" and we may as well take it to be n_{11} . If the cups are presented to the lady in a random order, the relevant null distribution is hypergeometric:

$$P(N_{11} = n_{11}) = \begin{bmatrix} n_{1\cdot} \\ n_{11} \end{bmatrix} \begin{bmatrix} n_{2\cdot} \\ n_{21} \end{bmatrix} \div \begin{bmatrix} n_{1\cdot} + n_{2\cdot} \\ n_{1\cdot} \end{bmatrix} .$$

Now N_{11} , appropriately standardized from

$$E[N_{11}] = n_{1.}n_{.1}/n\ ,\ Var(N_{11}) = n_{1.}n_{.1}n_{2.}n_{.2}/n^2(n-1)\ ,$$

will be asymptotically normal; then

$$t = (n_{11} - E[N_{11}])/\sqrt{Var(N_{11})}$$
$$= (n_{11}n_{22} - n_{21}n_{12})((n-1)/n_{1.}\,n_{.1}\,n_{2.}\,n_{.2})^{1/2}\ .$$

But in the present situation, the distribution of N_{11} is completely known; it is not necessary to use asymptotics to get probabilities. In particular for the lady tasting tea, we have

n_{11}	0	1	2	3	4
$P(n_{11})$	1/70	16/70	36/70	16/70	1/70

If the lady is just guessing, the two classifications are essentially independent and the observed n_{11} is "arbitrary"; otherwise, her n_{11} ought to be "large". In this example, there is only one reasonable p–value:

$$P(N_{11} \geq 4) = 1/70 \approx .014\ .$$

If she classifies all 8 cups correctly, we hire her as a taster; otherwise, her success rate is "not unusual".

Exercise 6: Find the distribution of N_{11} for a taste test with the results:

Identified as		Butter	Margarine
Presented	Butter	5	20
as	Margarine	20	5

Intuitively, Case II follows Case I by conditioning on $n_{1.}$ and $n_{2.}$ and Case III follows Case II by further conditioning on $n_{.1}$ and $n_{.2}$. This can be made rigorous by taking note of the exponential form of all the densities and applying the results of Lesson 3.

For Case I, the parametric portion of $L(\theta)$ takes the form

$$e^{n_{1.}\ln p_1}e^{n_{.1}\ln p_2}((1-\theta_{1.})/(1-\theta_{.1}))^n$$

where $p_1 = \theta_{1.}/(1 - \theta_{1.})$ and $p_2 = \theta_{.1}/(1 - \theta_{.1})$. Here $n_{1.}$

and $n_{.1}$ are jointly complete sufficient statistics for p_1 and p_2 so that UMPU tests are based on n_{11} given $n_{1.}$ and $n_{.1}$. This is just the hypergeometric of Case III.

For Case II, $L(\theta)$ has the form

$$\begin{bmatrix} n_{1\cdot} \\ n_{11} \end{bmatrix} \begin{bmatrix} n_{2\cdot} \\ n_{21} \end{bmatrix} e^{n_{.1} \ln p} \frac{}{(1-p)^n}$$

where $p = \theta_c/(1 - \theta_c)$. Now $n_{.1}$ is the complete sufficient statistic for p so the UMPU test is based on n_{11} given $n_{.1}$ which is again the hypergeometric.

Indeed, you should have noticed that the original statistic in all three cases is a function of the difference

$$n_{11} n_{22} - n_{12} n_{21} .$$

This discussion means that any "one–sided" test based on the difference is equivalent to one based on n_{11} alone. Unfortunately, this is not true for "two–sided" tests. For further discussion, see Davis, 1986.

PART VII: SPECIAL TOPICS

Overview

The topics herein are special because each is only the merest introduction to some mode of statistical inference: decision theory, linear models, sequential analysis, non–parametrics, goodness–of–fit, classified data.

The material in the first three lessons is in the spirit of decision theory. In particular, "Bayesians" look upon parameters as not really fixed but (changing with the seasons?) random and treat $f(x;\theta)$ as the conditional density of X given θ. They then use distributions on Θ to help in "minimizing risk", etc. One way to reduce the class of estiamtors under consideration is to impose "invariance" under certain transformations. A more modern phrase is "equivariance" and further studies along this line would bring other algebraic structures into statistical theory.

Lesson 4 contains an introduction to linear regression with very minimal assumptions and Lesson 5 is concerned with estimation and testing under the classical assumption of "normal errors". Of course, most real problems are not "linear" but this is the simplest non–trivial starting point.

Historically, this began with "fitting curves to data" in two dimensions: $(y_1,x_1,)\ (y_2,x_2)\ \cdots\ (y_n,x_n)$. Gauss was quick to admit that his choice to minimize $\sum_{i=1}^{n}(y_i - f(x_i))^2$ was totally arbitrary. Yet, it could be that as much as 75% of the techniques used in statistical applications are an outgrowth of material reviewed in these two lessons. With some modest assumptions of "smoothness",

$$f(x) = f(x_o) + f'(x_o)(x - x_o) + R(emainder)\ .$$

In strictly dynamic processes, this remainder usually involves higher order derivatives, while in "first order" stochastic processes, this remainder is taken to be random. (Just because a cricket chirps 20 times per 10 second does not mean that the temperature is exactly 91 degrees. See the interesting work of Pierce, 1949.) Then the Taylor form above can be written as

$$y = f(x) = \beta_0 + \beta_1 \cdot x + \varepsilon$$

where $\beta_0 = f(x_o) - f'(x_o)\cdot x_o$ and $\beta_1 = f'(x_o)$.

In the late 1880's, Galton observed that the distribution of the heights of the sons of tall[short] fathers tended to be less spread out than that of their fathers. This was a "turn toward the mean", a re–turn, a re–gression; this name is still used for the process but it no longer has that particular sense. In fact, Galton was observing (what he likened to) a bivariate normal distribution and consequently found the linear relation:

$$E[Y \mid X = x] = \mu_Y + \rho\sigma_Y(x - \mu_X)/\sigma_X .$$

This means that the nice results of Lessons 4 and 5 can be used to analyze "Y" when "X" is fixed (or indeed, visa–versa).

Lessons 6 and 7 are suggestions of other areas of research which have been found to be informative and, particularly for sequential techniques, very practical. It has been shown that his latter methodology leads (almost surely) to the use of samples of smaller size than in the (fixed size) techniques of Parts V and VI.

Lessons 8 and 9 contain an introduction to "non–parametric or distribution–free" methods of testing. Neither of these terms is quite correct and some explanation of this terminology is included in the lessons. Historically, these tests were invented to treat questions about means without the assumptions of normality as in Lesson 12, Part VI.

There may be almost as many "goodness–of–fit" tests as there are statisticians but essentially only one is presented in Lesson 10. Similar remarks apply to the extensions from 2 by 2 to r by c tables presented in Lesson 11.

LESSON *1. MINIMAX AND BAYES ESTIMATORS – I

In some previous lessons, we restricted discussion to unbiased estimators and sought those with minimum variance. Under this guideline, we ignored biased estimators which might be "better" than the UMVUE in terms of some other risk. To pursue this possibility, we turn to the *mean square error of* Lesson 1, Part V.

Let the model be $\{f(x;\theta), x \in R^m, \theta \in \Theta \subset R^d\}$. From a random sample $X_1, \cdots, X_n$ of this model, we use a statistic $T(X_1,\cdots,X_n)$ to estimate a function $\varphi(\theta)$. In order to compare different estimators of $\varphi(\theta)$, we take the *risk* as MSE:

$$R(\theta,T) = E_\theta[\| T - \varphi(\theta)\|^2]$$

(henceforth assuming existence of risk whenever it appears).

Example: Let X be the normal $N(0, \sigma^2)$, $\theta = \sigma^2 > 0$.

a) By the Lehmann–Scheffe' theorem, the sample variance $S^2 = \frac{1}{n-1}\sum_{i=1}^{n}(X_1 - \bar{X})^2$ is an UMVUE of θ. The risk $R(\theta,S^2) = \mathrm{Var}_\theta(S^2) = \frac{2\theta^2}{n-1}$.

b) Now $T(X_1,\cdots,X_n) = \frac{1}{n+1}\sum_{i=1}^{n}(X_i - \bar{X})^2$ is a biased estimator with mean $(n-1)\sigma^2/(n+1)$ and $R(\theta,T)$

$$= R(\theta, \frac{n-1}{n+1}S^2) = E_\theta\left[\left[\frac{n-1}{n+1}S^2 - \sigma^2\right]^2\right]$$

$$= E_\theta\left[\left[\frac{n-1}{n+1}(S^2 - \sigma^2) + \frac{n-1}{n+1}\sigma^2 - \sigma^2\right]^2\right]$$

$$= E_\theta\left[\left[\frac{n-1}{n+1}\right]^2(S^2 - \sigma^2)^2\right] + \frac{4}{n+1}\sigma^4$$

because $E[S^2 - \sigma^2] = 0$. Thus,

$$R(\theta,T) = \frac{(n-1)^2}{(n+1)^2}\text{Var}(S^2) + \frac{4}{(n+1)^2}\sigma^4$$

$$= \frac{(n-1)^2}{(n+1)^2}\,\frac{2\sigma^4}{n-1} + \frac{4}{(n+1)^2}\sigma^4$$

$$= \frac{2\sigma^4}{n+1} = \frac{2\theta^2}{n+1}.$$

c) It follows that $R(\theta,T) < R(\theta,S^2)$ for all $\theta \in \Theta$. In geometric terms, the average (squared) distance from T to φ is less than that from S^2 to φ; one might prefer T to S^2. In fancier language, the unbiased estimator S is inadmissible within the class of all estimators (with second moments).

This situation is perfectly general; when allowing all possible estimators into the competition, a "best" estimator within a subclass might be *inadmissible* within the larger class. This suggests that additional criteria be employed.

First we consider the pessimistic attitude that a "good" estimator should minimize the risk at the least favorable parameter point. More precisely,

Definition: *The estimator* T^* *is a minimax estimator of* $\varphi(\theta)$ *if* $\sup_{\theta\in\Theta} R(\theta,T^*) = \inf_T \sup_{\theta\in\Theta} R(\theta,T)$.

Here, the inf is taken over all estimators T so that

$$\sup_{\theta\in\Theta} R(\theta,T^*) \le \sup_{\theta\in\Theta} R(\theta,T) \text{ for all } T.$$

Thus the minimax estimator is the one which minimizes the maximum risk (and gives it its name).

Now we introduce another principle of estimation which will provide minimax estimators and *admissible* estimators; this principle requires a "good" estimator to minimize an averaged risk. More precisely, taking the so-called *Bayesian* viewpoint, we regard the unknown parameter θ as a random variable (or vector) with a known prior *probability density* on Θ, say, $\Lambda(\theta)$

(including σ–fields and all that). Here $f(x;\theta)$ is now the conditional density of X given θ.

Definition: *The Bayes risk of the estimator T with respect to the prior Λ is defined to be*

$$R_\Lambda(T) = E_\Lambda[R(\theta,T)] = \int_\Theta R(\theta,T)\Lambda(\theta)\, d\theta \,.$$

T^* is called a Bayes estimator of $\varphi(\theta)$ with respect to Λ if

$$R_\Lambda(T^*) = \inf_T R_\Lambda(T)$$

with the inf taken over all estimators T of $\varphi(\theta)$.

For simplicity, we will consider one dimensional parameters. The following theorem gives a method for finding Bayes estimators.

Theorem: *The Bayes estimator of θ is given by*

$$T(x_1,\cdots,x_n) = E[\theta\,|\,(X_1,\cdots,X_n) = (x_1,\cdots,x_n)]$$

Proof: $R_\Lambda(T) = \int_\Theta R(\theta,T)\Lambda(\theta)\, d\theta$

$$= \int_\Theta \Lambda(\theta)\, d\theta \int_{\mathscr{X}^n} [\theta - T(x_1,\cdots,x_n)]^2 \prod_{i=1}^n f(x_i;\theta)\, dx_i \,.$$

Now observe that $\Lambda(\theta) \prod_{i=1}^n f(x_i;\theta)$ is the joint density of the variables $\theta, X_1, \cdots, X_n$, and thus the marginal density of $(X_1,\cdots,X_n)$ is, say,

$$g(x_1,\cdots,x_n) = \int_\Theta \Lambda(\theta) \prod_{i=1}^n f(x_i;\theta)\, d\theta \,.$$

It follows that the conditional density of θ given $(x_1,\cdots,x_n)$ is

$$h(\theta\,|\,x_1,\cdots,x_n) = \Lambda(\theta) \prod_{i=1}^n f(x_i;\theta)/g(x_1,\cdots,x_n) \,.$$

Then, $R_\Lambda(T)$ is

$$\int_{\mathscr{B}^n}\left[\int_{\Theta}(\theta - T)^2\, h(\theta\,|\,x_1,\cdots,x_n)\, d\theta\right] g(x_1,\cdots,x_n)\, dx_1\cdots dx_n$$

and this will be minimal when the inner integral

$$\int_{\Theta}[\theta - T(x_1,\cdots,x_n)]^2\, h(\theta\,|\,x_1,\cdots,x_n)\, d\theta \qquad (*)$$

is minimal.

To shorten notation, write $\tilde{x} = (x_1,\cdots,x_n)$. Let $\alpha(\tilde{x}) = \int_{\Box} \theta h(\theta\,|\,\tilde{x})\, d\theta$. Since $\int_{\Theta} h(\theta\,|\,\tilde{x})\, d\theta = 1$,

$$\int_{\Theta}(\theta - \alpha(\tilde{x}))h(\theta\,|\,\tilde{x})\, d\theta = 0 .$$

From this, we get $\int_{\Theta}[\theta - T(\tilde{x})]^2 h(\theta\,|\,\tilde{x})\, d\theta$

$$= \int_{\Theta}[\theta - \alpha(\tilde{x}) + \alpha(\tilde{x}) - T(\tilde{x})]^2 h(\theta\,|\,\tilde{x})\, d\theta$$

$$= \int_{\Theta}(\theta - \alpha(\tilde{x}))^2 h(\theta\,|\,\tilde{x})\, d\theta + (\alpha(\tilde{x}) - T(\tilde{x}))^2 .$$

The right–hand side and hence also (*) is minimal when

$$T(\tilde{x}) = \alpha(\tilde{x}) = E[\theta\,|\,\tilde{x}] .$$

Exercise 1: (The technique used in the proof is quite general.) Show that for real random variables Y and g(X),

$E[\{Y - g(X)\}^2]$ is minimal when $g(X) = E[Y\,|\,X]$.

Corollary: *When the parameter of interest is* $\varphi(\theta)$ *, the Bayes estimator is* $E[\varphi(\theta)\,|\,\tilde{x}]$

$$= \int_{\Theta}\varphi(\theta)[\prod_{i=1}^{n} f(x_i;\theta)]\Lambda(\theta)\, d\theta \Big/ \int_{\Theta}[\prod_{i=1}^{n} f(x_i;\theta)]\Lambda(\theta)\, d\theta .$$

Example: Let $X_1, \cdots, X_n$ be a random sample from $N(\theta,1)$,

$\theta \in \Theta = R$. Let $\Lambda(\theta) = (2\pi)^{-1/2}\exp\{-\frac{1}{2}\theta^2\}$. Then

$$g(\tilde{x}) = \int_0^\infty \Lambda(\theta)f(\tilde{x};\theta)\,d\theta$$

$$= (2\pi)^{-n/2}(n+1)^{-1/2}\exp\{-\frac{1}{2}\sum_{i=1}^n x_i^2 + \frac{n^2(\bar{x})^2}{2(n+1)}\}$$

and $h(\theta\,|\,\tilde{x}) = \Lambda(\theta)\,f(\tilde{x};\theta)/g(\tilde{x})$

$$= (2\pi)^{=1/2}(n+1)^{1/2}\exp\{-\frac{n+1}{2}[\theta - \bar{x}(1+\tfrac{1}{n})^{-1}]^2\}$$

which is the density for $N(\bar{x}\,n/(n+1),\ 1/(n+1))$. Thus the Bayes estimator of θ is

$$\theta^*(x_1,\cdots,x_n) = E[\theta\,|\,\tilde{x}] = \int_{-\infty}^\infty \theta h(\theta\,|\,\tilde{x})d\theta = \bar{x}n/(n+1)\,.$$

The associated Bayes risk is

$$R_\Lambda(\theta^*) = \int_\Theta \Lambda(\theta)\int_{\mathscr{X}^n}[\theta^* - \theta]^2 f(\tilde{x};\theta)\,d\tilde{x}\,d\theta = 1/(n+1).$$

Exercise 2: Write the integration details of $R_\Lambda(\theta^*)$.

Exercise 3: a) Let X be uniformly distributed over $(0,\theta)$, $\theta > 0$, and $\Lambda(\theta) = e^{-\theta}I(\theta > 0)$. Find the Bayes estimator of θ and its Bayes risk.
b) Let X be Bernouilli, $\theta \in (0,1)$, and let

$$\Lambda(\theta) = I(0 < \theta < 1).$$

Find the Bayes estimator of θ and its Bayes risk.

The following result shows that Bayes estimators are often admissible. As usual, "uniqueness" is almost surely, that is, except for a set N with $P_\theta(N) = 0$ for each $\theta \in \Theta$.

Theorem: *If T is the unique Bayes estimator with respect to a prior density Λ of θ, then T is admissible.*

Proof: It suffices to show that if S is an estimator such that

$$R(\theta,S) \le R(\theta,T) \text{ for all } \theta \in \Theta ,$$

then there is no θ_0 such that $R(\theta_0,S) < R(\theta_0,T)$.

When $R(\theta,S) \le R(\theta,T)$ for all $\theta \in \Theta$,

$$E_\Lambda[R(\theta,S)] \le E_\Lambda[R(\theta,T)] \text{ , say, } R_\Lambda(S) \le R_\Lambda(T) .$$

By definition of T, it must be that $R_\Lambda(S) = R_\Lambda(T)$ and by hypothesis, S = T a.s. Hence for all $\theta \in \Theta$,

$$R(\theta,S) = \int_{\mathscr{X}^n} |\theta - T(\tilde{x})|^2 f(\tilde{x};\theta)\, d\tilde{x} = R(\theta,T) .$$

The random variable $E[\varphi(\theta)|\tilde{x}]$ is unique up to an equivalence of random variables. Thus in the above theorem, the uniqueness of the Bayes estimator will follow if "G–a.s. implies P_θ–a.s. for $\theta \in \Theta$", where P_θ is the conditional distribution of $\tilde{X}$ given θ and G is the marginal distribution of $\tilde{X}$. In the following, this implication does not hold.

Example: X given $\theta \in [0,1]$ is binomial.

$$\begin{aligned} \Lambda(\theta) &= 1/2 \text{ if } \theta = 0 \text{ or } 1 \\ &= 0 \quad \text{otherwise} \end{aligned} .$$

$$P(X = 0, \theta = 0) = P(X = 0 \mid \theta = 0) \cdot P(\theta = 0) = 1/2 .$$

$$P(X = n, \theta = 1) = 1/2 .$$

The marginal distribution

$$\begin{aligned} g(x) = P(X = x) &= 1/2 \text{ for } x = 0 \text{ or } n \\ &= 0 \quad \text{otherwise} \end{aligned} .$$

For $A = \{1, 2, \cdots, n-1\}$, $P(X \in A) = 0$ but for $\theta \ne 0$ or 1,

$$P_\theta(X \in A|\theta) = \sum_{x=1}^{n-1} \binom{n}{x} \theta^x (1-\theta)^{n-x} \ne 0 .$$

Thus G–a.s. does not imply P_θ–a.s.

LESSON *2. MINIMAX AND BAYES ESTIMATORS – II

As promised, we now show that Bayes estimators may be minimax.

Theorem: *If T is a Bayes estimator of $\varphi(\theta)$ with respect to a prior Λ, such that the risk $R(\theta,T)$ is independent of θ, then T is minimax.*

Proof: Since by hypothesis, $R(\theta,T) = c$, a constant independent of θ,

$$R(\theta,T) = \sup_{\theta \in \Theta} R(\theta,T).$$

But $\int_\Theta \Lambda(\theta)d\theta = 1$ and $R(\theta,T) = c$ imply,

$$R(\theta,T) = \int_\Theta R(\theta,T)\Lambda(\theta)\,d\theta = R_\Lambda(T).$$

Also, since T is Bayes, $R_\Lambda(T) \le R_\Lambda(S)$, for any S. But from

$$R_\Lambda(S) = \int_\Theta R(\theta,S)\Lambda(\theta)d\theta \le \sup_{\theta \in \Theta} R(\theta,S),$$

we get $\sup_{\theta \in \Theta} R(\theta, T) \le \sup_{\theta \in \Theta} R(\theta, S)$ for all S; T is minimax.

In the next example we use the following

Definition: *The family of Beta distributions (indexed by the positive parameters θ_1, θ_2) has density*

$$x^{\theta_1 - 1}(1-x)^{\theta_2 - 1} I(0 < x < 1)/B(\theta_1,\theta_2)$$

where $B(\theta_1,\theta_2) = \int_0^1 x^{\theta_1 - 1}(1-x)^{\theta_2 - 1}\,dx = \dfrac{\Gamma(\theta_1)\cdot\Gamma(\theta_2)}{\Gamma(\theta_1+\theta_2)}$

and Γ is the gamma function. This distribution may be symbolized as $\mathcal{B}(\theta_1,\theta_2)$.

Exercise 1: The conditional density of X given θ is binomial with parameters n and θ ; the marginal distribution of θ is $\mathscr{B}(\frac{\sqrt{n}}{2}, \frac{\sqrt{n}}{2})$. Show that the conditional of distributon of θ given x is $\mathscr{B}(x + \frac{\sqrt{n}}{2}, n - x + \frac{\sqrt{n}}{2})$.

Exercise 2: Show that the mean of the $\mathscr{B}(\theta_1, \theta_2)$ distribution is $\frac{\theta_1}{\theta_1 + \theta_2}$. Hint: make use of the properties of $B(\theta_1, \theta_2)$.

Example: Let X be binomial with parameters n , θ , θ ∈ (0,1). The MLE of θ is $\frac{X}{n}$ which is an UMVUE with $R(\theta,\frac{X}{n}) = \theta(1 - \theta)/n$. We will look for a minimax estimator of θ by using the previous theorem.

a) Suppose that the prior Λ(θ) of θ on (0,1) is $\mathscr{B}(\frac{\sqrt{n}}{2}, \frac{\sqrt{n}}{2})$. By Exercises 1 and 2, we have

$$E(\theta|X) = (X + \frac{\sqrt{n}}{2})(n + \sqrt{n})^{-1} .$$

The Bayes estimator $T(X) = (X + \frac{\sqrt{n}}{2})(n + \sqrt{n})^{-1}$ has the risk $R(\theta,T) = \frac{1}{4}(1 + \sqrt{n})^{-2}$ which is independent of θ , and hence is minimax.

For $|\theta - \frac{1}{2}| < \{1 + 2\sqrt{n}\}^{\frac{1}{2}}\{2(1 + \sqrt{n})\}^{-1}$, we have $R(\theta,\frac{X}{n}) > R(\theta,T)$.

b) Note that T has the form $\frac{\alpha}{n}X + \beta$ with $\alpha \neq 0$.

$$R(\theta,\frac{\alpha}{n}X + \beta) = E_\theta[(\frac{\alpha}{n}X + \beta - \theta)^2]$$

$$= \beta^2 + \frac{\theta}{n}\left[1 - 2(1 - \alpha) + (1 - \alpha)^2 - 2n\beta(1 - \alpha)\right]$$

$$-\frac{\theta^2}{n}\left[1 - 2(1-\alpha) + (1-\alpha)^2(1-n)\right].$$

This risk will be independent of θ iff the two factors in large brackets are both zero. This requires

$$\alpha = \frac{\sqrt{n}}{1+\sqrt{n}}, \quad \beta = \tfrac{1}{2}(1+\sqrt{n})^{-1}$$

and at these points, $\frac{\alpha}{n}X + \beta = T$, so that

$$R(\theta,T) = \beta^2 = \tfrac{1}{4}(1+\sqrt{n})^{-2}.$$

We look again at the sample mean from a normal population and explain two important properties:

In the uni/bi–variate case, the sample mean is an admissible and minimax estimator which can be related to the information inequality of regular models (e.g., exponential families);

In higher dimensions, by contrast, the sample mean is not admissible but an appropriate function of it, the *James-Stein estimator*, is. First let X be distributed as $N(\theta, 1)$ with a RS $X_1, \cdots, X_n$. The sample mean $\overline{X} = \sum_{i=1}^{n} X_i/n$ is an UMVUE of θ, and $\mathrm{Var}_\theta(\overline{X}) = \frac{1}{n} = R(\theta,\overline{X})$. Also, the model $N(\theta, 1)$ is regular with Fisher information $I(\theta) = 1$, and $\overline{X}$ is efficient. In the discussion of the admissibility of $\overline{X}$, it makes sense to consider only estimators $T(X_1, \cdots, X_n)$ such that $R(\theta,T) \le R(\theta,\overline{X}) = \frac{1}{n}$ for all $\theta \in \Theta$, more generally, estimators T with bounded risk. We use the following properties of a result in Ibragimov and Has'minski,1981.

Lemma: *If T is an estimator of θ with bounded risk and the underlying density $f(x;\theta)$ is regular with $I(\theta) > 0$ for all $\theta \in \Theta \subset R$, then the bias $b(\theta) = E_\theta[T] - \theta$ is continuously differentiable on R and*

$$R(\theta,T) \ge \frac{(1 + b'(\theta))^2}{n\, I(\theta)} + b^2(\theta) = \frac{(1+b'(\theta))^2}{n} + b^2(\theta). \qquad (*)$$

Exercise 3: Note that (*) above is the information inequality for a biased estimator. Show (*) directly for the normal case with $R(\theta,T) \le R(\theta,\overline{X}) = 1/n$.

To show that $\overline{X}$ is admissible, we use "reductio absurdum". So, suppose that $\overline{X}$ is inadmissible; then there exists $T(X_1,\cdots,X_n)$ such that $R(\theta,T) \le R(\theta,\overline{X})$ for all $\theta \in R$ and $R(\theta',T) < R(\theta',\overline{X})$ for some $\theta' \in R$. Note that such an estimator T must be biased since $\overline{X}$ is an UMVUE. By the previous observation, we have

$$\frac{(1 + b'(\theta))^2}{n} + b^2(\theta) \le R(\theta,T) \le R(\theta,\overline{X}) = \frac{1}{n}$$

which implies $b'(\theta)^2 + 2b'(\theta) \le -n\, b^2(\theta)$ and $2b'(\theta) \le -n\, b^2(\theta)$. Since T is biased, there is some θ_o such that $b(\theta_o) \ne 0$. But, then

$$\int_{\theta_o}^{\theta} -\frac{2b'(\alpha)}{b^2(\alpha)}\, d\alpha = 2\left[\frac{1}{b(\theta)} - \frac{1}{b(\theta_o)}\right] \ge n\,(\theta - \theta_o)$$

or

$$1/b(\theta) \ge n(\theta - \theta_o)/2 + 1/b(\theta_o) .$$

It follows that $\lim_{\theta \to +\infty} b(\theta) = 0$. Similarly, for $\theta < \theta_o$,

$$2\left[\frac{1}{b(\theta_o)} - \frac{1}{b(\theta)}\right] \ge n(\theta - \theta_o)$$

and hence $\lim_{\theta \to -\infty} b(\theta) = 0$. But, on the other hand,

$$2\, b'(\theta) \le -n\, b^2(\theta) \le 0 ,$$

so that $b(\theta)$ is a non–increasing function; this means that $b(\theta) = 0$ for all $\theta \in R$, a contradiction. Therefore, $\overline{X}$ is admissible.

In order to show that $\overline{X}$ is minimax, we first observe that if there exists a sequence of Bayes estimators T_k (with respect to prior Λ_k on Θ) such that

$$R(\theta,\overline{X}) \leq \limsup_{k\to+\infty} R_{\Lambda_k}(T_k),$$

then $\overline{X}$ is minimax. It suffices to show that, for any estimator S, we have:

$$R(\theta,\overline{X}) \leq \sup_{\theta\in\Theta} R(\theta',S), \text{ for all } \theta .$$

Now, $\sup_{\theta\in\Theta} R(\theta, S) \geq \int_{\Theta} R(\theta, S)d\,\Lambda_k(\theta) = R_{\Lambda_k}(S) \geq R_{\Lambda_k}(T_k)$

since T_k is Bayes with respect to Λ_k for all k. Hence,

$\sup_{\theta\in\Theta} R(\theta,S) \geq \limsup_{k\to+\infty} R_{\Lambda_k}(T_k) \geq R(\theta,\overline{X})$ by hypothesis.

Exercise 4: Let the model be $N(\theta, 1)$. Take Λ_k to be $N(0, \sigma^2 = k)$. Show that the corresponding Bayes estimator $T_k = [\frac{nk}{nk+1}]\overline{X}$ and that the Bayes risk is $R_{\Lambda_k}(T_k) = \frac{k}{nk+1}$.

With the result of the Exercise 4, we see that

$$\lim_{k\to+\infty} R_{\Lambda_k}(T_k) = \lim_{k\to+\infty} \left(\frac{k}{nk+1}\right) = \frac{1}{n} = R(\theta,\overline{X})$$

for all $\theta \in R$. Thus $\overline{X}$ is minimax.

A much shorter proof of this fact is as follows! Since $\overline{X}$ is admissible and has a constant risk $\frac{1}{n}$ (independent of θ), it is minimax. This is true in general: let S be an admissible estimator with constant risk. Suppose there is an estimator S^* such that

$$\sup_{\theta\in\Theta} R(\theta,S^*) < \sup_{\theta\in\Theta} R(\theta,S) ;$$

then for any θ, $R(\theta,S^*) < \sup_{\theta'\in\Theta} R(\theta',S) = R(\theta,S)$ contradicting the hypothesis that S is admissible.

Is $\bar{X}$ a Bayes estimator? To answer this question, we need the following necessary condition (Lehmann, 1986).

Lemma: *Let* Λ *be a prior on* Θ *and* $f(x|\theta)$ *(as the conditional density of* X *given* θ*) be the model. If* T *is an unbiased Bayes estimator of* $\varphi(\theta)$*, then necessarily* $E[(T - \varphi(\theta))^2] = 0$.

Proof: For $T(X) = E[\varphi(\theta)|X]$,

$$
\begin{aligned}
E[T(X)\varphi(\theta)] &= E[\,E[T(X)\varphi(\theta)|X]\,] = E[\,T(X)E[\varphi(\theta)|X]\,] \\
&= E[T^2(X)], \\
&= E[\,E[T(X)\varphi(\theta)|\theta]\,] = [\varphi(\theta)E[T(X)|\theta]\,] \\
&= E[\varphi^2(\theta)]
\end{aligned}
$$

since $E[T(X)|\theta] = \varphi(\theta) = E_\theta[T(X)]$ by hypothesis.

Thus $E[(T - \varphi)^2] = E[T^2] - 2E[T\varphi] + E[\varphi^2] = 0$ as prescribed.

Now in the normal model, $\bar{X}$ is unbiased for $\varphi(\theta) = \theta$, and $E[(\bar{X} - \theta)^2] = \frac{1}{n} \neq 0$ so that $\bar{X}$ is *not* a Bayes estimator.

It can be shown that when X is normal in R^2 (the density with variance 1 is

$$f(x;\theta) = (2\pi)^{-1} \exp\{-\tfrac{1}{2}\|x - \theta\|^2\}, \ x \in R^2, \theta \in R^2)$$

then $\bar{X}$ still has these same properties. However, in higher dimensions, the situation is different. We examine the case R^3.

Let $X = (X_1, X_2, X_3)'$ have independent components, normally distributed with

$$\text{means } \theta = (\theta_1, \theta_2, \theta_3)' \text{ and variances all } 1.$$

For i = 1,2,3, let $g_i : R^3 \to R$ so that $g = (g_1, g_2, g_3)'$ maps R^3 into R^3. For a random sample of X of size n,

$\overline{X} = (\overline{X}_1, \overline{X}_2, \overline{X}_3)'$ is normally distributed with independent components, mean θ, and variances $1/n$. The Stein (or James–Stein) estimators are of the form

$$T = \overline{X} + g(\overline{X})/n .$$

The difference in the risks is

$$R(\theta,\overline{X})-R(\theta,T) = E_\theta[\|\overline{X}-\theta\|^2] - E_\theta[\|\overline{X}+g(\overline{X})/n-\theta\|^2]$$

$$= -2E_\theta[(\overline{X}-\theta)'g(\overline{X})]/n - E_\theta[\|g(\overline{X})\|^2]/n$$

$$= -2E_\theta\left[\sum_{i=1}^{3}(\overline{X}_i-\theta_i)g_i(\overline{X})\right]/n^2 - E_\theta[\|g(\overline{X})\|^2]/n^2 .$$

Exercise 5: Use integration by parts and add conditions on g to establish the following

Lemma: $$\int \frac{\partial g_i(\bar{x})}{\partial \bar{x}_i} e^{-n(\bar{x}_i-\theta_i)^2/2}\, d\bar{x}_i / \sqrt{(2\pi/n)}$$

$$= n\int g_i(\bar{x})(\bar{x}_i-\theta_i)e^{-n(\bar{x}_i-\theta_i)^2/2}\, d\bar{x}_i / \sqrt{(2\pi/n)} .$$

Using the result of this lemma makes

$$R(\theta,\overline{X})-R(\theta,T) = -2E_\theta[\Sigma\partial g_i(\overline{X})/\partial\overline{x}_i]/n^2 - E_\theta[\|g(\overline{X})\|^2]/n^2 .$$

Now find φ so that $g(x) = \nabla\log\varphi(x) = \dfrac{\partial\varphi(x)}{\partial x}/\varphi(x)$. Then,

$$g_i(x) = \frac{\partial\varphi(x)}{\partial x_i}/\varphi(x)$$

and

$$\frac{\partial g_i(x)}{\partial x_i} = (\varphi(x)\cdot\partial^2\varphi/\partial x_i^2 - (\partial\varphi/\partial x_i)^2)/\varphi(x)^2$$

$$= (\partial^2\varphi/\partial x_i^2)/\varphi(x) - \left[\frac{\partial\varphi/\partial x_i}{\varphi(x)}\right]^2$$

$$= (\partial^2\varphi/\partial x_i^2)/\varphi(x) - \|g(x)\|^2 \ .$$

This makes $R(\theta,\overline{X}) - R(\theta,T)$

$$= -\,2E_\theta[\sum_{i=1}^{3} \frac{\partial^2\varphi/\partial x_i^2}{\varphi(\overline{X})} - \|g(\overline{X})\|^2\,]/n^2$$

$$= E_\theta[\ \|g(\overline{X})\|^2\,]/n^2 - 2E_\theta[\sum_{i=1}^{3} \frac{\partial^2\varphi/\partial x_i^2}{\varphi(\overline{X})}]/n^2 \ .$$

Exercise 6: Recall that $\|x\| = (x'x)^{1/2}$. Let

$$\varphi(x) = 1/\|x\| \ \text{ if } \|x\| \ge 1$$

$$= e^{(1 - \|x\|)/2} \ \text{ if } \|x\| \le 1 \ .$$

Verify the following derivatives including continuity as appropriate.

a) For $\|x\| \ge 1$, $\partial\varphi/\partial x = -x/|x\|^3$

$\partial^2\varphi/\partial x_i^2 = -(x'x - 3x_i^2)(x'x)^{-5/2}$.

b) For $\|x\| \le 1$, $\partial\varphi/\partial x = -\varphi(x)\cdot x$

$\partial^2\varphi/\partial x_i^2 = -\varphi(x)(1 - x_i^2)$.

To evaluate $E_\theta[\sum_{i=1}^{3} \dfrac{\partial^2\varphi(\overline{X})/\partial x_i^2}{\varphi(\overline{X})}\,]/n^2$, (*)

we split the integration into two pieces:

for $\|x\| \ge 1$, the sum is $-(3\bar{x}'\bar{x} - 3\bar{x}'\bar{x}\,)(x'x)^{-5/2}(x'x)^{1/2} = 0$;

for $\|x\| \le 1$, the sum is $-(3 - \bar{x}'\bar{x}) < 0$.

Therefore, (*) is non–positive.

With these choices of φ and g ,

$$T = \overline{X} + g(\overline{X})/n = \overline{X} + \nabla\log\varphi(\overline{X})/n$$

$$= (1 - 1/n\|\overline{X}\|^2)\overline{X} \qquad \text{if } \|\overline{X}\| \geq 1$$

$$= (1 - 1/n)\overline{X} \qquad \text{if } \|\overline{X}\| < 1 .$$

And, $R(\theta,\overline{X}) - R(\theta,T) = E_\theta[\ \|g(\overline{X})\|^2\ ]/n^2 - (*) \geq 0$. Thus means that $\overline{X}$ has greater risk than T and is inadmissible.

INE–Exercise: Continuing with this normal model, let $\varphi(x) = \|x\|$, $g(x) = \nabla \log \varphi(x)$, $S = (1 - 1/n\overline{X}'\overline{X})\overline{X}$. Show that $R(\theta,\overline{X}) - R(\theta,S) = E_\theta[\ \|g(\overline{X})\|^2\ ]/n^2 > 0$.

LESSON *3. EQUIVARIANT ESTIMATORS

In a previous lesson, we obtained some uniformly minimum risk estimators by restricting consideration to a special class of estimators, namely that of unbiased estimators. In the same spirit, we now consider another class of estimators which will be called *equivariant estimators*. The invariance of estimators can be treated in a general framework of groups of transformations but here we will consider only a special case of the so-called location parameters for continuous type real X. We continue to use quadratic loss and, of course, random samples.

Consider the problem of estimating the mean θ of a population X which is distributed as $N(\theta,1)$ based upon a random sample $X_1, \cdots, X_n$. The density of X is

$$f(x;\theta) = (2\pi)^{-1/2}\exp\{-\tfrac{1}{2}(x-\theta)^2\} = \phi(x-\theta)$$

where $\phi(x) = (2\pi)^{-1/2}\exp\{-\frac{1}{2}x^2\}$. This leads to

Definition: *If for X real, the density $f(x;\theta) = g(x - \theta)$ for some function g, then θ is referred to as a location parameter.*

Exercise 1: Show that θ is a location parameter in the model

$$f(x;\theta) = \tfrac{1}{2}\, I(\theta - 1 < x < \theta + 1)\ ,\ \theta \in R .$$

Now, in our normal example, let $\overline{X}$ be the sample mean. We see that, for any $c \in R$, $T(X_1,\cdots,X_n) = \overline{X}$ is such that

$$T(X_1 + c,\cdots,X_n + c) = \overline{X} + c = T(X_1,\cdots,X_n) + c .$$

This leads to

Definition: *An estimator $T(X_1,\cdots,X_n)$ such that*

$$T(X_1 + c,\cdots,X_n + c) = T(X_1,\cdots,X_n) + c$$

for all real c is called equivariant.

Exercise 2: Show that T is equivariant if and only if

$$T(X_1, \cdots, X_n) = X_1 + S(X_2-X_1, \cdots, X_n-X_1) \text{ for some } S.$$

Hint: for "only if" consider

$$T(x_1-x_1+x_1, x_2-x_1+x_1, \cdots, x_n-x_1+x_1).$$

The following is an important property of equivariant estimators.

Theorem: *If T is an equivariant estimator of a location parameter θ, then its risk is constant (i.e. independent of θ).*

Proof:
$$\begin{aligned} R(\theta,T) &= E_\theta[|T-\theta|^2] \\ &= \int_{R^n} |T(x_1, \cdots, x_n) - \theta|^2 \prod_{i=1}^n g(x_i - \theta)dx_i \\ &= \int_{R^n} |T(x_1-\theta, \cdots, x_n-\theta)|^2 \prod_{i=1}^n g(x_i - \theta)dx_i \\ &= \int_{R^n} |T(x_1, \cdots, x_n)|^2 \prod_{i=1}^n g(x_i)dx_i \end{aligned}$$

which is free of θ.

When $f(x;\theta) = g(x-\theta)$, we have $f(x;0) = g(x)$, and

$R(\theta,T) = E_0[T(X_1,\cdots,X_n)]^2 = R(0,T)$ for all $\theta \in \Theta$. As a consequence, within the class of equivariant estimators T, the "best" one will be T^* such that $R(0,T^*) = \min_T R(0,T)$. Such a T^* is called *minimum risk equivariant estimator.*

Exercise 3: Let T be an equivariant estimator of a location parameter θ. Show that the bias and the variance of T are independent of θ.

We now proceed to find the minimum risk equivariant estimator. For simplicity, we are considering the case $\theta \in R$ but the multi-dimensional case can be treated in the same fashion.

Definition: *If* $\psi(y) = \psi(y,(x_1,\cdots,x_n))$

$$= \int_R (y-\theta)^2 \prod_{i=1}^{n} g(x_i-\theta)d\theta$$

is well defined for $y \in R$ *, and* $\hat{\theta}(x_1,\cdots,x_n)$ *is such that* $\psi(\hat{\theta}) = \min_{y \in R} \psi(y)$ *, then* $\hat{\theta}$ *is called the Pitman estimator.*

Theorem: *The Pitman estimator* $\hat{\theta}$ *is a minimum risk equivariant estimator.*

Proof: First we will actually compute $\hat{\theta}$; then we will show that $\hat{\theta}$ is an equivariant estimator; finally we will demonstrate that $\hat{\theta}$ is in fact a minimum risk equivariant estimator.

Solving $\dfrac{\partial\psi(y)}{\partial y} = \int_R 2(y-\theta)\prod_{i=1}^{n} g(x_i-\theta)\,d\theta = 0$ for y ,

we get $y = \hat{\theta}(x_1,\cdots,x_n) = [\int_R \theta \prod_{i=1}^{n} g(x_i-\theta)d\theta]/G(x_1,\cdots,x_n)$

where $G(x_1,\cdots,x_n) = \int_R \prod_{i=1}^{n} g(x_i-\theta)d\theta$.

Since $\partial^2\psi/\partial y^2 = \int_R \prod_{i=1}^{n} g(x_i-\theta)\,d\theta > 0$, this is a unique maximal point.

Using the substitution $\theta = \tilde{\theta} + x_1$ in the integral for $\psi(y)$ yields $\psi(y) = \int_R (y-x_1-\tilde{\theta})^2 \prod_{i=1}^{n} g(x_i-x_1-\tilde{\theta})d\tilde{\theta}$. Then $\partial\psi/\partial y = 0$ yields

$$y - x_1 = [\int_R \tilde{\theta} \prod_{i=1}^{n} g(x_i-x_1-\tilde{\theta})d\tilde{\theta}]/G(0, x_2-x_1, \cdots, x_n-x_1) .$$

The ratio of the integrals for y and $y - x_1$ is a function of the differences $\{x_i - x_1\}$. Therefore,

$$\hat{\theta} = y = x_1 + S(x_2 - x_1, \cdots, x_n - x_1)$$

and equivariance follows by exercise 2.

Let T be an equivariant estimator of θ. By the first theorem, T has a constant risk

$$R(0,T) = E_0[T^2] = E_0[E_0[T^2 \mid X_2 - X_1, \cdots, X_n - X_1]] .$$

We need a little

Lemma: $E_0[T^2(X_1, \cdots, X_n) | X_2 - X_1, \cdots, X_n - X_1]$

$$= \int_R T^2(x_1 - u, \cdots, x_n - u)[\prod_{i=1}^{n} g(x_i - u)][\int_R \prod_{j=1}^{n} g(x_j - v)dv]^{-1}\, du .$$

Partial proof: Consider n = 2. First note that

$$Y = \frac{\int_R T^2(x_1-u, x_2-u)[g(x_1-u)g(x_2-u)]du}{\int_R g(x_1-v)g(x_2-v)\, dv}$$

$$= \frac{\int_R T^2(w, x_2-x_1+w)g(w)g(x_2-x_1+w)\, dw}{\int_R g(w)g(x_2-x_1+w)\, dw}$$

is a function of $x_2 - x_1$ alone. Let $V = v(X_2 - X_1)$ be any function of $X_2 - X_1$ alone (with appropriate moments). Then with some obvious substitutions, $E_0[VY]$ is

$$\iint v\left(\frac{\int T^2(x_1-u,\ x_2-u)g(x_1-u)g(x_2-u)du}{\int g(x_1-z)g(x_2-z)dz}\right) g(x_1)g(x_2)\, dx_1\, dx_2$$

$$= \iiint (v(x_2-x_1)\left[\frac{T^2(x_1-u,x_2-u)g(x_1-u)g(x_2-u)}{\int g(x_1-z)g(x_2-z)\ dz}\right]\times$$
$$g(x_1)g(x_2)dx_1dx_2du$$

$$= \iint v(y_2-y_1)T^2(y_1,y_2)\ g(y_1)\ g(y_2)$$
$$\times\left[\int g(y_1+u)g(y_2+u) + \int g(y_1+u-z)g(y_2+u-z)dz\right] du\ dy_1\ dy_2$$

$$= \iint v(y_2-y_1)T^2(y_1,y_2)\ g(y_1)\ g(y_2)$$
$$\times\left[\int g(y_1+u)g(y_2+u)du + \int g(y_1+z)g(y_2+z)dz\right] dy_1\ dy_2$$

$$= \iint v(y_2-y_1)T^2(y_1,y_2)\ g(y_1)\ g(y_2)\ dy_1\ dy_2$$
$$= E_0[VT^2]\ .$$

By properties of conditional expectation, the conclusion follows.

Continuing the proof of the theorem, we have

$$R(0,T) = E_0[E_0[T^2 \mid X_2-X_1, \cdots, X_n-X_1]]$$
$$= E_0\left[\int T^2(X_1-u, \cdots, X_n-u) \prod_{i=1}^n g(X_i-u)\ du/G(X_1, \cdots, X_n)\right].$$

But since T is equivariant, T is of the form

$$T(x_1-u,\cdots,x_n-u) = x_1 - u + S(x_2-x_1,\cdots,x_n-x_1)\ ,$$

and hence $R(0,T)$

$$= E_0\left[\int \{X_1 + S(X_2-X_1, \cdots, X_n-X_1) - u\}^2\right.$$
$$\left.\times \prod_{i=1}^n g(X_i-u)\ du/G(X_1, \cdots, X_n)\right]$$

$$\geq E_0\left[\min_y \int (y-u)^2 \prod_{i=1}^{n} g(X_i - u)\, du/G(X_1, \cdots, X_n)\right]$$

$$= E_0\left[\int (\hat{\theta} - u)^2 \prod_{i=1}^{n} g(X_i - u)\, du/G(X_1, \cdots, X_n)\right]$$

by the definition of $\hat{\theta}$. From

$$E_0[\, E_0[\hat{\theta}^2 \,|\, X_2 - X_1, \cdots, X_n - X_1]\,] = E_0(\hat{\theta}^2) = R(0,\hat{\theta})$$

it follows that $R(0,T) \geq R(0,\hat{\theta})$ and so $\hat{\theta}$ is equivariant .

Since $\hat{\theta}$ is equivariant, we have

$$E_0[\hat{\theta}^2 \,|\, X_2 - X_1, \cdots, X_n - X_1]$$

$$= \int \hat{\theta}^2(x_1 - u, \cdots, x_n - u) \prod_{i=1}^{n} g(x_i - u)\, du/G(x_1, \cdots, x_n)$$

$$= \int [\hat{\theta}(x_1, \cdots, x_n) - u]^2 \prod_{i=1}^{n} g(x_i - u)\, du/G(x_1, \cdots, x_n)\,.$$

Example: Let X be $N(\theta, 1)$. The Pitman estimator of θ is $\overline{X}$.

Exercise 4: Let X be uniform on $(\theta - 1/2 , \theta + 1/2)$. Show that the Pitman estimator of θ is $[X_{(1)} + X_{(n)}]/2$.

LESSON *4. SIMPLE LINEAR REGRESSION–I

As noted in the overview, "fitting curves to data" begins with straight lines. In the spirit of Gauss, we consider the "best line" $\beta_0 + \beta_1 x$ for a set of observations $(y_1, x_1), \cdots, (y_n, x_n)$ to be that which minimizes $G = \sum_{i=1}^{n} (y_i - \beta_0 - \beta_1 x_i)^2$. In the sketch,

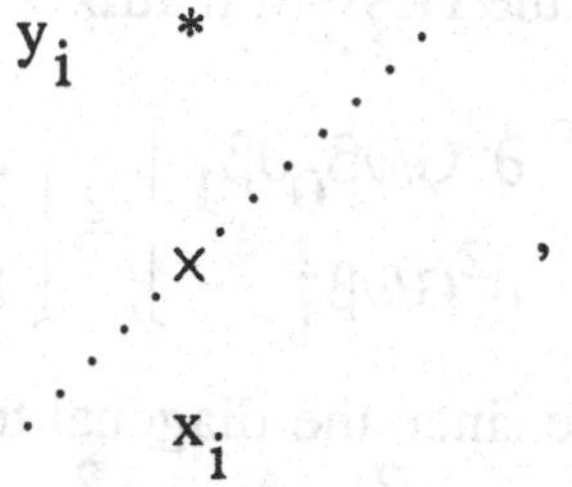

we can see that we are going to minimize the sum of the squares of the vertical distances from observed points * to corresponding points × on the line. In an algebraic view of the geometry, for the vectors

$$y = (y_1, \cdots, y_n)' \text{ and } \tilde{y} = (\beta_0+\beta_1 x_1, \cdots, \beta_0+\beta_1 x_n)' ,$$

$\sqrt{G}$ is just the Euclidean distance (norm) $\|y - \tilde{y}\|$. (In the rest of this lesson, we maintain the index $i = 1(1)n$ but do not always write it out.)

Formally, we have

$$\frac{\partial G}{\partial \beta_0} = \Sigma 2(y - \beta_0 - \beta_1 x)(-1) , \quad \frac{\partial G}{\partial \beta_1} = \Sigma 2(y - \beta_0 - \beta_1 x)(-x) .$$

The stationary point, $(\hat{\beta}_0, \hat{\beta}_1)$, is the solution of the so-called

normal equations $\quad \begin{matrix} \partial G/\partial \beta_0 = 0 \\ \partial G/\partial \beta_1 = 0 \end{matrix}$;

here these are equivalent to: $\quad \begin{matrix} \Sigma y = \hat{\beta}_0 n + \hat{\beta}_1 \Sigma x \\ \Sigma yx = \hat{\beta}_0 \Sigma x + \hat{\beta}_1 \Sigma x^2 \end{matrix}$.

Exercise 1: Show that the solution of these normal equations is:

$$\hat{\beta}_1 = \frac{n\cdot\Sigma xy - \Sigma x\cdot\Sigma y}{n\cdot\Sigma x^2 - (\Sigma x)^2},\ \hat{\beta}_0 = \Sigma y/n - \hat{\beta}_1\Sigma x/n .$$

Note immediately, that $\hat{\beta}_1$ is undefined unless there are at least two distinct values among the x's ; henceforth, this is assumed. Technically, to complete the minimization, we should check the positivity of the Hessian matrix:

$$\begin{bmatrix} \partial^2 G/\partial\beta_0^2 & \partial^2 G/\partial\beta_0\partial\beta_1 \\ \partial^2 G/\partial\beta_1\partial\beta_0 & \partial^2 G/\partial\beta_1^2 \end{bmatrix} = \begin{bmatrix} \Sigma 2 & \Sigma 2x \\ \Sigma 2x & \Sigma 2x^2 \end{bmatrix}.$$

This is positive definite since the diagonal terms are positive and

$$\Sigma 2\cdot\Sigma 2x^2 - (\Sigma 2x)^2 = 4(n\cdot\Sigma x^2 - (\Sigma x)^2) > 0$$

whenever there are at least two distinct values among x's . The *O(rdinary) L(east) S(quares) R(egression) L(ine) of y on x* is:

$$\hat{y} = \hat{\beta}_0 + \hat{\beta}_1 x$$

with *estimated = predicted* points $\hat{y}_i = \hat{\beta}_0 + \hat{\beta}_1 x_i$ and *error = residual* values $y_i - \hat{y}_i$. Also,

$$\min_{\beta_0,\beta_1} \Sigma(y-\beta_0-\beta_1 x)^2 = \Sigma(y_i - \hat{y}_i)^2 = \Sigma(y_i-\hat{\beta}_0-\hat{\beta}_1 x_i)^2 .$$

Exercise 2: For $y_1, \cdots, y_n$, let $\bar{y} = \Sigma y_i/n$. Show that

$$\min_{\gamma} \Sigma(y_i - \gamma)^2 = \Sigma(y_i - \bar{y})^2 .$$

The result of exercise 2 shows that we have used the minimization principle indirectly when dealing with the sample variance. We can relate these two minimal sums of squares as follows:

$$\Sigma(y - \bar{y})^2 = \Sigma(y-\hat{y} + \hat{y}-\bar{y})^2 = \Sigma(y-\hat{y})^2 + \Sigma(\hat{y}-\bar{y})^2 \qquad (*)$$

because

$$\Sigma(y - \hat{y})(\hat{y} - \bar{y}) = \Sigma(y - \hat{\beta}_0 - \hat{\beta}_1 x)(\hat{y} - \bar{y}) = 0(\hat{y} - \bar{y}) = 0$$

is just another form of one of the normal equations. Of course, (*) is just a Pythagorean form in n dimensions. The left-hand side of (*) is the "total sum of squares"; the right-hand side is "sum of squares for error" plus "sum of squares for regression". Symbolically, we write

$$SST = SSE + SSR .$$

Then the *coefficient of determination*

$$R^2 = SSR/SST = 1 - SSE/SST$$

is the proportion of variation in y "explained by the regression (line)" and SSE/SST is the proportion "unexplained". If R^2 is "close" to 1 , the unexplained variation or SSE must be close to 0 and the line is a "good fit". (The number of significant digits to be retained in any calculations depends upon the subject matter and the statistical technique under use. "At the end", one often retains one more digit than determined in the original data; we shall not be precise herein.)

Example: Here are some calculations for cricket data (patterned after Pierce, 1949). "x" is cricket chirps per five seconds and "y" is temperature in degrees F .

	x	y	x^2	y^2	xy
	20	88.6	400	7849.96	1772
	16	71.6	256	5126.56	1145.6
	20	93.3	400	8704.89	1866
	18	84.3	324	7106.49	1517.4
	17	80.6	289	6496.36	1370.2
	16	75.2	256	5655.04	1203.2
	15	69.7	225	4858.09	1045.5
	17	82.0	289	6724	1344
	15	69.4	225	4816.36	1041
	16	83.3	256	6938.89	1332.8
	15	79.6	225	6336.16	1144
	17	82.6	289	6822.76	1404.2
SUMS	202	960.2	3434	77435.56	16285.9

From exercise 1, we get

$$\hat{\beta}_1 = \frac{12(16285.9) - 202\times 960.2}{12(3434) - 202\times 202} = \frac{1470.4}{404} \approx 3.6396 ,$$

$$\hat{\beta}_0 = 960.2/12 - \hat{\beta}_1(202/12) \approx 18.75 \,.\ \ \bar{y} \approx 80.0167 \,.$$

It follows that $\hat{y} \approx 18.75 + 3.6396 \cdot x$.
If $x = 20$, the estimated = predicted value of y is
$18.75 + 3.6396(20) = 91.5421$; etc.

We calculate the sums of squares in another table:

x_i	y_i	$\hat{y}_i$	$(y_i - \bar{y})^2$	$(y_i - \hat{y}_i)^2$	$(\hat{y}_i - \bar{y})^2$
20	88.6	91.5421	8.5833^2	$(-2.9421)^2$	11.5254^2
16	71.6	76.9837	$(-8.4167)^2$	$(-5.3837)^2$	$(-3.0330)^2$
20	93.3	91.5421	13.2833^2	1.7579^2	11.5254^2
18	84.3	84.2629	4.2833^2	$.0371^2$	4.2462^2
17	80.6	80.6233	$.5833^2$	$(-.0233)^2$	$.6066^2$
16	75.2	76.9837	$(-4.8167)^2$	$(-1.7837)^2$	$(-3.0330)^2$
15	69.7	73.3441	$(-10.3167)^2$	$(-3.6441)^2$	$(-6.6726)^2$
17	82.0	80.6233	1.9833^2	1.3767^2	$.6066^2$
15	69.4	73.3441	$(-10.6167)^2$	$(-3.9441)^2$	$(-6.6726)^2$
16	83.3	76.9837	3.2833^2	6.3163^2	$(-3.0330)^2$
15	79.6	73.3441	$(-.4167)^2$	6.2559^2	$(-6.6726)^2$
17	82.6	80.6233	2.5833^2	1.97767^2	$.6066^2$
		SUMS	603.5567	= 157.5839	+ 445.9728

Exercise 3: (One of the authors bakes a certain kind of bread.) Suppose that the following data are collected for the baking of a standard mix . "x" is time of baking and "y" is the loaf weight.

x	16	16	17	17	18	18	19	19	20	20 .
y	12.0	12.2	11.9	11.8	12.2	12.2	11.7	11.8	11.9	12.0

Find the OLSRL of y on x and its $SSE = \Sigma(y_i - \hat{y}_i)^2$.

To go beyond this strictly descriptive format, to introduce inferential techniques, we need to make some assumptions about the distributions of the variables X and Y. The first model is to have $x_1, \cdots, x_n$ given (as in the baking times above) with the errors as the only random components. Then,

$$Y_i - \beta_o - \beta_1 x_i = \varepsilon_i$$

are to be uncorrelated with common means 0 and common finite variance $\sigma^2 > 0$. While this is a rather weak asuumption (not specifying even a limit distribution), it is strong enough to prove a remarkable theorem involving minimum variance which will be presented after the

Definition: *Let $\{Y_1, \cdots, Y_n\}$ have a joint distribtion F; let $\theta \in R$ be a parameter of F. Suppose that the set of linear unbiased estimators,*

$$\mathscr{C} = \{\gamma = (c_1, \cdots, c_n) : c_i \in R,\ E[\sum_{i=1}^{n} c_i \cdot Y_i] = \theta\},$$

is not empty. Then the B(est) L(inear) U(nbiased) E(stimator) of θ (if it exists) is given by

$$\sum_{i=1}^{n} c_i^* \cdot Y_i \text{ for some } \gamma^* \text{ in } \mathscr{C}$$

when $Var(\sum_{i=1}^{n} c_i^* Y_i) \le Var(\sum_{i=1}^{n} c_i Y_i)$ *for all* $\gamma \in \mathscr{C}$.

Theorem: *(Gauss-Markov) Let $x_1, \cdots, x_n$ be given; suppose that β_0 and β_1 are non-random (real parameters). For $i = 1(1)n$, let*

$$Y_i - \beta_0 - \beta_1 x_i = \varepsilon_i$$

with $E[\varepsilon_i] = 0$ *and* $E[\varepsilon_i \varepsilon_j] = \begin{cases} 0 & \text{for } i \neq j \\ \sigma^2 & \text{for } i = j \end{cases}$. *Then the BLUEs of β_0, β_1 are the OLS estimators $\hat{\beta}_0, \hat{\beta}_1$.*

Partial proof: It is convenient to have some other notation:

$$\bar{x} = \Sigma x/n,\ S_{xx} = n\Sigma x^2 - (\Sigma x)^2 = n\Sigma(x_i - \bar{x})^2 = n\Sigma(x_i - \bar{x})x_i$$

$$S_{yy} = n\Sigma y^2 - (\Sigma y)^2,\ S_{xy} = n\Sigma xy - (\Sigma x)\Sigma y.$$

For the RV form, y is replaced by Y .

$$\text{Then, } \hat{\beta}_0 = \bar{Y} - \hat{\beta}_1\bar{x} = \frac{\Sigma Y_i}{n} - \frac{n\Sigma x_i Y_i - \Sigma x_i \Sigma Y_i}{S_{xx}}\,\bar{x}$$

$$= \sum_{i=1}^{n}\left[\frac{1}{n} - \frac{nx_i - n\bar{x}}{S_{xx}}\,\bar{x}\right]Y_i .$$

$$\text{Hence, } E[\hat{\beta}_0] = \sum_{i=1}^{n}\left[\frac{1}{n} - \frac{n\bar{x}(x_i - \bar{x})}{S_{xx}}\right]\cdot(\beta_0 + \beta_1 x_i)$$

$$= \sum_{i=1}^{n}\left[\frac{1}{n} - \frac{n\bar{x}(x_i - \bar{x})}{S_{xx}}\right]\beta_0$$

$$+ \beta_1\sum_{i=1}^{n}\left[\frac{x_i}{n} - \frac{n\bar{x}(x_i - \bar{x})x_i}{S_{xx}}\right]$$

$$= (1 - 0)\beta_0 + \beta_1(\bar{x} - \bar{x}) = \beta_0 .$$

Thus $\hat{\beta}_0$ is a linear unbiased estimator with

$$c_i = 1/n - n(x_i - \bar{x})\bar{x}/S_{xx} = \frac{S_{xx} - n^2 x_i\bar{x} - n^2\bar{x}^2}{nS_{xx}}$$

$$= \frac{\Sigma x^2 - (\Sigma x)x_i}{S_{xx}} .$$

Since the $\{\beta_0 + \beta_1 x_i\}$ are non–random, the covariances of $\{Y_i = \beta_0 + \beta_1 x_i + \varepsilon_i\}$ are the same as those of $\{\varepsilon_i\}$. Hence for any $\{c_i\}$, $\text{Var}(\Sigma c_i Y_i) = \Sigma c_i^2 \cdot \sigma^2$; we want to minimize this sum subject to $E[\Sigma c_i Y_i] = \Sigma c_i(\beta_0 + \beta_1 x_i) = \beta_0$.

This is equivalent to

$$\text{minimize } \Sigma c_i^2 \text{ such that } \Sigma c_i = 1 \text{ and } \Sigma c_i x_i = 0$$

and is accomplished most conveniently by using the method of Lagrange multipliers (Fleming, 1977): minimize

$$h(c_1,\cdots,c_n,\lambda,\mu) = \Sigma c_i^2 + \lambda(\Sigma c_i - 1) + \mu(\Sigma c_i x_i - 0)$$

as a function of $c_1, \cdots, c_n, \lambda, \mu$.

$$\partial h/\partial c_i = 2c_i + \lambda + \mu x_i$$

$$\partial h/\partial \lambda = \Sigma c_i - 1$$

$$\partial h/\partial \mu = \Sigma c_i x_i \ .$$

The stationary points are the zeroes of these derivatives. From the first we get

$$\hat{c}_i = -(\hat{\lambda} + \hat{\mu} x_i)/2 \ .$$

Substituting this in the second two forms yields

$$-n\hat{\lambda}/2 - \hat{\mu}\Sigma x_i/2 - 1 = 0$$

$$-\hat{\lambda}\Sigma x_i/2 - \hat{\mu}\Sigma x_i^2/2 = 0 \ .$$

From these equations, we get

$$\hat{\lambda} = -2\Sigma x_i^2/S_{xx} \qquad \hat{\mu} = 2(\Sigma x)/S_{xx}$$

whence $\hat{c}_i = \dfrac{n\bar{x}\cdot\Sigma x^2}{S_{xx}\ \Sigma x} - \dfrac{2n\bar{x}}{S_{xx}} x_i = \dfrac{\Sigma x^2 - n\bar{x}\cdot x_i}{S_{xx}}$

which is the same c_i as above.

$$\mathrm{Var}(\hat{\beta}_0) = \sum_{i=1}^{n} c_i^2\cdot\sigma^2 = \sum_{i=1}^{n}\left[\frac{\Sigma x^2 - n\bar{x}x_i}{S_{xx}}\right]^2 \sigma^2$$

$$= \frac{\sigma^2}{S_{xx}}\sum_{i=1}^{n}\left[(\Sigma x^2)^2 - 2n\,\bar{x}\,\Sigma x^2\, x_i + n^2\,\bar{x}^2\, x_i^2\right]$$

$$= \frac{\sigma^2}{S_{xx}} \left[n(\Sigma x^2)^2 - 2n^2 \bar{x}^2 \Sigma x^2 + n^2 \bar{x}^2 \Sigma x^2 \right]$$

$$= \frac{\sigma^2}{S_{xx}} (\Sigma x^2)(n\Sigma x^2 - n^2 \bar{x}^2) = \sigma^2 \Sigma x^2 / S_{xx} .$$

Exercise 4: Following the path for $\hat{\beta}_0$, complete the proof of the theorem by showing that:

a) $\hat{\beta}_1$ is a linear estimator ;

b) $E[\hat{\beta}_1] = \beta_1$;

c) $Var(\hat{\beta}_1) = \sigma^2 n / S_{xx}$ is "minimal".

Perhaps not surprizingly, we can get an unbiased estimator for σ^2 and hence also for the variances of $\hat{\beta}_0$ and $\hat{\beta}_1$. Recall that the second moment (of a real RV) is the variance plus the square of the first moment.

Since $E[Y_i] = \beta_0 + \beta_1 x_i$, $Var(Y_i) = \sigma^2$,

$$E[\bar{Y}] = \beta_0 + \beta_1 \bar{x} , \ Var(\bar{Y}) = \sigma^2/n ,$$

the expectation of $\Sigma(Y_i - \bar{Y})^2 = \Sigma Y_i^2 - n\bar{Y}^2$ is

$$\Sigma(\sigma^2 + (\beta_0 + \beta_1 x_i)^2) - n(\sigma^2/n + (\beta_0 + \beta_1 \bar{x})^2)$$

$$= (n-1)\sigma^2 + \Sigma(\beta_0 + \beta_1 x_i)^2 - n(\beta_0 + \beta_1 \bar{x})^2$$

$$= (n-1)\sigma^2 + \Sigma(\beta_1(x_i - \bar{x}))^2$$

$$= (n-1)\sigma^2 + \beta_1^2 \cdot \Sigma(x_i - \bar{x})^2 .$$

Since $\Sigma(Y_i - \bar{Y})^2 = \Sigma(\hat{\beta}_0 + \hat{\beta}_1 x_i - \bar{Y})^2$

$$= \Sigma(\bar{Y} - \hat{\beta}_1 \bar{x} + \hat{\beta}_1 x_i - \bar{Y})^2 = \hat{\beta}_1^2 \cdot \Sigma(x_i - \bar{x})^2 ,$$

its expectation $E[\hat{\beta}_1{}^2]\cdot\Sigma(x_i - \bar{x})^2$

$$= (\sigma^2 n/S_{xx} + \beta_1{}^2)S_{xx}/n = \sigma^2 + \beta_1{}^2 S_{xx}/n .$$

By use of (*), we get

$$E[\Sigma(Y_i - \hat{Y}_i)^2] = E[\Sigma(Y_i - \bar{Y})^2] - E[\Sigma(\hat{Y}_i - \bar{Y})^2]$$

$$= (n-1)\sigma^2 + \beta_1{}^2\Sigma(x_i - \bar{x})^2 - \sigma^2 - \beta_1{}^2 S_{xx}/n = (n-2)\sigma^2 .$$

Exercise 5: Show that the unbiased estimate of σ^2,

$$s_e^2 = \sum_{i=1}^{n}(y_i - \hat{y}_i)^2/(n-2) ,$$

can be written as $(S_{xx}S_{yy} - S_{xy}{}^2)/n(n-2)S_{xx}$.

It follows that $s_e^2\ \Sigma x^2/S_{xx}$ and $s_e^2\ n/S_{xx}$ are unbiased estimators for the variances $\sigma^2_{\hat{\beta}_0}$ and $\sigma^2_{\hat{\beta}_1}$ respectively. The square roots of those estimators are denoted by $s_{\hat{\beta}_0}$ and $s_{\hat{\beta}_1}$ and are called *standard errors* .

LESSON *5. SIMPLE LINEAR REGRESSION–II

The vector/matrix version of this material has the property that when extending from simple (one "x") to multiple (several "x") regression, the algebraic forms of estimators, sums of squares, etc., are preserved; we will not make this extension but we will complete the inference for S(imple) L(inear) R(egression) by switching to vector–matrix notation.

Our model can be written as $Y = X\cdot\beta + \varepsilon$ if:

Y is the n by 1 vector $(Y_1, \cdots, Y_n)'$;

X is the n by 2 matrix $\begin{bmatrix} 1 & 1 & \cdots & 1 \\ x_1 & x_2 & \cdots & x_n \end{bmatrix}'$;

β is the 2 by 1 vector $(\beta_0, \beta_1)'$;

ε is the n by 1 vector $(\varepsilon_1, \cdots, \varepsilon_n)'$.

As in Lesson 9, Part VI, we take J to be the n by 1 vector all of whose elements are 1 ; then,

$$\sum_{i=1}^{n} Y_i = J'Y .$$

The vector all of whose elements are the sample mean is

$$J\cdot(\sum_{i=1}^{n} Y_i/n) = JJ'Y/n = K\cdot Y$$

where K is the n by n matrix all of whose elements are 1/n . Also, I_n denotes the n by n identity matrix.

Exercise 1: Show that in the matrix notation just introduced, the normal equations (from the previous lesson) can be written as $X'Y = X'X\cdot\beta$.

From exercise 1, it follows that

$$\hat{\beta} = (X'X)^{-1}X'Y = \frac{\begin{bmatrix} \Sigma x^2 & -\Sigma x \\ -\Sigma x & n \end{bmatrix}}{S_{xx}} \begin{bmatrix} 1 & 1 & \cdots & 1 \\ x_1 & x_2 & \cdots & x_n \end{bmatrix} \begin{bmatrix} Y_1 \\ \vdots \\ Y_n \end{bmatrix}$$

which shows the linearity of the estimators. Expanding this gives

$$\hat{\beta} = \frac{\begin{bmatrix} \Sigma x^2 & -\Sigma x \\ -\Sigma x & n \end{bmatrix}}{S_{xx}} \begin{bmatrix} \Sigma Y_i \\ \Sigma x_i Y_i \end{bmatrix} = \begin{bmatrix} \Sigma x^2 \cdot \Sigma Y - \Sigma x \cdot \Sigma x Y \\ -\Sigma x \cdot \Sigma Y + n \cdot \Sigma x Y \end{bmatrix} / S_{xx}$$

so that as before,

$$\hat{\beta}_0 = (\Sigma x^2 \cdot \Sigma Y - \Sigma x \cdot \Sigma x Y)/S_{xx} \quad \text{and} \quad \hat{\beta}_1 = S_{xY}/S_{xx} .$$

$$\begin{aligned} \text{Now, } E[\hat{\beta}] &= E[(X'X)^{-1}X'Y] \\ &= (X'X)^{-1}X' \cdot E[Y] = (X'X)^{-1}X'X \cdot \beta = \beta \end{aligned}$$

which confirms the unbiasedness of $\hat{\beta}$. The "predicted values" are also "unbiased": $\hat{Y} = X \cdot \hat{\beta} = X(X'X)^{-1}X'Y$ has

$$E[\hat{Y}] = X(X'X)^{-1}X'X \cdot \beta = X \cdot \beta .$$

The covariances are:

$$Cov(Y) = Cov(\varepsilon) = E[\varepsilon\varepsilon'] = \left[E[\varepsilon_i \varepsilon_j]\right] = \sigma^2 \cdot I_n ;$$

$$\begin{aligned} Cov(\hat{\beta}) &= Cov((X'X)^{-1}X'Y) = (X'X)^{-1}X' \cdot Cov(Y) \cdot X(X'X)^{-1} \\ &= (X'X)^{-1}X' \cdot \sigma^2 \cdot I_n \cdot X(X'X)^{-1} = \sigma^2 \cdot (X'X)^{-1} ; \end{aligned}$$

$$Cov(\hat{Y}) = Cov(X \cdot \hat{\beta}) = X \cdot \sigma^2 \cdot (X'X)^{-1} \cdot X' = \sigma^2 \cdot X(X'X)^{-1}X' .$$

$$\begin{aligned} \text{Now } Y - \hat{Y} &= Y - X \cdot \hat{\beta} = Y - X(X'X)^{-1}X'Y \\ &= (I_n - X(X'X)^{-1}X')Y . \end{aligned}$$

Also,the random vector for the observations

$$(y_1 - \bar{y}, \cdots, y_n - \bar{y})'$$

is

$$Y - K \cdot Y = (I_n - K)Y .$$

With this notation, the Pythagorean relation

$$SST = SSE + SSR$$

becomes $(Y - K \cdot Y)'(Y - K \cdot Y)$

$$= (Y - \hat{Y})'(Y - \hat{Y}) + (\hat{Y} - K \cdot Y)'(\hat{Y} - K \cdot Y) \qquad (*)$$

equivalently,

$$Y'(I_n - K)Y = Y'(I_n - X(X'X)^{-1}X')Y + Y'(X(X'X)^{-1}X' - K)Y .$$

It follows that $\hat{\sigma}^2 = Y'(I_n - X(X'X)^{-1}X')Y/(n-2)$.

Exercise 2: Substitute $\hat{Y}$ in (*) and perform the algebra to derive its equivalent form.

We are now ready to add the traditional assumption that the errors $\{\varepsilon_i\}$ are normally distributed. This will enable us to answer some questions about tests of hypotheses and confidence intervals for the parameters β and σ^2 . For this, we use the results of Lessons 7 and 9, Part VI, which are stated in terms of ε/σ^2 , an n–dimensional normal vector with mean 0 and covariance I_n . It will be convenient to write

$$Z = Y - X \cdot \beta \text{ and } C = I_n - X(X'X)^{-1}X' .$$

Since $\hat{\beta} - \beta = (X'X)^{-1}X'(Y - X \cdot \beta) = (X'X)^{-1}X'Z$ is a linear function of the normal Z , it is also normally distributed with mean 0 and covariance $\sigma^2(X'X)^{-1}$. In particular,

$$\hat{\beta}_0 - \beta_0 \text{ is } N(0, \sigma^2 \Sigma x^2/S_{xx})$$

and

$$\hat{\beta}_1 - \beta_1 \text{ is } N(0, \sigma^2 n/S_{xx}) .$$

Since $X'C = X'(I_n - X(X'X)^{-1}X') = 0$,

$$Z'CZ = (Y - X\cdot\beta)'(I_n - X(X'X)^{-1}X')(Y - X\cdot\beta)$$

$$= Y'CY = \hat{\sigma}^2(n-2) .$$

Since $C = C' = C^2$, $Z'CZ/\sigma^2 = (\varepsilon/\sigma)'C(\varepsilon/\sigma)$ is distributed as chisquare with n–2 dof; that is,

$$(n-2)\hat{\sigma}^2/\sigma^2 \text{ is } \chi^2(n-2) .$$

We can now find a CI for σ^2 (or σ) just as was done in Lesson 12, Part VI.

Exercise 3: Verify the properties of the matrix C above.

We are continuing the discussion of the distribution. Since $(X'X)^{-1}X'C = 0$, $\hat{\beta}$ and $\hat{\sigma}^2$ are independent. Actually, to use the theorem (Lesson 9, Part VI), we must look at the two quadratic forms $(\hat{\beta} - \beta)'(\hat{\beta} - \beta) = Z'X(X'X)^{-1}(X'X)^{-1}X'Z$,

$$Z'(I_n - X(X'X)^{-1}X')Z$$

and check that the product of their matrices is 0 . That theorem then gives the independence of

$$(\hat{\beta} - \beta)'(\hat{\beta} - \beta)/\sigma^2 \text{ and } \hat{\sigma}^2/\sigma^2$$

from which the conclusion here follows.

In particular, $\hat{\beta}_0$ and $\hat{\sigma}^2$ are independent. Let

$$T = \frac{\hat{\beta}_0 - \beta_0}{\hat{\sigma}\sqrt{\Sigma x^2/S_{xx}}} = \frac{\hat{\beta}_0 - \beta_0}{\sqrt{\sigma^2\Sigma x^2/S_{xx}}} \div \sqrt{\hat{\sigma}^2/\sigma^2}$$

$$= \frac{\hat{\beta}_0 - \beta_0}{\sqrt{\sigma^2_{\hat{\beta}_0}}} \div \sqrt{\frac{Z'CZ/\sigma^2}{n-2}} .$$

In this expression, the numerator involving $\hat{\beta}_0$ is a standard normal RV ; the quantity under the radical in the denominator is an independent chisquare RV divided by its dof. It follows that T has the (central) Student distribution with n–2 dof.

Example: a) For the Student–T with n–2 dof, let $P(T \geq t_{\alpha/2}) = \alpha/2$. Then a $100(1-\alpha)\%$ CI for β_0 is given by

$$\hat{\beta}_0 \pm t_{\alpha/2} \cdot s_{\hat{\beta}_0} = \hat{\beta}_0 \pm t_{\alpha/2} \cdot \hat{\sigma} \cdot \sqrt{\Sigma x^2/S_{xx}} \ .$$

b) For the cricket data of the previous lesson,

$$n = 12 \quad \Sigma x = 202 \quad \Sigma y = 960.2 \quad \Sigma x^2 = 3434$$

$$\Sigma xy = 16285.9 \quad S_{xy} = 404$$

$$\hat{\beta}_0 \approx 18.75 \quad SSE = 157.5839 \quad \hat{\beta}_1 \approx 3.6396$$

$$s_{\hat{\beta}_0} = \sqrt{\frac{157.5839}{10}} \cdot \sqrt{3434/404}$$

$$\approx 3.9697(2.9155) = 11.5737 \ .$$

For $1 - \alpha = .95$, (using the table at the end of this volume), we find $t_{.025} = 2.228$ and the CI for β_0 is

$$18.75 \pm 2.228 \cdot 11.5737$$

which is

$$(7.04, 44.54) \ .$$

Exercise 4: a) Following the pattern for β_0 above, develop a general CI for β_1 .

b) Find a 95% CI for β_1 using the cricket data.

Exercise 5: Find 99% CIs for β_0 and β_1 in the exercise on baking bread in the previous lesson.

Naturally, the first question to be asked is whether or not

"x and Y" are really linearly related. This was answered hueristically by suggesting a "good fit" when

$$R^2 = 1 - SSE/SST$$

was close to 1 . Making use of the last exercise in the previous lesson, we get

$$R^2 = 1 - \frac{(S_{xx}S_{YY} - S_{xY}{}^2)/nS_{xx}}{S_{YY}/n}$$

$$= S_{xY}{}^2/S_{xx}S_{YY} = \hat{\beta}_1^2 \cdot (S_{xx}/S_{YY}) .$$

A "lack of fit" can then mean that R^2 or $\hat{\beta}_1^2$ is close to 0 . This is usually rephrased as testing $\beta_1 = 0$ but, strictly speaking, we have: the independent observables

$$(Y_1, x_1), \cdots, (Y_n, x_n)$$

have $\{Y_i\}$ or $\{\varepsilon_i\}$ normally distributed with common variance $0 < \sigma^2 < \infty$. A test of

$$H_o: Y_i = \beta_0 + \varepsilon_i \text{ vs } H_a: Y = \beta_0 + \beta_1 x_i + \varepsilon_i$$

has critical region $|\hat{\beta}_1/s_{\hat{\beta}_1}| > t_{\alpha/2}$ with $P(|T| > t_{\alpha/2}) = \alpha$ for the Student–T with n–2 dof.

Example: Recall the cricket data. The assumptions, hypotheses and test are as just outlined. With n = 12 and $\alpha = .01$, $t_{.005} = 3.169$. The data yield

$$\hat{\beta}_1 \approx 3.6396 , \; s_{\hat{\beta}_1}{}^2 \approx \frac{157.56839}{10} \cdot \frac{12}{404} \approx .46808 .$$

Hence, $\hat{\beta}_1 / s_{\hat{\beta}_1} \approx 3.6396/.6841 \approx 5.320 > 3.169$. We reject H_o and conclude that temperature is (at least) linearly related to these numbers of cricket chirps. The "p–value" here is $P(|T| > 5.320)$ which is effectively 0 .

Exercise 6: Test for linearity in the bread baking data. Include the p–value.

Exercise 7: Write out a procedure like that for $\beta_1 = 0$ but for testing that $\beta_0 = 0$ in the SLR model.

It should be obvious that one can also test one–sided hypotheses so that, for example, the critical region for

$$H_o: \beta_1 = .6 \text{ vs } H_a: \beta_1 > .6 \text{ is } (\hat{\beta}_1 - .6)/s_{\hat{\beta}_1} > t_\alpha .$$

Another parameter of substantial interest is the mean

$\mu_v = \beta_0 + \beta_1 \cdot x_v$ for an arbitrary (usually new) value x_v. Since $\hat{\beta}$ is unbiased,

$$\hat{\mu}_v = \hat{\beta}_0 + \hat{\beta}_1 \cdot x_v = (1, x_v)\hat{\beta}$$

is an unbiased estimator for μ_v. The variance of $\hat{\mu}_v$ is

$$(1, x_v)\sigma^2(X'X)^{-1}\begin{bmatrix}1\\ x_v\end{bmatrix} = \sigma^2(1, x_v)\frac{\begin{bmatrix}\Sigma x^2 & -\Sigma x\\ -\Sigma x & n\end{bmatrix}}{S_{xx}}\begin{bmatrix}1\\ x_v\end{bmatrix}$$

$$= \sigma^2\frac{\Sigma x^2 - 2(\Sigma x)x_v + nx_v^2}{S_{xx}} = \sigma^2\frac{n\Sigma x^2 - 2n(\Sigma x)x_v + n^2x_v^2}{nS_{xx}}$$

$$= \sigma^2\frac{S_{xx} + (\Sigma x)^2 - 2n(\Sigma x)x_v + n^2x_v^2}{nS_{xx}}$$

$$= \sigma^2(1/n + n(x_v - \bar{x})^2/S_{xx}) .$$

A $100(1 - \alpha)\%$ CI for $\beta_0 + \beta_1 \cdot x_v$ is

$$\hat{\beta}_0 + \hat{\beta}_1 \cdot x_v \pm t_{\alpha/2}\,\hat{\sigma}\cdot[1/n + n(x_v - \bar{x})^2/S_{xx}]^{1/2} .$$

Example: For $1-\alpha = .90$, and $x_v = 19$ chirps per second, a CI for the average temperature is

$$18.75 + 3.6396(19)$$

$$\pm 1.812(3.9697)(1/12 + 12(19 - 16.83)^2/404)^{1/2}$$

$$87.902 \pm 3.398 \text{ or } (84.50, 91.30) .$$

That is, 90% of the subpopulations with $x_v = 19$ will have *mean* temperatures between 84.50 and 91.30 .

Note that the length of the confidence interval above is very highly dependent on the distance from x_v to $\bar{x}$; when this distance is large, $(x_v - \bar{x})^2$ is large and the CI is wide. This may be interpreted as having the precision be small. Interpolation (with values of x_v between the max and min of the observed $\{x_i\}$) is reasonable; extrapolation (using values of x_v outside this interval) is unreliable and may be dangerous. Similar remarks apply to the estimation = prediction of the value of Y_v rather than only its mean.

Let $Y_v = \beta_0 + \beta_1 \cdot x_v + \varepsilon_v$ be independent of $Y = X \cdot \beta + \varepsilon$ in a SLR model with estimator $\hat{\beta}$ as above. Then,

$$Y_v - \hat{\beta}_0 - \hat{\beta}_1 \cdot x_v$$

is $N(0, \sigma^2[1 + 1/n + n(x_v - \bar{x})^2/S_{xx}])$ and is independent of $(n-2)$ $\hat{\sigma}^2/\sigma^2$ which is still distributed as chisquare. Hence,

$$\frac{Y_v - \hat{\beta}_0 - \hat{\beta}_1 \cdot x_v}{\hat{\sigma}\sqrt{1 + 1/n + n(x_v - \bar{x})^2/S_{xx}}}$$

is distributed as Student–T with $n-2$ dof. It follows that the inequality

$$\hat{\beta}_0 + \hat{\beta}_1 \cdot x_v - t_{\alpha/2}\hat{\sigma}\sqrt{1 + 1/n + n(x_v - \bar{x})^2/S_{xx}} < Y_v$$

$$< \hat{\beta}_0 + \hat{\beta}_1 \cdot x_v + t_{\alpha/2}\hat{\sigma}\sqrt{1 + 1/n + n(x_v - \bar{x})^2/S_{xx}}$$

has probability $1 - \alpha$ and thereby determines a $100(1 - \alpha)\%$ prediction interval for Y_v. By essentially the same arguments, we get

$$\hat{\beta}_0 + \hat{\beta}_1 \cdot x_v \pm t_{\alpha/2}\hat{\sigma}\sqrt{1/k + 1/n + n(x_v - \bar{x})^2/S_{xx}}$$

as a $100(1 - \alpha)\%$ prediction interval for the mean $\sum_{j=1}^{k} Y_j/k$ of a an independent RS from the subpopulation determined by the fixed value x_v.

Exercise 7: Prove (by a verbal argument) that

$$Y_v - \hat{\beta}_0 - \hat{\beta}_1 x_v \text{ is independent of } (n-2)\hat{\sigma}^2/\sigma^2.$$

Example: The following are the 90% prediction limits for the temperature when $x_v = 19$ cricket chirps are heard.

$$18.75 + 3.6396(19)$$

$$\pm\, 1.812(3.9697)(1 + 1/12 + 12(19 - 16.83)^2/404)^{1/2}$$

$$87.902 \pm 7.193(1.223)^{1/2} \text{ or } (79.95, 95.86)\,.$$

When $x_v = 19$, 90% of the corresponding temperatures will be between 79.95 and 95.86 .

Exercise 8: Find a 95% CI for $\beta_0 + \beta_1 \cdot x_v$ and a 95% prediction interval for Y_v when $x_v = 17$ minutes baking a loaf of bread.

Obviously, these distributional results can also be used to set up some tests of hypotheses.

LESSON *6. SUFFICIENT STATISTICS AND UNIFORMLY MOST POWERFUL TESTS

This lesson is based on a paper by Neyman and Pearson, 1936, with basically the same title; the idea is to show some inter–relationships of these two statistical concepts. Of course, the original paper contains other discussion and rather complete details; their comments on Fisher's work may not be relevant today. First, N–P defined two kinds of sufficient statistics; our previous discussions concenred only one of these and we did not use the modifier "specific".

Definition: *Let* $\{P_\theta : \theta \in \Theta \subset \mathbb{R}^p\}$ *be a family of probability measures on* $\mathcal{V} = R^n$, $\mathcal{B} = \mathcal{B}_n$. *A statistic is a measurable function* $t: \mathcal{V} \to R^m$, $m \geq n\text{-}1$. *Then* $T = t(V)$ *is a specific [shared] sufficient statistic for* $\theta_1 \in R$ $[(\theta_1, \cdots, \theta_q), 1 < q < p]$ *if, for any other statistic* T_2 , *the conditional distribution of* T_2 *given* T *is free of* θ_1 $[(\theta_1, \cdots, \theta_q)]$ *but may depend on other components of* θ .

Any general reference to "θ" or "statistic" will be with respect to this definition. You should recall that a necessary and sufficient condition that T be a (specific) sufficient statistic for θ is that, almost surely, the likelihood

$$L_\theta(v) = g(t(v);\theta)h(v)$$

where h is a function of v, free of θ both in its domain and in its range. A N–P generalization has a similar proof: $T = t(V)$ is a shared sufficient statistic for $(\theta_1, \cdots, \theta_q)$ iff

$$\text{a.s. } L_\theta(v) = g(t(v);\theta_1, \cdots, \theta_q)h(v;\theta_{q+1}, \cdots, \theta_p)$$

with h free of $(\theta_1, \cdots, \theta_q)$.

Although the delayed exponential family (indexed by parameters θ_1 real and $\theta_2 > 0$) is used herein for particular reasons, it is a common in modeling survival data; the PDF is

$$f(x;\theta) = \theta_2 \cdot [\exp(-\theta_2(x - \theta_1)] \cdot I\{\theta_1 \le x\} .$$

Exercise 1: a) Show that f given just above is indeed a density of a RV X .
b) Find the density of $Y = (X - \theta_1)/\theta_2$.
c) Let $Y_1, \cdots, Y_n$ be IID as Y ; find the density of $\Sigma_{i=1}^n Y_i$.

Example: Let $X_{(1)}$ be the minimum of the IID observables $X_1, \cdots, X_n$ with this delayed exponential distribution; then,

$$L_\theta(v) = \theta_2^{\,n} \cdot [\exp(-\theta_2 \sum_{i=1}^{n} (x_i - \theta_1))] \cdot I\{\theta_1 \le x_{(1)}\} .$$

For $H_o: \theta_1 = \theta_{1o}, \theta_2 = \theta_{2o}$ vs $H_a: \theta_1 < \theta_{1o}, \theta_2 > \theta_{2o}$, each BCR B_α will be defined by

$$L_{\theta_a}(v) \ge kL_{\theta_o}(v) .$$

Obviously, B_α will contain all of

$$\mathscr{V}_0 = \{v : L_{\theta o}(v) = 0\}$$

and a subset of $\mathscr{V}_+ = \{v : L_{\theta o}(v) > 0\}$ such that

$$\theta_{2a}^{\,n}[\exp(-n\theta_{2a}(\Sigma x_i - n\theta_{1a}))] \ge k\theta_{2o}^{\,n}[\exp(-n\theta_{2o}(\Sigma x_i - n\theta_{1o}))]$$

equivalently,

$$\Sigma x_i \le k \text{ (generic)} .$$

Since k will be determined under H_o , this B_α is a UMPCR. Also, under H_o, the parameters are known so that $\Sigma x_i \le k$ is also equivalent to

$$\Sigma y_i/\theta_{2o} + n\theta_{1o} \le k$$

for which probabilities can be determined via the gamma distribution.

Exercise 2: Continue the example above.

a) Determine the UMPCR for

$$H_o \text{ vs } H_a: \theta_1 = \theta_{1o}, \theta_2 > \theta_{2o}.$$

(Note the gamma distribution.)

b) When H_a contains $\theta_2 = \theta_{2o}$, the condition in $\mathcal{V}_+$ reduces to $0 \le k$ which is no restriction at all. Describe some part of $\mathcal{V}_+$ that could be included in B_α.

Let $T = t(V)$ be a shared sufficient statistic for θ_1 and θ_2; then

$$L_\theta(v) = g(t(v); \theta_1,\theta_2)h(v)$$

where h is free of θ_1 and θ_2. Take $\theta_{1a} < \theta_{1o}$ and $\theta_{2a} > \theta_{2o}$; when $t(v_1) = t(v_2)$, the ratios $(j = 1,2)$ $L_{\theta_o}(v_j)/L_{\theta_a}(v_j) = g(t(v_j); \theta_{1o},\theta_{2o})/g(t(v_j); \theta_{1a},\theta_{2a})$ are equal. In $\mathcal{V}_+$ of the example at hand, each of these ratios contains the corresponding

$$\bar{x}(j): (\theta_{2o}/\theta_{2a})^n[\exp(-n(\theta_{2o} - \theta_{2a})\bar{x}(j) + \theta_{1o}\theta_{2o} - \theta_{1a}\theta_{2a})].$$

These ratios will be equal iff $\bar{x}$ is constant. There are two cases:
1) If the orbits of $\bar{X}$ and T are the same, each will be a function of the other so that $\bar{X}$ will also be a shared sufficient statistic. But

$$L_\theta(v) = \theta_2^n \cdot [\exp(-\theta_2 n(\bar{x} - \theta_1))] \cdot I\{\theta_1 \le x_{(1)}\}$$

and the indicator function cannot be factored so that $\bar{X}$ cannot be "shared"; it follows that T cannot be "shared".

2) If the orbits of T are contained in the orbits of $\bar{X}$, let v' and v'' be such that $t(v') = t(v'') = t_1$; also, let $x'_{(1)}$ and $x''_{(2)}$ be the corresponding minima. For $x'_{(1)}$ and $x''_{(2)}$ greater than θ_1, we have

$$L_\theta(v') = g(t_1;\theta)h(v') > 0$$

and

$$L_\theta(v'') = g(t_1;\theta)h(v'') > 0 .$$

If $x'_{(1)} < x''_{(1)}$, we can find $x'_{(1)} < \theta'_1 < x''_{(1)}$.

Then $L_{\theta'}(v') = 0$ implies $g(t_1;\theta') = 0$
while $L_{\theta'}(v'') > 0$ implies $g(t_1;\theta') > 0$.

This contradiction means that $x'_{(1)} = x''_{(1)}$. In other words,

$$t(v') = t(v'') \text{ implies } \bar{x}' = \bar{x}'' \text{ and } x'_{(1)} = x''_{(1)} ;$$

then $T \in R^{n-2}$ and is no longer a statistic.

Exercise 3: Continue the example. Let $\theta_2 = 1 + \theta_1^{\,2}$. Consider $H_o: \theta_1 = 0, \theta_2 = 1$ vs $H_a: \theta_1 < 0, \theta_2 > 1$.

a) Show that a UMPCR exists.

b) Show that, in fact, $\bar{X}$, $X_{(1)}$ are sufficient for θ_1.

This discussion shows that one can have a UMPCR not based on any sufficient statistics. (Incidentally, this last exercise presents an example where the parameter space for θ_1 is real but the sufficient statistic is defined on a "half–plane".) The next example reverses the condition: existence of a sufficient statistic does not imply a UMPCR.

Example: The PDF of $V = (X_1, \cdots, X_n)$ is given by

$$L_\theta(v) = e^{-Q/2}/(2\pi)^{n/2}\sqrt{\theta_2}$$

where $Q = (x_1 - n\theta_1)^2/\theta_2 + \sum_{i=2}^{n} (1 + 1/\theta_2)x_i^{\,2}$

$$+ 2\sum_{i=2}^{n} (x_1 - n\theta_1)x_i/\theta_2 + 2\sum_{i\neq j>1} x_i x_j/\theta_2$$

θ_1 is real and $\theta_2 > 0$. We have one observation of an n–variate normal distribution with

$$\text{mean } (n\theta_1, 0, \cdots, 0)'$$

and covariance matrix $\Sigma = \begin{bmatrix} \theta_2 + n-1 & -1' \\ -1 & I \end{bmatrix}$

where $1'$ is a 1 by n–1 vector of all ones, I is an n–1 by n–1 identity matrix. Let Σ^{-1} be the matrix inverse of Σ; then Q is the quadratic form

$$(x_1 - n\theta_1, x_2, \cdots, x_n)\Sigma^{-1} \begin{bmatrix} x_1 - n\theta_1 \\ x_2 \\ \vdots \\ x_n \end{bmatrix}.$$

Exercise 4: a) From the information (notation) given in this normal example, write out Σ and Σ^{-1} for the special case $n = 4$ and verify that $\Sigma \cdot \Sigma^{-1}$ is the 4 by 4 identity.
b) Show that

$$\theta_2 \cdot Q = (\Sigma_{i=1}^{n} x_i - n\theta_1)^2 - \Sigma_{i=2}^{n} x_i^2 .$$

Example: Continue with the normal introduced in the last example above. For $H_0: \theta_1 = \theta_{1o}, \theta_2 = \theta_{2o}$ and any simple alternative,

$$L_{\theta a}(v) > kL_{\theta o}(v) \text{ iff}$$

$$[\exp(-\sum_{i=1}^{n} x_i - n\theta_{1a})^2/2\theta_{2a} - \sum_{i=2}^{n} x_i^2/2)]/(\sqrt{\theta_{2a}})(2\pi)^{n/2}$$

is greater than

$$k[\exp(-\sum_{i=1}^{n} x_i - n\theta_{1o})^2/2\theta_{2o} - \sum_{i=2}^{n} x_i^2/2]/(\sqrt{\theta_{2o}})(2\pi)^{n/2}$$

which reduces to

$$-(\bar{x} - \theta_{1a})^2/\theta_{2a} > k - (\bar{x} - \theta_{1o})^2/\theta_{2o}$$

or

$$\bar{x}^2(\theta_{2a} - \theta_{2o}) - 2\bar{x}(\theta_{2a}\theta_{1o} - \theta_{2o}\theta_{1a}) > k\,.$$

Obviously, the BCR depends on the alternative θ_{2a}, θ_{1a} so that there is no UMPCR for this arbitrary H_a.

Exercise 5: Continuing with this normal likelihood, find the UMPCR for $H_a: \theta_2 > \theta_{2o}, \theta_1 > \theta_{1o}$.

Example: Let us modify this normal likelihood by restricting Θ to the subset $\sqrt{\theta_2} = \theta_1 > 0$. Then,

$$L_\theta(v) = L_{\theta_1}(v)$$

$$= [\exp(-n^2(\bar{x}-\theta_1)^2/2\theta_1^{\,2} - \Sigma_{i=2}^{n} x_i^{\,2}/2]/\theta_1(2\pi)^{n/2}$$

and a BCR is determined by

$$\bar{x}^2(\theta_{1a}^{\,2} - \theta_{1o}^{\,2}) - 2\bar{x}(\theta_{1a} - \theta_{1o})\theta_{1a}\theta_{1o} \geq k\,.$$

Again, this BCR depends on θ_{1a} and so there is no UMPCR.

However, it is seen from $L_{\theta_1}(v)$ that $\bar{X}$ is a (specific) sufficient statistic for θ_1.

Starting with the existence of a sufficient statistic, essentially with the exponential family, addition of the MLR condition led to the existence of certain uniformly most powerful tests. The discussion above implies that some additional condition will be needed to move from UMPCR to the existence of sufficient statistics. Before presenting the N–P discovery in this regard, we need some more terminology.

Definition: *For a point* $v_o \in \mathscr{V}$ *and a positive number* δ, *a* δ*-neighborhood of* v_o *is the set*

$$\mathscr{N}_\delta(v_o) = \{\, v : |v - v_o| < \delta\} \,;$$

$|\cdot|$ *is the ordinary Euclidean distance. The point* v_o *is a boundary point of a subset* S *of* $\mathscr{V}$ *iff for each* $\delta > 0$ *, there is a* $v_1 \neq v_o$ *and a* $v_2 \neq v_o$ *such that*

$$v_1 \in \mathscr{N}_\delta(v_o) \cap S \text{ and } v_2 \in \mathscr{N}_\delta(v_o) \cap S^c .$$

The boundary of S is the set of all its boundary points.

Example: a) The sphere itself, $x^2 + y^2 + z^2 = 1$, is the boundary of its interior, $x^2 + y^2 + z^2 < 1$, and of its exterior, $x^2 + y^2 + z^2 > 1$, and of the solid disk,

$$x^2 + y^2 + z^2 \leq 1 .$$

b) $S = [0,1) \subset \mathbb{R}$ has one boundary point, 0 , in S and one boundary point 1 not in S; its boundary is $\{0,1\}$.

Definition: *Let* B_α *be a UMPCR for the simple hypothesis* $H_o: \theta = \theta_o$*. The point v belongs to the positive boundary of* B_α *if v is an ordinary boundary point of* B_α *and*

$$L_{\theta a}(v) \geq k \cdot L_{\theta o}(v) > 0 \text{ for all } \theta_a .$$

The next little trick requires that we restrict the rest of the discussion to V with continuous distribution functions.
If $\alpha_2 < \alpha_1$, then there is a subregion of B_{α_1} with size α_2 obtained by changing "k" in the defining inequality. In this sense, B_α is "decreasing" with α . Here is the final result, the existence of a sufficient statistic.

Theorem: *Let each* P_θ *have a continuous CDF. Suppose that for each* α *with* $0 < \alpha \leq \alpha_o \leq 1$ *, there is a UMPCR* B_α *for the simple hypothesis* H_o *. Suppose that each point (Lebesgue almost all) of* $\mathscr{V}$ *belongs to the positive boundary of some* B_α*. Then there is a (shared) sufficient statistic for* θ *.*

Proof: If v is a positive boundary point of only one B_α , let

$$t(v) = \alpha .$$

If v is a positive boundary point of more than one B_α, let

$$t(v) = (\alpha_1+\alpha_2)/2 ,$$

α_1 [α_2] is the inf [sup] of the sizes of all its B_α. By the remark preceding the theorem, v is a positive boundary point of B_α for all α between α_1 and α_2. If $t(v_1) = t(v_2)$, then v_1 and v_2 belong to the same boundary; but then

$$L_{\theta o}(v_i) \text{ and } L_{\theta a}(v_i) , i = 1 \text{ and } 2,$$

are both positive. For i = 1 or 2, there is

$$\text{a } v_i' \in \mathcal{N}_\delta(v_i) \cap B_\alpha \text{ and a } v_i'' \in N_\delta(v_i) \cap B_\alpha{}^c$$

such that $L_{\theta a}(v_i') \geq k_\alpha \cdot L_{\theta o}(v_i')$ and

$$L_{\theta a}(v_i'') \leq k_\alpha \cdot L_{\theta o}(v_i'') .$$

Letting $\delta \downarrow 0$, yields $L_{\theta a}(v_i) = k_\alpha \cdot L_{\theta o}(v_i)$ and $k_\alpha = L_{\theta a}(v_i(t))/L_{\theta o}(v_i(t))$ whence $L_\theta(v) = k(t(v);\theta) \cdot L_{\theta o}(v)$ as required for sufficiency.

LESSON *7. SEQUENTIAL PROBABILITY RATIO TESTS

All of our discussions about "best tests" involved fixed sample sizes; even the likelihood ratio tests had fixed sample sizes although some were required to be "large". As the title above suggests, this lesson treats another variation of the ratio; the classical results are in Wald, 1947. While explaining his procedure, we will also point out some of its advantages and contrary to earlier admonitions, here we will use the phraseology "accept/reject".

Let $X_1, X_2, X_3, \cdots$ be IID observable RVs. Suppose it is known that their density is either p_0 or p_1. Usually these will be written as if they are members of a family $\{f(x;\theta)\}$, but, generally, p_0 can be any specific distribution (say standard normal) and p_1 any other specific distribution (say Student with 9 dof).

A *sequential probability ratio test* (SPRT) of

$$H_0\text{: the density is } p_0 \text{ vs } H_a\text{: the density is } p_1$$

is conducted as follows:
fix $A_0 < A_1$; observe x_1 ;

$$\text{if } p_1(x_1)/p_0(x_1) \le A_0, \text{ accept } H_0 ;$$

$$\text{if } p_1(x_1)/p_0(x_1) \ge A_1, \text{ reject } H_0 ;$$

otherwise, observe x_2 ;

$$\text{if } p_1(x_1)p_1(x_2)/p_0(x_1)p_0(x_2) \le A_0, \text{ accept } H_0 ;$$

$$\text{if } p_1(x_1)p_1(x_2)/p_0(x_1)p_0(x_2) \ge A_1, \text{ reject } H_0 ;$$

otherwise, observe x_3 ; $\cdots$
otherwise, observe x_n ;

$$\text{if } \Pi_{i=1}^{n} p_1(x_i)/p_0(x_i) \le A_0, \text{ accept } H_0 ;$$

$$\text{if } \Pi_{i=1}^{n} p_1(x_i)/p_0(x_i) \ge A_1, \text{ reject } H_0 ;$$

otherwise, observe x_{n+1} ; · · ·

Note that we are still using the likelihood ratio but the number of items observed is a random variable. An immediate question is, "What if the sampling goes on forever?". We will see that the probability that this happens is 0 . The following exercise is crucial.

Exercise 1: Let $W_1, W_2, W_3, \cdots$ be a sequence of real numbers. Fix $0 < A_0 < A_1$. Show that if for all positive integers m,

$$\log A_0 < W_1 + \cdots + W_m < \log A_1 ,$$

then for all positive integers m,

$$|W_m| < \log A_1 - \log A_0 .$$

Theorem: *Suppose that for* $i = 1(1)\infty$,

$$Z_i = \log p_1(X_i)/p_0(X_i)$$

are IID non-degenerate RVs with $P(Z_i = 0) < 1$ *. Then the SPRT terminates with probability 1.*

Proof: This test terminates at the smallest n for which the following inequality fails to hold:

$$\log A_0 < Z_1 + Z_2 + \cdots + Z_n < \log A_1 . \qquad (*)$$

Partition the sequence $\{Z_i\}$ into segments of length r ; if

$$W_1 = Z_1 + \cdots + Z_r , W_2 = Z_{r+1} + \cdots + Z_{2r} , \cdots ,$$
$$W_m = Z_{(m-1)r+1} + \cdots + Z_{mr} , \text{ etc.,}$$

then $W_1, W_2, W_3, \cdots$ are also IID. Let

$$D = \log A_1 - \log A_0 .$$

If the test does not terminate, the inequality (*) holds for all positive integers n so by Exercise 1,

$$|W_1| < D, |W_2| < D, \cdots, |W_m| < D$$

for all positive integers m. Then for each $m \geq 1$ and $p = P(W_i^2 < D^2)$, the probability of no termination is

$$P(\cap_{i=1}^{\infty}\{W_i^2 < D^2\}) \leq \Pi_{i=1}^{m} P(W_i^2 < D^2) = p^m .$$

If the variance of Z_i is finite, then

$$E[W_i^2] = Var(W_i) + (E[W_i])^2 = rVar(Z_i) + (rE[Z_i])^2$$

can be made arbitrarily large by choosing r large; if $Var(Z_i) = \infty$, so is $E[W_i^2]$. In either case, p will be less than 1 so that $p^m \to 0$ as $m \to \infty$. The conclusion follows.

Exercise 2: Show that when $p = P(W_i^2 < D^2) = 1$, $E[W_i^2] < D^2$. Where does this fit in the proof above?

The accept/reject phraseology should remind you of an acceptance sampling plan (Lesson 16, Part I) which we will now rewrite in terms of a normal distribution.

Example: Suppose we are sampling cracking temperature of the seals on certain fuel rockets. Assume the outcomes $X_1, \cdots, X_n$ are IID normal RVs with mean θ and (for simplicity) known variance $\sigma^2 = 81$. The standard normal (Gauss) RV will be denoted by G. For

$$H_o : \theta \leq 25 \text{ vs } H_a : \theta > 25 ,$$

we base the test on the sample mean $\bar{X}$. Producer's risk

$$\alpha = P(\text{Reject } H_o \mid H_o \text{ true})$$
$$= P(\bar{X} > c \mid \theta = 25) = P(G > (c - 25)\sqrt{(n/81)}) ;$$

for $\alpha = .01$, this leads to

$$(c - 25)\sqrt{(n/81)} = 2.326. \quad (1)$$

Consumer's risk

$$\beta = P(\text{Accept } H_o \mid H_o \text{ false}) = P(\bar{X} \leq c \mid \theta = 30)$$

$= P(G \le (c - 30)\surd(n/81))$;

for $\beta = .05$, this leads to

$$(c - 30)\surd(n/81) = -1.645. \qquad (2)$$

Solving (1) and (2) simultaneously yields

$$n = 7.148^2 ,\ c = 32.071 ;$$

of course, n is rounded off to 52 .

Exercise 3: Continue this example. Show that a general solution for testing $\theta = \theta_o$ vs $\theta = \theta_a > \theta_o$ with $P(G > g_\alpha) = \alpha$, has

$$\surd n = (g_\alpha + g_\beta)\sigma/(\theta_a - \theta_o) ,\ c = \theta_o + g_\alpha\sigma/\surd n .$$

Exercise 4: Let X and Y be independent RVs, X normal with mean 0 and variance 1 , Y Poisson with mean 2; let $g : R^2 \to R$. Write out the formalities to illustrate

$$E[g(X,Y)] = E[\ E[g(X,Y)\,|\,Y]\ ] .$$

The proof of the following theorem is elementary except for the manipulation of the conditional distribution function:

$$F(z\,|\,n) = P(Z \le z, N = n)/P(N = n) ;$$

the "natural processes" we use can (but won't) be justified formally. Note that the dependence of N on $Z_1, \cdots, Z_{N-1}$ is not specific so that the result includes that for the SPRT.

Theorem: *Let $Z_1, Z_2, Z_3, \cdots$ be IID RVs; let N be a positive integer valued random variable whose value n depends only on $Z_1, \cdots, Z_{n-1}$. Suppose that E[N] and $E[Z_1]$ are both finite. Then, $E[Z_1 + \cdots + Z_N] = E[Z_1]\cdot E[N]$.*

Proof: $E[Z_1 + \cdots + Z_N] = E[\ E[Z_1 + \cdots + Z_N\,|\,N]\ ]$

$$= \sum_{n=1}^{\infty} E[Z_1 + \cdots + Z_n \mid N = n]\cdot P(N = n)$$

$$= \sum_{n=1}^{\infty} \sum_{i=1}^{n} E[Z_i \mid N = n] \cdot P(N = n)$$

$$= \sum_{i=1}^{\infty} \sum_{n=i}^{\infty} E[Z_i \mid N = n] \cdot P(N = n)$$

$$= \sum_{i=1}^{\infty} \sum_{n=i}^{\infty} \int_R z_i \, dF(z_i \mid n) \cdot P(n = n)$$

$$= \sum_{i=1}^{\infty} \sum_{n=i}^{\infty} \int_R z_i \, dF(z_i;n)$$

$$= \sum_{i=1}^{\infty} \int_R z_i \cdot \sum_{n=i}^{\infty} dF(z_i;n)$$

$$= \sum_{i=1}^{\infty} E[Z_i] \cdot P(N \geq i)$$

since the condition on N makes $\{N \geq i\}$ independent of Z_i. As the Z_i are IID, $E[Z_i] = E[Z_1]$ can be factored outside the summation sign. By an example in Lesson 11, Part II,

$$\sum_{i=1}^{\infty} P(N \geq i) = E[N] .$$

The conclusion follows.

Exercise 5: One should justify the interchange of order of summation in the above proof by showing that the convergence of the series is absolute. Do this by showing first that

$$E[|Z_1 + \cdots + Z_N|] \leq E[|Z_1|] \cdot E[N] .$$

Now we need to get some information about A_o and A_1. It turns out that a certain approximation is easy to obtain while the actual values are elusive. Remember that with probability 1, we either accept or reject H_o. The set which leads to rejection of H_o is the union of the sets

$$S_1 = \{k_1 = \log A_1 \le Z_1\} ,$$

$$S_2 = \{k_1 \le Z_1 + Z_2 , k_o = \log A_o < Z_1 < k_1\} ,$$

$$S_3 = \{k_1 \le Z_1+Z_2+Z_3, k_o < Z_1+Z_2<k_1, k_o<Z_1< k_1\}, \text{ etc.}$$

In S_n, $\log A_1 \le Z_1 + \cdots + Z_n$ is equivalent to

$$\Pi_{i=1}^n f(x_i;\theta_1) \ge A_1 \cdot \Pi_{i=1}^n f(x_i;\theta_o) .$$

Hence, $\alpha_o = P(\text{Reject } H_o \mid H_o \text{ true}) = \sum_{n=1}^{\infty} P(S_n \mid \theta_o)$

$$= \sum_{n=1}^{\infty} \int_{S_n} \Pi_{i=1}^n f(x_i;\theta_o)$$

$$\le \sum_{n=1}^{\infty} \int_{S_n} \Pi_{i=1}^n f(x_i;\theta_1)/A_1 = (1-\alpha_1)/A_1$$

where $\alpha_1 = P(\text{Do not reject } H_o \mid H_o \text{ false})$. Similarily for the probability of accepting H_o, the probability of the union of the T_n, which are the sets

$$\{k_o \ge Z_1+\cdots+Z_n, k_o < Z_1 < k_1,\cdots,k_o < Z_1+\cdots+Z_{n-1} < k_1\} ,$$

leads to $1 - \alpha_o = \sum_{n=1}^{\infty} \int_{T_n} \Pi_{i=1}^n f(x_i;\theta_o) \ge \alpha_1/A_o$. Then these two inequalities involving the alphas can be written as

$$A_o \ge \alpha_1/(1-\alpha_o) , \ A_1 \le (1-\alpha_1)/\alpha_o .$$

If we take these boundary values as initial values, that is,

$$A_o' = \alpha_1/(1-\alpha_o) , A_1' = (1-\alpha_1)/\alpha_o ,$$

the the same process leads to

$$A_o' \ge \alpha_1'/(1-\alpha_o'), \ A_1' \le (1-\alpha_1')/\alpha_o' .$$

From $\alpha_1/(1-\alpha_0) \geq \alpha_1'/(1-\alpha_0')$, we get

$$\alpha_1'(1-\alpha_0) \leq \alpha_1(1-\alpha_0') ; \qquad (**)$$

since $\alpha_1(1-\alpha_0') \leq \alpha_1$, this implies

$$\alpha_1' \leq \alpha_1/(1-\alpha_0) .$$

From $(1-\alpha_1)/\alpha_0 \leq (1-\alpha_1')\alpha_0'$, we get

$$\alpha_0'(1-\alpha_1) \leq (1-\alpha_1')\alpha_0 ; \qquad (***)$$

since $(1-\alpha_1')\alpha_0 \leq \alpha_0$, this implies

$$\alpha_0' \leq \alpha_0/(1-\alpha_1) .$$

By addition of the second inequalities in (**) and (***), we get

$$\alpha_0' + \alpha_1' \leq \alpha_0 + \alpha_1 .$$

In applications, the desired α_0 and α_1 are generally less than 1/10 ; since the sum of the α_0' and α_1' values attained by using the boundary values for A_0 and A_1 is then less than 2/10 , at least one of α_0', α_1' will be less than 1/10 . The following theorem suggests that the SPRT may lead to "smaller" sample sizes in a manner to be illustrated forthwith.

Theorem: *Among all tests for which the error probabilities are* $P(\text{Reject } H_o \mid \theta_o) \leq \alpha_o$ *and* $P(\text{Accept } H_o \mid \theta_1) \leq \alpha_1$ *and for which the expected sample sizes are finite, the SPRT minimizes* $E[N \mid \theta_o]$ *and* $E[N \mid \theta_1]$ *. Moreover,*

$$E[N \mid \theta_o] \geq \frac{((1-\alpha_o)\log \alpha_1/(1-\alpha_o) + \alpha_o \log(1-\alpha_1)/\alpha_o)}{E[Z_1 \mid \theta_o]} ,$$

$$E[N \mid \theta_1] \geq \frac{(\alpha_1 \log \alpha_1/(1-\alpha_o) + (1-\alpha_1)\log(1-\alpha_1)/\alpha_o)}{E[Z_1 \mid \theta_1]} .$$

Proof: See Wald, 1947.

The following heuristic argument gives another approximation which is consistent with these bounds. We note first that when

a) the test leads to rejection of H_o, then

$$E[Z_1 + \cdots + Z_N] \approx \log A_1 ;$$

b) the test leads to acceptance of H_o, then

$$E[Z_1 + \cdots + Z_N] \approx \log A_o .$$

It follows that $E[N \mid H_o \text{ true}]$ can be approximated by

"The value of N accepting H_o when true"$\cdot$P(Accept $H_o | \theta_o$)

+ "The value of N rejecting H_o when true"$\cdot$P(Reject $H_o | \theta_o$)

$$\text{or } \frac{E[Z_1 + \cdots + Z_N \mid \text{Accept } H_o, \theta_o]}{E[Z_1 \mid \theta_o]} \cdot (1-\alpha_o)$$

$$+ \frac{E[Z_1 + \cdots + Z_N \mid \text{Reject } H_o, \theta_o]}{E[Z_1 \mid \theta_o]} \cdot \alpha_o$$

$$\approx ((1-\alpha_o)\log A_o + \alpha_o \log A_1) \div E[Z_1 \mid \theta_o]$$

$$\approx ((1-\alpha_o)\log \alpha_1/(1-\alpha_o) + \alpha_o \log(1-\alpha_1)/\alpha_o) \div E[Z_1 \mid \theta_o] .$$

Similarly, $E[N \mid H_o \text{ false}]$ is approximately

$$(\alpha_1 \log \alpha_1/(1-\alpha_o) + (1-\alpha_1)\log (1-\alpha_1)/\alpha_o) \div E[Z_1 \mid \theta_1] .$$

Exercise 6: Verify the heurisitic approximation of

$$E[N|H_o \text{ false}] .$$

Example: Let us use these approximations in the example on the normal with $H_o: \theta \le 25$ vs $H_a: \theta > 25$,

$$\alpha_0 = .01 , \theta_o = 25 , \alpha_1 = .05 , \theta_1 = 30 .$$

Since $Z_1 = \log f(X_1;30)/f(X_1;25) = (10X - 275)/162$,

$$E[Z_1 | \theta_o = 25] = -25/162 \text{ and } E[Z_1 | \theta_1 = 30] = 25/162 .$$

$$E[N | H_o \text{ true}] \approx (.99 \cdot \log(.05/.99) + .01 \cdot \log(.95/.01)) \div (-25/162) = 18.8565 ;$$

$$E[N | H_o \text{ false}] \approx (.05 \cdot \log(.05/.99) + .95 \cdot \log(.95/.01)) \div (25/162) = 27.0663 .$$

Thus if H_o is true, it will be accepted in about 19 observations while if H_o is false, it will be rejected in about 28 observations. On the other hand, the fixed sample size test requires 52 observations to reach a conclusion. In practice, a saving of about half of the observations is certainly economical.

Exercise 7: Consider sampling from a normal population with mean θ and variance 100 . Find the fixed sample size and the approximate average sample sizes for the SPRT with

$$\theta_o = 50 , \alpha_o = .05 , \theta_1 = 53 , \alpha_1 = .01 .$$

Exercise 8: Consider sampling from a Poisson population with mean θ . Find the fixed sample size and the approximate average sample sizes for the SPRT with

$$\theta_o = 2 , \alpha_o = .04 , \theta_1 = 4 , \alpha_1 = .02 .$$

LESSON *8. A TEST BY MANN, WHITNEY, WILCOXON

In some previous lessons, we considered as a null hypothesis the equality of means, say H_o: $\mu_1 = \mu_2$. With the additional assumption that the underlying population was normal, it was found that a test could be based on Student T ; or, with large samples, the criterion involved Gauss Z . A next question is "What procedure can be used when neither of these assumptions is tenable ?" The test in the title (briefly, MWW) was invented for this purpose. We have a random sample $X_1, X_2, \cdots, X_n$ from one population and a RS $Y_1, Y_2, \cdots, Y_m$ from a second population. As before, the populations are independent and of the same kind of variable: weight, time, mileage, proficiency, $\cdots$.

Suppose that the populations are known to be essentially the same except for a possible difference between their means. Specifically, consider

$$H_o: \mu_X = \mu_Y \text{ vs } H_a: \mu_X > \mu_Y .$$

Under H_a , we might have population densities like:

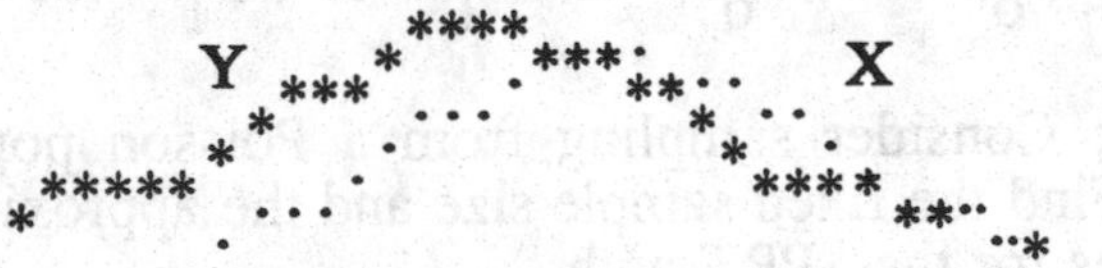

Then when all the observations $x_1, \cdots, x_n, y_1, \cdots, y_m$ are arranged in order, "more" of the X values will be "on the right" and "more" of the Y values will be "on the left". Under H_o, the X values and the Y values should be "well–mixed". This intuitive analysis can be formalized as follows:

for $i = 1(1)n$, $j = 1(1)m$, let $U_{ij} = \begin{matrix} 1 \text{ if } X_i > Y_j \\ 0 \text{ otherwise} \end{matrix}$; if the observed value of $U = \sum_{i=1}^{n} \sum_{j=1}^{m} U_{ij}$ is "large", reject H_o.

Note: U_{ij} is the indicator function of the set

$$\{X_i > Y_j\}$$

and U is the total number of times that an "X" is greater than a "Y" or a "Y" is less than an "X" .

This analysis does not focus strictly on the difference of the means but on differences in "location". The null hypothesis is the equivalence of the distributions and the alternative is that "X is bigger than Y", or, properly, "X is stochastically larger than Y" .

Definition: *The RV X is stochastically larger than the RV Y iff $P(Y \le a) \ge P(X \le a)$ for all real a . For CDFs F_X and F_Y, we may write $F_Y \ge F_X$.*

The inequalities above may appear to be backwards but if, for example, $P(Y \le 3) \ge P(X \le 3)$,then indeed, "more" X values are greater than 3 : $P(Y > 3) \le P(X > 3)$. The first exercise shows that, at least in cases like the normal, the condition does generalize that of the ordered means.

Exercise 1: Let X and Y be independent normal RVs with equal variances σ^2 . Show that X is stochastically larger than Y when $\mu_X > \mu_Y$.

This gives us a third way to test a hypothesis about the difference in means of independent normal populations but, of course, we need to determine the distribution of this Mann–Whitney U . It happens that U is equivalent to a Wilcoxon rank sum so that we will discuss rank first.

Example: Let the Y's be 23 24 29 26 22 and let the X's be 21 25 33 31 30 34 . Arrange all 11 numbers in order and assign *ranks* from smallest to largest.

	X:	21				25			30	31	33	34
	Y:		22	23	24		26	29				
	U =	0				+ 3			+5	+5	+5	+5
X ranks	S:	1				5			8	9	10	11
Y ranks	T:		2	3	4		6	7				

The pertinent sums are $U = 23$, $S_X = 44$, $T_Y = 22$.

Exercise 2: Find U , the ranks of the X's , and the Y's , their sums S_X , T_Y for Y: 11 21 15 19 22 25 16 18
X: 10 17 23 26 14 27 28 .

Lemma: *Let distinct numbers $x_1, \cdots, x_n, y_1, \cdots y_m$ be arranged in order. Let $s_1, \cdots, s_n$ be the ranks of the x's and let $t_1, \cdots, t_m$ be the ranks of the y's . Let $N = n + m$. Let*

$$u = \sum_{i=1}^{n} \sum_{j=1}^{m} I\{x_i > y_j\}$$

where $I\{A\}$ is the indicator function of the set A. Then,

$$s_1 + \cdots + s_n + t_1 + \cdots + t_m = N(N + 1)/2 ,$$

$$u = s_1 + \cdots + s_n - n(n + 1)/2 .$$

Proof: Let the ordered values of the x's be arrayed with their

ranks: $\begin{matrix} x_{(1)} & x_{(2)} & x_{(n-1)} & x_{(n)} \\ s_1 & s_2 & s_{n-1} & s_n \end{matrix}$. Then there are:

$s_1 - 1$ y's less than $x_{(1)}$;

$s_2 - 2$ y 's less than $x_{(2)}$;

. . .

$s_n - n$ y's less than $x_{(n)}$.

Hence, u = the number of y's less than some x

$= s_1 + \cdots + s_n - 1 - 2 - \cdots - n$

$= s_1 + \cdots + s_n - n(n + 1)/2$.

Since $s_1, \cdots, s_n, t_1, \cdots, t_m$ are just the numbers $1, 2, \cdots, N$ in some other order, their sum is $N(N + 1)/2$ and the conclusion follows.

Now the criterion $U \geq c$ can be taken as

$$S_X = S_1 + \cdots + S_n \geq c_1 \text{ or}$$

$$T_Y = T_1 + \cdots + T_m \le c_2 .$$

The significance level (p–value) is

$$\alpha = P(U \ge c \mid H_o) = P(S_X \ge c_1 \mid H_o) = P(T_Y \le c_2 \mid H_o) .$$

Under H_o, which is now equality of the distributions, all the observables are IID and the separation of the ranks into an X group and a Y group is equivalent to the separation of N (positive) integers into groups of size n and m . Thus

$$P(S_1 = s_1, \cdots, S_n = s_n | H_o)$$
$$= P(T_1 = t_1, \cdots, T_m = t_m | H_o) = 1/\begin{bmatrix} N \\ n \end{bmatrix} .$$

The significance probability or, in general the distribution, of U or S_X or T_Y can be obtained by listing the cases and counting. This may be a formidable task since, for example, $\begin{bmatrix} 15 \\ 7 \end{bmatrix} = 6435$.

Example: For m = 2 , n = 3 , the ranks are 1 2 3 4 5 and there are $\begin{bmatrix} 5 \\ 3 \end{bmatrix}$ = 10 cases. The distributions are:

X	Y	S_X	T_Y
1 2 3	4 5	6	9
1 2 4	3 5	7	8
1 2 5	3 4	8	7
1 3 4	2 5	8	7
1 3 5	2 4	9	6
1 4 5	2 3	10	5
2 3 4	1 5	9	6
2 3 5	1 4	10	5
2 4 5	1 3	11	4
3 4 5	1 2	12	3

S_X	P	T_Y
6	1/10	9
7	1/10	8
8	2/10	7
9	2/10	6
10	2/10	5
11	1/10	4
12	1/10	3

For the criterion $S_X \ge c_1$, the values of α which can actually be attained, equivalently, the p–values, are

c_1	6	7	8	9	10	11	12
p	1	9/10	8/10	6/10	4/10	2/10	1/10

Exercise 3: a) Find the distribution of S_X when n = m = 3 .

b) Is X stochastically larger than Y when the data are

Y: 12.6 18.2 9.3
X: 16.1 17.8 9.6 ?

Hint: find the p–value for $S_X \geq 11$.

For $n,m \leq 10$, tables of the S_X distribution or a centralized form $S_X - n(n+1)/2$, or U itself are available in various texts; for example, Lehmann, 1975, has a distribution for W_{XY} which counts the number of X's less than a Y . This text also includes details for the following theorem; the asymptotic distribution is fairly accurate for $n,m \geq 7$.

Theorem: *For the independent random samples* $X_1, \cdots, X_n$, $Y_1, \cdots, Y_m$ *of continuous type RVs, let* S_X *be the sum of the ranks of the X's when all* $N = n+m$ *values are ordered together. Under the null hypothesis of the equality of the distributions,* $E[S_X] = n(N+1)/2$, $Var(S_X) = mn(N+1)/12$ *and the asymptotic distribution of the standardized RV* $(S_X - E[S_X])/\sqrt{Var(S_x)}$ *is the standard normal.*

Example: In exercise 2, we find $S_X = 63$. For $n = 7$ and $m = 8$, $E[S_X] = 7(16)/2 = 56$, $Var(S_X) = 8 \cdot 7(16)/12 = 224/3$.

As in the normal approximation of a binomial distribution, the correction for continuity improves the accuracy. Hence for the p–value, we calculate $P(S_X \geq 63 \mid H_o)$

$$\approx P(Z \geq (62.5 - 56)/\sqrt{224/3}) \approx P(Z \geq .75) = .7734 .$$

We do not reject H_o ; this sample does not suggest that X is stochastically larger than Y .

Exercise 4: Let F_X and F_Y be the CDFs of the corresponding distributions; find the appropriate p–value for testing

$H_o: F_X = F_Y$ vs $H_a: F_X \le F_Y$ when the data is

X: 25 17 28 30 19 28 23 27 26 31 29
Y: 15 12 13 14 11 21 22 20 16 8 9 .

Of course, one can interchange the rolls of "X and Y" and use an alternative that makes Y stochastically larger than X . It is also possible to consider a two–sided alternative which does not specify the direction of the shift; then the rejection region is

$$S_X \ge c_1 \text{ or } S_X \le c_2 .$$

Because of the symmetry of the null distribution (see Lehmann, 1975), one usually considers

$$P(S_X \ge c_1 \,|H_o) \approx \alpha/2 \approx P(S_X \le c_2 \,|H_o) .$$

The following "experimental design" also suggests use of this MWW test.

There are N experimental units (subjects) available for comparison of two treatments. In many applications, one treatment is a "control" with which the effects of the "new treatment" will be compared; we have all heard of placebos. The subjects must be selected so as to be "homogeneous" with respect to characteristics that might interact with the treatments. For example, if the treatment is for a disease, subjects are selected to have the "same" levels of infection, the "same" sex, the "same" age group, etc. Each subject is assigned "at random" to be one of n "X's" or m "Y's".

If the effects of the treatments are not different, the reaction of each subject will be the same no matter which treatment it receives. It follows that the random assignment of subjects to treatment groups will be equivalent to the random separation of N independent objects into groups of sizes n and $m = N - n$.

If reaction to one treatment (X) is expected to be greater than that of the other (Y), then the sum of those ranks, S_X , will tend to be "large". We reject H_o: the effects of X and Y are similar in favor of H_a: the effects of X are greater than those of Y when S_X is "large".

Exercise 5: a) Find a critical value (at $\alpha = .05$) for testing that a new vaccine (Y) induces less (than the old vaccine X) inflammation at the injection sight when $n = m = 5$. Hint: There are $\begin{bmatrix}10\\5\end{bmatrix} = 252$ cases. Since $252(.05) = 12.6$, you want to find the 12 or 13 cases in which S_X is less than its

$$\max = 10 + 9 + 8 + 7 + 6 = 40 .$$

b) What is the "α value" closest to .05 ?

You should note that since the criterion is based on ranks, it could be used in any experiment in which data can be "ranked" even when the outcome is not "measured". This kind of data often results in ties; for example, if we classify teachers as good, indifferent, bad, there has to be a tie as soon as there are more than three teachers! Ties can also occur with "interval data" partly due to the limitations of our measuring tools which are not infinitely precise. The principles discussed above still apply and we get a MWW test but the distributions are different; for example, 1 2 2 has only three distinguishable arrangements not 6 . The normal approximation must be modified to take account of ties. (Again see Lehmann, 1975).

This MWW is one of the first in a large class of "non-parametric" or "distribution-free" procedures which have been developed. Neither name is quite appropriate although both suggest that the criterion used does not "depend" on the true underlying population parameters or distribution function. Of course, if you calculate a probability, you are using *some* distribution; and in the next lesson, we will see how *other* parameters become involved. In many of the procedures, the distribution (of U $\cdots$) under the null hypothesis is the same for all underlying continuous CDF F_X, F_Y ; the distributions under H_a do not enjoy the same uniformity. This makes general consideration of the power of such tests intractable.

However, if we restrict our model slightly we can make some comparison of the MWW test with the Student T test. The rest of this lesson is a paraphrase of some results from Lehmann, 1975, Chapter 2.

Theorem: *Let* $X_1, \cdots, X_n$ *be IID with continuous CDF* F *; let* $Y_1, \cdots, Y_m$ *be IID with CDF* G *such that*

$$\text{for } \Delta \geq 0 ,\ G(y-\Delta) = F(y) \text{ for all } y .$$

Let $S_1 < \cdots < S_n$ *denote the ranks of the* X*'s among all*

N = n+m observations.

a) *When* $\Delta = 0$, $P(S_1 = s_1, \cdots, S_n = s_n) = 1/\binom{N}{n}$;

b) *The power function* $P(S_X \geq c_1 \mid \Delta)$ *is a non-decreasing function of* Δ.

Exercise 6: Show that the MWW test in this theorem is unbiased for alternatives $\Delta > 0$.

Under the hypotheses of the theorem, let

$$p_1 = P(X_1 > Y_1),$$
$$p_2 = P(\{X_1 > Y_1\} \cap \{X_1 > Y_2\}),$$
$$p_3 = P(\{X_1 > Y_1\} \cap \{X_2 > Y_1\}).$$

Then $Var(S_X)$

$$= mnp_1(1-p_1) + mn(n-1)(p_2 - p_1^2) + nm(m-1)(p_3 - p_1^2)$$

and for large m and n, the power function has a normal approximation: $P(S_X \geq c_1 \mid \Delta)$

$$\approx P(Z \geq (c_1 - n(n+1)/2 - .5 - nmp_1)/\sqrt{Var(S_X)}).$$

Now let the sample means and variances be

$$\bar{X},\ \bar{Y},\ \Sigma_{i=1}^{n}(X_i - \bar{X})^2/(n-1),\ \Sigma_{j=1}^{m}(Y_j - \bar{Y})^2/(m-1).$$

The traditonal Student T criterion for testing

$$\Delta = 0 \text{ vs } \Delta > 0$$

is $(\bar{X}-\bar{Y})/\sqrt{(\Sigma_i(X_i - \bar{X})^2 + \Sigma_j(Y_j - \bar{Y})^2)/(n+m-2)} \geq t_\alpha$.

When F is actually normal and the sample sizes are equal, this Student test requires only about 95% as many observations to obtain the same limiting power as the MWW test. This is called the *Pitman efficiency* of MWW with respect to T.

When F is not normal, particularly when the density has "heavier tails" than the normal, MWW may be considerably more (Pitman) efficient than T.

LESSON *9. TESTS FOR PAIRED COMPARISONS

As in the previous lesson, we first recall earlier procedures. We have independent pairs of observations

$$(X_1,Y_1),\cdots,(X_n,Y_n)$$

of two similar variables: before–after weight, morning– afternoon rainfall, generally, treatment1–treatment2. A test of

$$H_o: \mu_1 = \mu_2$$

can be based on the sample mean $\bar{D} = \Sigma_{i=1}^n(x_i - y_i)/n$. When the $D_i = X_i - Y_i$ are normally distributed, significance probabilities can be calcluted using Student T; when n is "large", Gauss Z can be used. In the rest of this lesson, we suppose that neither of these distributions is tenable.

In a bit different vein, we start with n pairs of subjects drawn at random from some population with X_i assigned to "treatment" and Y_i assigned to "control". Under the null hypothesis,

$$H_o\text{: the treatment has no effect,}$$

the differences $D_i = X_i - Y_i$ have the same distribution as $-D_i = Y_i - X_i$; in other words, the distribution of the D's is symmetric about 0 . Suppose that the alternative is

$$H_a\text{: treatment increases response ;}$$

then "X is larger than Y" (X >> Y) so that the D's tend to be positive. This verbiage can be translated to testing

$$H_o: P(D_i > 0) = 1/2 \quad \text{vs} \quad H_a: P(D_i > 0) > 1/2 \,.$$

Let $I_i = 1$ if $D_i > 0$, $= 0$ otherwise. Then $S_n = \Sigma_{i=1}^n I_i$

is the number of "successes" in n independent Bernoulli trials with $\theta = P(D_i > 0)$. Thus we can look at the whole thing as testing

$$H_o: \theta = 1/2 \text{ vs } H_a: \theta > 1/2$$

in a Bernoulli experiment. We reject H_o when

$$S_n \geq c \text{ with } P(S_n \geq c \mid \theta = 1/2) \approx \alpha .$$

The power function $P(S_n \geq c \mid \theta)$ is an increasing function of θ so that for these alternatives, this sign test is unbiased. For large n, this power function has the normal approximation

$$P(Z \geq (c - .5 - n\theta)/\sqrt{n\theta(1-\theta)}) .$$

Actually, to get such a *sign test*, we only have to be able to decide whether or not "X dominates Y" ; in this way, qualitative outcomes can be considered. For example, each ith pair might arise by having the same individual tasting candy bars with peanut butter (X) and coconut (Y) ; preference for peanut butter gives I_i the value 1 .

Exercise 1: Find the p–value for

$$H_o: \theta = 1/2 \text{ vs } H_a: \theta > 1/2$$

when $S_n = 10$ and $n = 16$.

Another variation in the experimental design is to consider the median, say ν , of the distribution of $D = X-Y$ without symmetry. Then

$$H_o: \nu = \nu_o \text{ means } P(D > \nu_o) = 1/2$$

and $$H_a: \nu > \nu_o \text{ means } P(D > \nu_o) > 1/2 .$$

A "success" is $D_i > \nu_o$ and the total S_n is again binomial.

Without saying so, we have been thinking that the probability of a tie, $X_i = Y_i$, is zero. Let us now take $P(D_i < 0) = p_1$, $P(D_i = 0) = p_2$, $P(D_i > 0) = p_3$; then we have a 3–cell Bernoulli distribution. Let N_1 be the number of negative signs, N_2 be the number of zeroes, and N_3 be the number of positive signs in $n = N_1 + N_2 + N_3$ independent trials. This generates a multinomial distribution:

$$P(N_1 = n_1, N_2 = n_2, N_3 = n_3)$$
$$= P(N_1 = n_1, N_2 = n_2)$$
$$= (n!/n_1!\cdot n_2!\cdot n_3!)p_1^{n_1}\cdot p_2^{n_2}\cdot p_3^{n_3} .$$

Inconveniently, the distributions of N_1, $N_1 + N_2/2, \cdots$ all depend on the unknown p_1.

One "cure" is to consider the conditional distribution given $N_2 = n_2 : P(N_1 = n_1 \mid N_2 = n_2)$

$$= \frac{(n!/n_1!\cdot n_2!\cdot n_3!)p_1^{n_1}\cdot p_2^{n_2}\cdot p_3^{n_3}}{(n!/n_2!\cdot(n-n_2)!)p_2^{n_2}\cdot(1-p_2)^{n-n_2}}$$

$$= \frac{(n_1 + n_3)!}{n_1!\ n_3!}\left[\frac{p_1}{p_1 + p_3}\right]^{n_1}\left[\frac{p_3}{p_1 + p_3}\right]^{n_3}$$

which is binomial. The null hypothesis becomes

$$p_1/(p_1 + p_3) = 1/2 .$$

Thus one solution to the problem of what to do with ties (zeroes) is to discard them and use a sign test with only n_1+n_3 trials.

Exercise 2: Use the sign test, discarding ties, with α close to .05 , for H_o: X and Y are equivalent vs H_a: X exceeds Y . The data are

X :	2	6.1	3	9.4	7	5	2	6.9	4	8
Y :	2	5	3	9	6	7	3	6.6	4	6 .

Obviously, the sign test is "non–parametric" and uses only the signs of the differences $D_i = X_i - Y_i$. Also, obviously, some information is lost when the magnitude of the differences is ignored; 2–6 must mean something other than 2–9 . If we continue to look for a "non–parametric" procedure, it is natural to examine combinations of signs and ranks. One such answer here is the Wilcoxon signed–ranks test.

Example: If the differences are

$$-11 \quad 12 \quad 15 \quad -14 \quad -13 \quad 7 \quad 9 \quad -10 \quad 6\,,$$

the ranks of the absolute values are

$$5 \quad 6 \quad 9 \quad 8 \quad 7 \quad 2 \quad 3 \quad 4 \quad 1\,.$$

The positive–signed ranks are

$$6 \quad 9 \quad 2 \quad 3 \quad 1 \quad \text{with a sum of } 21\,;$$

the negative–signed ranks are

$$5 \quad 8 \quad 7 \quad 4 \quad \text{with a sum of } 24\,.$$

Exercise 3: What are the sum of the positive–signed ranks and the negative–signed ranks in

$$\begin{matrix} X: & 3.8 & 5.9 & 6.3 & 5.4 & 8.7 & 6.4 & 7.9 & 5.7 & 6.6 \\ Y: & 2.7 & 6.8 & 5.5 & 4.9 & 7.8 & 6.6 & 6.8 & 4.9 & 5.5 \end{matrix}\ ?$$

In general, let the ranks of the absolute values of all n differences be separated into

$S_1 < \cdots < S_p$ corresponding to positive differences,

$T_1 < \cdots < T_q$ corresponding to negative differences.

Assuming no ties, p, 0, q are the values of the RVs N_1, N_2, N_3 discussed above. Under the hypothesis of no difference in the distributions of X and Y, the $n = p+q$ signs are independent and each of the 2^n n–tuples of $+,-$ has probability $1/2^n$. Each sign combination yields one value of

$$N_1 = p \text{ and } S_1 < \cdots < S_p\,.$$

It follows that the null distribution can be taken as

$$P(N_1 = p, S_1 = s_1, \cdots, S_p = s_p) = 1/2^{p+q}\,.$$

If the alternative hypothesis is that X is greater than Y, we reject H_o when p is large and $S_1,\cdots,S_p$ are "larger" than $T_1,\cdots,T_q$. One single statistic which combines these intuitive measures is

$$W = S_1 + \cdots + S_p .$$

Note that this (Wilcoxon sum W) differs from the (MWW) rank sum of the previous lesson especially because here the value p is the value of a RV while in the earlier sum $S_1 + \cdots + S_n$, n was fixed before the experiment was conducted.

Example: The following table is for the case $n = p + q = 4$.

Rank	All possible signs															
1	+	+	+	+	−	+	+	−	−	−	+	+	−	−	−	−
2	+	+	+	−	+	+	−	+	+	−	−	−	+	−	−	−
3	+	+	−	+	+	−	+	+	−	+	−	−	−	+	−	−
4	+	−	+	+	+	−	−	−	+	+	+	−	−	−	+	−
W	10	6	7	8	9	3	4	5	6	7	5	1	2	3	4	0

Under H_o, the relevant distribution is

W	0	1	2	3	4	5	6	7	8	9	10
P	$\frac{1}{16}$	$\frac{1}{16}$	$\frac{1}{16}$	$\frac{2}{16}$	$\frac{2}{16}$	$\frac{2}{16}$	$\frac{2}{16}$	$\frac{2}{16}$	$\frac{1}{16}$	$\frac{1}{16}$	$\frac{1}{16}$

Exercise 4: Find the null distribution for W when $n = 5$.

Even though N_1 and N_3 are random, we have (without ties) $S_1 + \cdots + S_{N_1} + T_1 + \cdots + T_{N_3} = n(n+1)/2$ so that a rejection region could be based on small values of $T_1 + \cdots + T_q$; when q is smaller than p, this may simplify some arithmetic.

Example: Suppose the signed–ranks are

$$6 \;\; 10 \;\; -2 \;\; -8 \;\; 5 \;\; 4 \;\; 9 \;\; 3 \;\; 7 \;\; 1 .$$

Now $q = 2$ and $T_1 + T_2 = 10$. If we reject H_o when this sum is "small", we find the siginificance probability (p–value) by counting cases with $T_1 + \cdots + T_{N_3} \le 10$.

N_3	Possible $T_1, \cdots, T_{N_3}$	number of cases
0	empty	1
1	1 2 3 4 5 6 7 8 9 10	10
2	1,2 1,3 1,4 1,5 1,6 1,7 1,8 1,9	
	2,3 2,4 2,5 2,6 2,7 2,8	
	3,4 3,5 3,6 3,7	
	4,5 4,6	20
3	1,2,3 1,2,4 1,2,5 1,2,6 1,2,7	
	1,3,4 1,3,5 1,3,6	
	1,4,5	
	2,3,4 2,3,5	11
4	1,2,3,4	1

The p–value associated with $T_1 + T_2 = 10$ is

$$P(T_1+\cdots+T_{N_3} \le 10) = 43/1024 \approx .042 .$$

Exercise 6: Compute the p–value for H_a: X >> Y when the signed ranks are -3 -5 7 6 9 4 8 -2 1 .

Tables of the distribution of W for $n \le 20$ are available in texts like Lehmann, 1975, to which we refer the reader for further details. Under the assumption of no possible ties and H_o,

$$E[W] = n(n+1)/4 \text{ and } Var(W) = n(n+1)(2n+1)/24$$

and the distribution of $(W - E[W])/\sqrt{Var(W)}$ is asymptotically that of the standard normal. Surprizingly, the approximation is satisfactory for n as small as 10 .

Example: In our last example, we had $n = 10$, $q = 2$ and $T_1 + T_2 = 10$. The p–value is $P(T_1+\cdots+T_{N_3} \le 10 \mid H_o)$

$$= P(W \ge 10(11)/2 - 10) = P(W \ge 45)$$

$$\approx P(Z \ge (44.5 - 55/2)/\sqrt{385/4}) \approx P(Z \ge 1.73) = .0418$$

which is practically the same as the exact value obtained in the example.

Exercise 7: Find the approximate p–value in exercise 6 .

The rest of this lesson is paraphrased from Lehmann, 1975, who uses the following result to invoke more rigor in the derivation of the null distribution of the Wilcoxon W . This also allows us to look at the *shift model* when $D = X - Y$ is "interval–valued" with CDF G :

$$H_o \text{ is } p = G(0) \text{ and } H_a \text{ is } p = G(\Delta) .$$

Lemma: *Let* $Z_1, \cdots, Z_N$ *be IID with continuous CDF* G. *Let* n *be the number of positive observations* $z_1, \cdots, z_n$ *and let* $s_1 < \cdots < s_n$ *be their ranks among the absolute values* $|z_1|, \cdots, |z_N|$. *Under the hypothesis that* G *is symmetric about* 0 , $P(N_1 = n, S_1 = s_1, \cdots, S_n = s_n) = 1/2^N$.

Proof: $P(Z_1 > 0, |Z_1| \le z)$

$= P(0 < Z_1 \le z) = P(-z \le Z_1 \le z)/2 = P(Z_1 > 0) \cdot P(|Z_1| \le z)$.

Similarly, $P(Z_1 < 0, |Z_1| \le z) = P(Z_1 < 0) \cdot P(|Z_1| \le z)$.

Therefore, the sign of Z_1 and its absolute value are independent. Since $Z_1, \cdots, Z_N$ are also independent, all the signs are independent of all the absolute values; hence all the signs are independent of all the ranks of the absolute values.

We now examine the power function of W in this shift model:

$$\mathcal{P}(\Delta) = P(W \ge c \mid \Delta) \text{ with } P(Z_1 > 0) = G(\Delta) .$$

Evaluation of $\mathcal{P}(\Delta)$ is difficult even when G is known. It is true that a normal approximation still holds but, as will be seen, this is no cureall.

For $p_1 = P(Z_1 > 0)$,

$$p_2 = P(Z_1 + Z_2 > 0) ,$$

$$p_3 = P(Z_1 + Z_2 > 0, Z_1 + Z_3 > 0 ,$$

$$E[W] = N(n-1)p_2/2 + Np_1 ,$$

$$\mathrm{Var}(W) = N(N-1)(N-2)(p_3-p_2^2) + N(N-1)[2(p_1-p_2)^2 + 3p_3(1-p_2)]/2 + Np_1(1-p_1),$$

and

$$\mathscr{P}(\Delta) \approx P(\mathscr{Z} \geq (c - .5 - E[W])/\sqrt{\mathrm{Var}(W)})$$

where this $\mathscr{Z}$ is the standard normal.

Example: We show some calculation of the little p's when the distribution of $Z_1,\cdots,Z_N$ is $N_1(\Delta,1)$. The values of

$$p_1 = P(Z_1 > 0) = P(\mathscr{Z} > -\Delta)$$

can be obtained from the usual table. Since $Z_1 + Z_2$ is also normal, the same is true of

$$p_2 = P(Z_1 + Z_2 > 0) = P(\mathscr{Z} > -2\Delta/\sqrt{2}) .$$

But for
$$p_3 = P(Z_1 + Z_2 > 0, Z_1 + Z_3 > 0)$$
$$= P(\mathscr{Z}_1 > -2\Delta/\sqrt{2}, \mathscr{Z}_2 > -2\Delta/\sqrt{2})$$

one needs a table of the bivariate normal $(\mathscr{Z}_1, \mathscr{Z}_2)$ with means 0, variances 1 and correlation $1/2$.

Exercise 8: Verify the correlation of $Z_1 + Z_2$, and $Z_1 + Z_3$ in the last example.

Now we suppose that the underlying distribution is really normal with mean Δ and variance σ^2. The power function for

$$H_o: \Delta = 0 \text{ vs } H_a: \Delta > 0$$

is approximated by:

$$P(\mathscr{Z} > z_\alpha - E[Z_1]\sqrt{(N/\sigma^2)}) \text{ with the T test;}$$

$$P(\mathscr{Z} > z_\alpha/2 - (p-1/2)\sqrt{(N/p(1-p))} \text{ with the sign test } S_N;$$

$$P(\mathscr{Z} > (c - 1/2 - E[W])/\sqrt{\mathrm{Var}(W)}) \text{ with the signed-ranks test } W.$$

The "Pitman relative efficiency" of a test1 to another test2 is basically the ratio of the sample sizes needed to have equal α

at $\Delta = 0$ and equal β or power $(1-\beta)$ at a given Δ_1. With the tests herein,

a) for W to T, the limiting value of this ratio is about .955 which means that, although the power functions will be essentially equal, we need about 5% fewer observations in the Student test than in the signed–ranks test;

b) for S_N to W, this limiting ratio is about 2/3 so that there is a considerable saving in sampling costs (1/3) when one uses the signed–ranks test instead of the simple sign test;

c) for S_N to T, this limit is about .64 which is consistent with the two previous results and also indicates a saving when the Student test is used.

LESSON *10. TESTS OF KOLMOGOROV, SMIRNOV TYPE

The Pearson chisquare test was developed for testing the goodness-of-fit of a sample to a multinomial distribution. In this lesson, we discuss some tests developed for goodness-of-fit to a continuous distribution; these are based on the (sample) or empirical distribution function.

Definition: *Let* $X_1, X_2, \cdots, X_n$ *be real RVs (of like measurements). For each real* x *and each* $i = 1(1)n$, *let* $I_x(X_i) = 1$ *when* $X_i \le x$ *and* $= 0$ *otherwise. Then, the sample distribution function (SDF) is* $F_n(x) = \sum_{i=1}^{n} I_x(X_i)/n$.

Some inspiration for the Kolmogorov test is contained in the following lemma (whose first part already occurred in Lesson 9, Part III).

Lemma: *Let* $X_1, X_2, X_3, \cdots$ *be independent real Rvs with a common CDF* F .

a) *For each real* x, $F_n(x) \overset{as}{\to} F(x)$.

b) *The distribution of* $D_n = \sup_x |F_n(x) - F(x)|$ *is the same for all continuous* F .

Proof: a) For each real x, $I_x(X_i)$ is a Bernoulli RV with $P(I_x(X_i) = 1) = P(X_i \le x) = F(x)$. Hence, the sample mean (proportion) $F_n(x)$ converges a.s. to the population mean (proportion) $F(x)$.

b) For the transformation determined by $Y_i = F(X_i)$,

$$I_y(Y_i) = 1 \text{ iff } Y_i \le y .$$

Now $F(X_i) \le y$ and $X_i \le F^{-1}(y) = \inf \{x : F(x) \ge y\}$

imply $Y_i \le y$. Hence for $F(x) = y$,

$$F_n(x) = k/n \text{ iff } F_n(F^{-1}(y)) = k/n . \qquad (*)$$

Then in

$$P(\sup_x |F_n(x) - F(x)| \le w) = P(\sup_y |F_n(F^{-1}(y)) - y| \le w)$$

the differences are all $|k/n - y|$ just as if the distribution were uniform and so do not depend on F .

Exercise 1: Let 2 3 4.1 6 8.7 9.3 10 be an observed sample from the exponential distribution with mean .5 . For

$$y = f(x) = 1 - e^{-.5x} , F^{-1}(y) = -2 \cdot \log(1-y) .$$

Verify the equality at (*) .

Now we formulate the testing problem. The simple null hypothesis H_o is that the observed sample $x_1, x_2, \cdots, x_n$ could have come from the population with the continuous CDF F_o ; the composite alternative hypothesis is that the sample might be from a population with some other continuous F . We reject H_o when the observed $D_n = \sup_x |F_n(x) - F_o(x)|$ is "too large". The significance probability is obtained from the distribution of D_n under H_o . Because of the strong convergence of F_n to the true CDF F , the D_n test is consistent; ie., the power tends to 1 if F_o is not correct.

For F continuous, Kolmogorov, 1933, gave recurrence relations for finite n that have been used to tabulate the distribution of D_n . In this article, he also established the limiting distribution as

$$\lim_{n\to\infty} P(\sqrt{n}D_n > z) = 2\sum_{r=1}^{\infty}(-1)^{r-1}\exp(-2r^2z^2) .$$

For example,

$$P(\sqrt{n}D_n > 1.36) \approx .05 \text{ and } P(\sqrt{n}D_n > 1.63) \approx .01 .$$

The following short table has been adapted from Miller, 1956.

$P(D_{max} > .) = \alpha$

Sample size	α = .10	.05	.02	.01
5	.50945	.56328	.62718	.66853
10	.36866	.40925	.45662	.44893
11	.35242	.39122	.43670	.46770
12	.33815	.37543	.41918	.44905
13	.32549	.36143	.41918	.43247
14	.31417	.34890	.38970	.41762
15	.30397	.33760	.37713	.40420
16	.29472	.32733	.36571	.39201
17	.28627	.31796	.35528	.38086
18	.27851	.30936	.34569	.37062
19	.27136	.30143	.33685	.36117
20	.26473	.29408	.32866	.35241
21	.25858	.28724	.32104	.34427
22	.25283	.28087	.31394	.33666
23	.24746	.27490	.30728	.32954
24	.24242	.26931	.30104	.32286
25	.23768	.26404	.29516	.31657
30	.21756	.24170	.27023	.28987
39	.19148	.21273	.23786	.25518
40	.18913	.21012	.23494	.25205
Approximation for n > 40	$1.22/\sqrt{n}$	$1.36/\sqrt{n}$	$1.52/\sqrt{n}$	$1.63/\sqrt{n}$

(This table has been reproduced with the permission of the American Statistical Association.)

For discrete RVs, it turns out that $P(\sqrt{n}D_n > 1.6276)$ is actually slightly less than .01 and the tables generated for continuous RVs are not satisfactory. Conover, 1972, has corrected this particular problem.

As with other "two-sided" tests, the procedure can be reversed to obtain a confidence interval. If d_α is the critical value of D_n, then from $P(\sup_x |F_n(x) - F(x)| > d_\alpha) \approx \alpha$, we get $P(F_n(x) - d_\alpha \le F(x) \le F_n(x) + d_\alpha) \approx 1-\alpha$ whence

$$F_n(x) - D_\alpha \, , \, F_n(x) + d_\alpha$$

is a(n approximate) $100(1-\alpha)\%$ confidence band for $F(x)$:

$100(1-\alpha)\%$ of the bands computed in this way will contain the true F.

Exercise 2: For $\alpha = .10$ and $n = 25$, $d_\alpha = .238$. Sketch $F_n(x) - .238$ and $F_n(x) + .238$ on the same set of axes. The sample is:

16 17 18 18 19 22 22 24 24 25 25 27
35 36 37 39 40 42 43 46 47 48 48 50 55.

Any CDF between these bounds could be that of the underlying population.

Next we show how to compute the observed D_n. Let the ordered values be

$$-\infty = x_o < x_{(1)} < x_{(2)} < \cdots < x_{(n)} < x_{(n+1)} = \infty .$$

As the following sketch shows, we need to examine *all* the differences $L_i = |F_n(x_{(i)}) - F_o(x_{(i)})|$ and

$$R_i = |F_o(x_{(i)}) - F_n(x_{(i-1)})| .$$

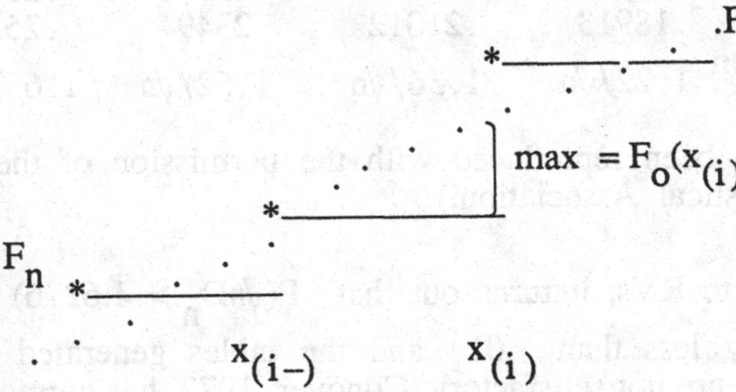

Example: Could the RS .355 .392 .612 .230 .530 .749 .701 .518 .545 .328 be from a uniform distribution on (0,1)? In this case, $F_o(x_{(i)}) = x_{(i)}$ and the arithmetic is summarized in:

i	$x_{(i)}$	$F_n(x_{(i)})$	$F_n(x_{(i-1)})$	L_i	R_i
1	.230	.1	0	.130	.230
2	.328	.2	.1	.128	.228
3	.355	.3	.2	.055	.155
4	.392	.4	.3	.008	.092
5	.518	.5	.4	.018	.118 .
6	.530	.6	.5	.070	.030
7	.545	.7	.6	.155	.055
8	.612	.8	.7	.188	.088
9	.701	.9	.8	.199	.099
10	.749	1.00	.9	.251	.151

This gives $D_{10} = .251$ and ,$P(D_{10} > .251) > .10$. The sample may be from the standard uniform distribution.

Exercise 3: Use the Kolmogorov D_n test for the hypothesis that the given sample could be from the indicated population:

a) 6 3 8 7 5 8 3 7 5 7 6 6 9 4 5 7 3 8, exponential distribution, mean 6;

b) 1.464 1.137 3.455 .677 .932 1.296 .812 2.298 1.241 .043 1.060 –1.526 .469 –.588 –.190 –.865, standard normal. Use the normal table to get $F_o(-1.526)$ etc.

One of the most common assumptions in statistical applications is that the underlying random error is essentially normally distributed and, in particular, that a given sample arose from a normal distribution but without knowledge of the population mean and/or variance. In such a case, we would be considering a family of F but not a specific F_o so that the Kolmogorov test illustrated above is not applicable. A solution using a modified form of this test was developed by Lilliefors, 1967. He uses

$$D_{max} = \sup_x |F_n(x) - F_*(x)|$$

where F_* is the cumulative normal distribution function using estimates of the parameters: $\hat{\mu} = \bar{x}$ and $\hat{\sigma}^2 = \Sigma(x_i - \bar{x})^2/(n-1)$.

Recently, Dallal and Wilkinson, 1986, presented a

"corrected" table of Upper Tail Percentiles

$$P(D_{max} > \cdot) = \alpha .$$

Sample size	"α" .20	.15	.10	.05	.01	.001
4	.303	.321	.346	.376	.413	.433
5	.289	.303	.319	.343	.397	.439
6	.269	.281	.297	.323	.371	.424
7	.252	.264	.280	.304	.351	.402
8	.239	.250	.265	.288	.333	.384
9	.227	.238	.252	.274	.317	.365
10	.217	.228	.241	.262	.304	.352
11	.208	.218	.231	.251	.291	.338
12	.200	.210	.222	.242	.281	.325
13	.193	.202	.215	.234	.271	.314
14	.187	.196	.208	.226	.262	.305
15	.181	.190	.201	.219	.254	.296
16	.176	.184	.195	.213	.247	.287
17	.171	.179	.190	.207	.240	.279
18	.167	.175	.185	.202	.234	.273
19	.163	.170	.181	.197	.228	.166
20	.159	.166	.176	.192	.223	.260
25	.143	.150	.159	.173	.201	.236
30	.131	.138	.146	.159	.185	.217
40	.115	.120	.128	.139	.162	.189
100	.074	.077	.082	.089	.104	.122
400	.037	.039	.041	.045	.052	.061
900	.025	.026	.028	.030	.035	.042

(This table has been reproduced with permission of the American Statistical Association.)

Perhaps the most interesting aspect of this form of the test is that the tables of the distribution of D_{max} have been obtained by simulation, the "Monte Carlo" technique. For example, take a random sample of size 4 from the standard normal distribution and compute D_{max} ; do this for a million or so independent samples and their histogram will approximate the distribution of D_{max} for $n = 4$. Etc.

An equivalent form of D_{max} is obtained by computing the SDF $F_n(x)$ for the "standardized values" $z_i = (x_i - \hat{\mu})/\hat{\sigma}$ and taking F_* as the standard normal CDF.

Example: Below are the last four digits of a sample of social security numbers. Could they be an observed sample of a normal distribution?

4055	3279	0903	8717	4060	5080	4017	5018	8814
5740	7878	9408	4034	3230	2731	4917	1219	6344
3092	4038	5067	4739	3417	6680	9135	2196	6129
4177	6446	9046	8541	5471	4030	8551	9789	2110
4524	5018	4035	0787	7625	9589	0312	3022	6123
1158	4040	7641	4424	2388	5474	3053	7883	0036
4457	3123	3822						

For $\hat{\mu} = 4923.37$, $\hat{\sigma} = 2546.15$, $z_i = (x_i - \hat{\mu})/\hat{\sigma}$:

−.3411	−.6458	−1.5790	1.4899	−.3391	.0615
−.3560	.0372	1.5280	.3207	1.1604	1.7613
−.3493	−.6651	−.8611	−.0025	−1.4549	.5580
−.7193	−.3477	.0654	−.0724	−.5916	.6899
1.6540	−1.0712	.4735	−.2931	.5980	1.6192
1.4208	.2151	−.3509	1.4248	1.9110	−1.1050
−.1569	−.0372	−.3489	−1.6246	1.0611	1.8324
−1.8111	−.7468	.4712	−1.4788	−.3469	1.0673
−.1961	−.9958	.2163	−.7346	1.1624	−1.9195
−.1832	−.7071	−.4326 .			

After looking up all $F_*(z_i)$ in the normal table, we find $D_{max} = .107$. Since the sample size 57 does not appear in our table, we interpolate to find $P(D_{max} > .107) \approx .105$ which is somewhat inconclusive.

Exercise 4: Try Lilliefors' test on the last *two* digits of the numbers in the example above.

As indicated in the table of the Kolmogorov statistics, it is possible to consider one–sided tests on a given F_o :

a) for H_o: $F(x) \geq F_o(x)$ vs H_a: $F(x) < F_o(x)$, we rejct H_o when $\sup_x (F_o(x) - F_n(x)) > d_{\alpha/2}$;

b) for H_o: $F(x) \leq F_o(x)$ vs H_a: $F(x) > F_o(x)$, we rejct H_o when $\sup_x (F_n(x) - F_o(x)) > d_{\alpha/2}$.

Recall that these inequalites may be expressed in terms of

"stochastically larger" and represent more general notions of location than statements about a mean or median. .

We close this lesson with a test of another assumption used in some applications:

two samples are from the same distribution.

The Smirnov test of this null hypothesis is a two sample version of the Kolmogorov test above (and some authors use both names on either test).

The observations will consist of two independent random samples: $X_1,X_2,\cdots,X_n$ with common CDF F_X

and $Y_1,Y_2,\cdots,Y_m$ with common CDF F_Y.

Let the corresponding SDFs be S_X and S_Y.

For H_o: $F_X(x) = F_Y(x)$ for all real x

vs H_a: $F_X(x) \neq F_Y(x)$ for some real x,

reject H_o when $W_1 = \sup_x |S_X(x) - S_Y(x)| > w_\alpha$.

For H_o: $F_X(x) \leq F_Y(x)$ for all real x vs

H_a: $F_X(x) > F_Y(x)$ for some real x,

reject H_o when $W_2 = \sup_x S_X(x) - S_Y(x) > w_{\alpha/2}$.

For H_o: $F_X(x) \geq F_Y(x)$ for all real x vs

H_a: $F_X(x) < F_Y(x)$ for some real x,

reject H_o when $W_3 = \sup_x S_Y(x) - S_X(x) > w_{\alpha/2}$.

Naturally, more extensive tables than appear here are available.

$P(W_1 > w_\alpha) = \alpha$ for the Smirnov test

n	m	α = .10	.05	.02	.01
2	5	4/5			
2	10	4/5	9/10		
3	10	7/10	8/10	9/10	9/10
4	10	13/20	14/20	16/20	16/20
5	10	4/6	4/6	5/6	5/6
5	20	11/20	12/20	14/20	15/20
7	14	7/14	8/14	9/14	10/14
10	20	9/20	10/20	11/20	12/20
12	12	5/12	6/12	7/12	7/12
15	20	24/60	26/60	29/60	31/60

For the one tail test, the p–values are halved.

Example: For a given set of data, we order the values and find the differences in their SDFs:

X	Y	$S_X(X) - S_Y(x)$	
6.1		1/12– 0	= 1/12
6.2		2/12– 0	= 2/12
6.3		3/12– 0	= 3/12
7.1		4/12– 0	= 4/12
	7.6	4/12–1/12	= 3/12
7.8		5/12–1/12	= 4/12
	7.9	5/12–2/12	= 3/12
8.1		6/12–2/12	= 4/12
8.2		7/12–2/12	= 5/12
8.3	8.3	8/12–3/12	= 5/12
	8.4	8/12– 4/12	= 4/12
	8.5	8/12– 5/12	= 3/12
8.6	8.6	9/12– 6/12	= 3/12
8.7		10/12– 6/12	= 4/12
	8.8	10/12– 7/12	= 5/12
	8.9	10/12– 8/12	= 4/12
	9.1	10/12– 9/12	= 3/12
9.2		11/12– 9/12	= 2/12
9.3		12/12–9/12	= 3/12
	9.4	12/12–10/12	= 2/12
	9.5	12/12–11/12	= 1/12
	9.6	12/12–12/12	= 0

The sup is 5/12 which leads to rejection of equivalent distributions if $\alpha = .10$ but not for smaller values.

Exercise 5: Apply the Smirnov test to:

X: 2.3 4.5 6.7 7.5 4.3 8.9 9.8 10.2 11.3 12.4 13.6 8.1
Y: 3.2 5.4 7.6 5.7 3.4 9.8 8.9 2.1 2.1 4.2 33.6 12.8.

We illustrate some of the theory for X and Y continuous so that, with probability 1, there are no ties. Then the values of W_1, W_2, W_3 will depend only on the order of the items in the sample and not their actual values (alà MWW). Under the null hypothesis each such arrangement will have the same probability, namely, $1/\binom{n+m}{m}$; the distributions can be tabulated.

Example: Suppose that the X sample is $a < b$ and the Y sample is $c < d < e$. For each arrangement, we list S_Y below S_X and the resultant W_1 :

a b c d e	1/2	1	1	1	1	1
	0	0	1/3	2/3	1	
a c b d e	1/2	1/2	1	1	1	2/3
	0	1/3	1/3	2/3	1	
a c d b e	1/2	1/2	1/2	1	1	1/2
	0	1/3	1/3	2/3	1	
a c d e b	1/2	1/2	1/2	1/2	1	1/2
	0	1/3	2/3	1	1	
c a d e b	0	1/2	1/2	1/2	1	1/2
	1/3	1/3	2/3	1	1	
c d a e b	0	0	1/2	1/2	1	2/3
	1/3	2/3	2/3	1	1	
c d e a b	0	0	0	1/2	1	1
	1/3	2/3	1	1	1	
c a d b e	0	1/2	1/2	1	1	1/3
	1/3	1/3	2/3	2/3	1	
c a b d e	0	1/2	1	1	1	2/3
	1/3	1/3	1/3	2/3	1	
c d a b e	0	0	1/2	1	1	2/3 .
	1/3	2/3	2/3	2/3	1	

Exercise 6: List the distributions of W_1, W_2, W_3 in case $n = 2$ and $m = 3$.

LESSON *11. CATEGORICAL DATA

In this lesson, we consider extensions of the Pearson Chisquare technique to classification of data in other forms. After giving some details for two–way tables, we indicate possibilities for multiway tables.

First n independent observations are classified by two criteria just as in the 2 by 2 tables but now we have r levels of one variable or category in the rows and c levels of the other in the columns. For the time being, the indices will be $i = 1(1)r$, $j = 1(1)c$; θ_{ij} is the probability that a single observation will fall (be classified) at level i of the row variable and level j of the column variable. The joint multi– nomial likelihood is

$$L(v) = n!\, \Pi_i\, \Pi_j\, \theta_{ij}^{\,n_{ij}}/n_{ij}!\ .$$

The data is usually presented in a table:

n_{11}	n_{12}	$\cdots$	n_{1c}	$n_{1\cdot}$
$\cdots$				
n_{r1}	n_{r2}	$\cdots$	n_{rc}	$n_{r\cdot}$
$n_{\cdot 1}$	$n_{\cdot 2}$	$\cdots$	$n_{\cdot c}$	n

where $n_{i\cdot} = \Sigma_j n_{ij}$, $n_{\cdot j} = \Sigma_i n_{ij}$, $n = \Sigma_i\, \Sigma_j n_{ij}$.

One application wherein this sort of thing occurs "automatically" is for census data; for example, income is limited to classes like "under 1000", "at least 1000 but under 2000", "at least 2000 but under 5000", etc. and another variable is ethnic group: Black, Hispanic, etc.

Now consider the null hypothesis of independence

$$H_o: \theta_{ij} = \theta_{i\cdot}\, \theta_{\cdot j} \quad \text{where } \theta_{i\cdot} = \Sigma_j \theta_{ij}\ , \theta_{\cdot j} = \Sigma_i \theta_{ij}\ .$$

The alternative contains no restrictions except, of course, $\Sigma_i\, \Sigma_j \theta_{ij} = 1$. Under H_a , the MLEs are

$$\hat{\theta}_{ij} = n_{ij}/n\ .$$

Under H_o , $L(v) = n!\, \Pi_j\, \Pi_i\, \theta_{i\cdot}^{\,n_{i\cdot}}\, \theta_{\cdot j}^{\,n_{\cdot j}}/n_{ij}!$ and the MLEs are

$$\hat{\theta}_{i\cdot} = n_{i\cdot}/n\,,\ \hat{\theta}_{\cdot j} = n_{\cdot j}/n\,.$$

Exercise 1: Verify the maximum likelihood estimates.

The LRT is to reject H_0 when $\lambda < \lambda_0$ where

$$\lambda = \Pi_i \Pi_j (n_{i\cdot}/n)^{n_{i\cdot}} (n_{\cdot j}/n)^{n_{\cdot j}} \div \Pi_i \Pi_j (n_{ij}/n)^{n_{ij}}\,.$$

Here the parameter space $\Theta = \{\theta_{ij} : \Sigma_i \Sigma_j \theta_{ij} = 1\}$ has dimension $rc-1$ and the null subspace

$$\omega = \{\theta_{ij} : \theta_{ij} = \theta_{i\cdot}\theta_{\cdot j}\,, \Sigma_i \theta_{i\cdot} = 1 = \Sigma_j \theta_{\cdot j}\}$$

has dimension $r-1 + c-1$ so that, by the general results of Lesson 10, Part VI, $-2\log\lambda$ has an asymptotic distribution which is chisquare with

$$rc - 1 - (r - 1 + c - 1) = (r - 1)(c - 1) \text{ dof.}$$

As in Lesson 9, Part VI, we get

$$-2\log\lambda \approx \Sigma_i \Sigma_j (n_{ij} - n_{i\cdot}n_{\cdot j}/n)^2/(n_{i\cdot}n_{\cdot j}/n)\,. \qquad (*)$$

Exercise 2: Supply some details for this discussion as follows. Let $w_{ij} = n_{i\cdot}n_{\cdot j}/n \cdot n_{ij}$. Show that:

a) $-2\log\lambda = -2\Sigma_i \Sigma_j n_{ij}(w_{ij} - 1 - (w_{ij} - 1)^2/2 + (w_{ij} - 1)^3/3(1 + \xi_{ij})^3$

where each ξ_{ij} is "between" w_{ij} and 1 ;

b) this is asymptotic to (*) by showing that the third order terms tend to 0 in probability.

Example: Consider "Final Grades in Statistics" vs "Undergraduate Major". (This school allows students to disenroll at midterm from courses in which their midterm grade is not at least C !)

	A	B	C	D	E
Math	33	61	95	8	2
EE	48	92	106	6	0
IE	39	49	73	7	1
ME	27	37	68	9	2
CS	16	29	83	4	0.

Since $\mathscr{X}^2 = \Sigma_i \Sigma_j (n_{ij} - n_{i.} n_{.j}/n)^2/(n_{i.} n_{.j}/n)$

$$= n(\Sigma_i \Sigma_j n_{ij}^2/n_{i.} n_{.j} - 1),$$

we first compute

$$33^2/163 \cdot 199 + 61^2/268 \cdot 199 + \cdots + 4^2/34 \cdot 132 \approx 1.0336$$

and then find the observed chisquare

$$895(1.0336 - 1) \approx 30.04 .$$

From tables of the (central) chisquare distribution, we get $P(\chi^2(16) > 30.04)$ close to .01 and reject H_o. The experimental conclusion is that grades and major are not independent.

Exercise 3: Apply the chisquare test of independence to the following table (which is used differently in Kendall and Stuart , 1974, page 595.)

	Unaided Distance Vision				
Right eye \ Left eye	High	Second	Third	Low	
High	821	112	85	35	1053
Second	116	494	145	27	782
Third	72	151	583	87	893
Low	43	34	106	331	514
	1052	791	919	480	3242

The following discussion is a generalization of the test for the equality of two simple proportions. At first it may appear a bit surprizing that the test criterion is the same as that for "independence" but note that with this hypothesis, the probability of a "success" does not depend on which population is being sampled. Here, we consider c independent populations (of some one variable) each classified into the same r cells. Now θ_{ij} is the probability that a single observation in population j will be classified into the ith cell. The corresponding multinomial likelihood is

$$L(v) = \Pi_j\, n_{\cdot j}!\, \Pi_i\, \theta_{ij}^{n_{ij}}/n_{ij}!\ .$$

The null hypothesis of *homogeneity* is

$$H_o : \begin{cases} \theta_{11} = \cdots = \theta_{1c} = \theta_1 \\ \cdots \\ \theta_{r1} = \cdots = \theta_{rc} = \theta_r \end{cases}$$

where $\theta_1, \cdots, \theta_r$ are not specified and $\Sigma_i \theta_i = 1$. The alternative H_a is unrestricted except for

$$\Sigma_i \theta_{i1} = \cdots = \Sigma_i \theta_{ic} = 1\ .$$

Under H_a , the MLEs are $\hat{\theta}_{ij} = n_{ij}/n_{\cdot j}$.

Under H_o, $L(v) = \Pi_j\, n_{\cdot j}!\, \Pi_i\, \theta_i^{n_{ij}}/n_{ij}! = \Pi_i\, \theta_i^{n_{i\cdot}}\ \Pi_j\, n_{\cdot j}!/n_{ij}!$

and the MLEs are $\hat{\theta}_i = n_{i\cdot}/n$.

Application of the LRT yields

$$\lambda = \Pi_i (n_{i\cdot}/n)^{n_{i\cdot}} / \Pi_j (n_{ij}/n_{\cdot j})^{n_{ij}}$$

which is the same as at (*). Here

$$\Theta = \{\theta_{ij} : \Sigma_i \theta_{i1} = \cdots = \Sigma_i \theta_{ic} = 1\}$$

has dimension $c(r-1)$ and $\omega = \{\theta_i : \Sigma_i \theta_i = 1\}$ has dimension $r-1$ so that $-2\log \lambda$ has

$$c(r-1) - (r-1) = (c-1)(r-1) \text{ dof}$$

as before but for different reasons. Again,

$$-2\log \lambda \approx \Sigma_i\, \Sigma_j (n_{ij} - n_{i\cdot}\, n_{\cdot j}/n)^2 / (n_{i\cdot}\, n_{\cdot j}/n)$$

and, under H_o , the asymptotic distribution is (central) chisquare with $(r-1)(c-1)$ dof.

When $r = 2$, this is a test for the equality of the proportions in c independent binomial distributions.

Exercise 4: Say that inspection of a certain part manufactured in different plants yields the following data.

Plant	A	B	C	D	E
Def	28	39	15	20	51
Non	172	161	85	80	149

Are the proportions of defectives equal in the five plants?

In the rest of this lesson, we concentrate on independence. A null hypothesis of independence usually does not specify $\theta_{i\cdot}$ and $\theta_{\cdot j}$ but we can see some interesting details when we proceed as if it did. Thus consider a LRT of

$$H_{oo}: \theta_{ij} = \theta_{i\cdot}\,\theta_{\cdot j} \text{ known vs } H_a \text{ as before.}$$

$$-2\log \lambda_o = 2\cdot\Sigma_i\,\Sigma_j\, n_{ij}\,\log(n_{ij}/n\theta_{i\cdot}\,\theta_{\cdot j})$$

can be rewritten as

$$2\Sigma_i\Sigma_j n_{ij}\,\log(n\cdot n_{ij}/n_{i\cdot}\,n_{\cdot j}) + 2\Sigma_j n_{\cdot j}\,\log(n_{\cdot j}/n\theta_{\cdot j})$$
$$+ 2\Sigma_i n_{i\cdot}\,\log(n_{i\cdot}/n\theta_{i\cdot})\,.$$

Exercise 5: Check the algebra in the last sentence above.

The first of the three sums above is $-2\log\lambda$ for the original test of independence.

The second is $-2\log\lambda_R$ for a test of a null hypothesis of independence with $\theta_{\cdot j}$ specified.

The third is $-2\log\lambda_C$ for a test of a null hypothesis specifiying independence and the $\theta_{i\cdot}$. Under H_{oo}, the asypmptotic distributions of each of these "information statistics" (see Kullbach, 1959) are central chisquare and

$-2\log\lambda_o$ has rc–1 dof

$-2\log\lambda$ has (r–1)(c–1) dof

$-2\log\lambda_R$ has r–1 dof

$-2\log\lambda_C$ has c–1 dof

As in earlier discussions, each is also (asymptotically) a Pearson–like sum so that we get the *asymptotic partition*

$$\Sigma_i \Sigma_j (n_{ij} - n\theta_{i.}\theta_{.j})^2/n\theta_{i.}\theta_{.j}$$

$$= \Sigma_i (n_{i.} - n\theta_{i.})^2/n\theta_{i.}$$

$$+ \Sigma_j (n_{.j} - n\theta_{.j})^2/n\theta_{.j}$$

$$+ \Sigma_i \Sigma_j (n_{ij} - n_{i.}n_{.j}/n)^2/(n_{i.}n_{.j}/n) .$$

Of course, under the original H_o , $\theta_{i.}$ and $\theta_{.j}$ must be estimated and then all

$$(n_{i.} - n\hat{\theta}_{i.})^2/n\hat{\theta}_{i.} = 0 = (n_{.j} - n\hat{\theta}_{.j})^2/n\hat{\theta}_{.j} .$$

More interesting partitions can be made when there are more variables (factors, categories) in the classification as we shall see.

A common application is the evaluation of manufactured parts where the criteria are different qualities of the part and the levels are the proportions of defective/non–defective. Then the indices i,j,k would have values 1,2 and $\{\theta_{kij}\}$ would represent the proportions of each of the eight kinds of parts.

Example: The following is a record (Pass/Fail comprehensive exam) of teaching assistants taking three exams per semester.

			Analysis P	Analysis F
P		Statistics P	16	3
	Algebra	F	8	4
F		Statistics P	12	6
		F	9	7

Symbolically, the table is

$$\begin{matrix} n_{111} & n_{112} \\ n_{121} & n_{122} \end{matrix} ;$$

$$\begin{matrix} n_{211} & n_{212} \\ n_{221} & n_{222} \end{matrix}$$

the marginal sums are indicated by

$$n_{ki\cdot} = \Sigma_j n_{kij}, \; n_{k\cdot\cdot} = \Sigma_i \Sigma_j n_{kij}, \text{ etc.}$$

A general three-way model with p "pages", r "rows", c "columns" and n observations has the multinomial likelihood

$$L(v) = n! \, \Pi_{k=1}^{p} \Pi_{i=1}^{r} \Pi_{j=1}^{c} \theta_{kij}^{n_{kij}} / n_{kij}! \, .$$

By analogy with the two-way table, we can get an asymptotic partition under the hypothesis of independence with

$\theta_{kij} = \theta_{k\cdot\cdot}\,\theta_{\cdot i\cdot}\,\theta_{\cdot\cdot j}$ known:

$$\Sigma\Sigma\Sigma\,(n_{kij} - n\theta_{k\cdot\cdot}\,\theta_{\cdot i\cdot}\,\theta_{\cdot\cdot j})^2 / n\theta_{k\cdot\cdot}\,\theta_{\cdot i\cdot}\,\theta_{\cdot\cdot j}$$

$$= \Sigma\Sigma\Sigma\,(n_{kij} - n_{k\cdot\cdot}\,n_{\cdot i\cdot}\,n_{\cdot\cdot j}/n)^2/(n_{k\cdot\cdot}\,n_{\cdot i\cdot}\,n_{\cdot\cdot j}/n)$$

$$+ \Sigma_i \Sigma_j\,(n_{\cdot ij} - n_{\cdot i\cdot}\,n_{\cdot\cdot j}/n)^2/(n_{\cdot i\cdot}\,n_{\cdot\cdot j}/n)$$

$$+ \Sigma_k \Sigma_j\,(n_{k\cdot j} - n_{k\cdot\cdot}\,n_{\cdot\cdot j}/n)^2/(n_{k\cdot\cdot}\,n_{\cdot\cdot j}/n)$$

$$+ \Sigma_k \Sigma_i\,(n_{ki\cdot} - n_{k\cdot\cdot}\,n_{\cdot i\cdot}/n)^2/(n_{k\cdot\cdot}\,n_{\cdot i\cdot}/n)$$

$$+ \Sigma_k\,(n_{k\cdot\cdot} - n\theta_{k\cdot\cdot})^2/n\theta_{k\cdot\cdot}$$

$$+ \Sigma_i\,(n_{\cdot i\cdot} - n\theta_{\cdot i\cdot})^2/n\theta_{k\cdot\cdot}$$

$$+ \Sigma_j\,(n_{\cdot\cdot j} - n\theta_{\cdot\cdot j})^2/n\theta_{\cdot\cdot j}$$

and the corresponding degrees of freedom are additive:

$$\begin{aligned} prc - 1 = {} & (p-1)(r-1)(c-1) \\ & +(r-1)(c-1) \\ & +(p-1)(c-1) \\ & +(p-1)(r-1) \\ & +p-1 \\ & +r-1 \\ & +c-1 \end{aligned} \, .$$

This suggests that one might test independence of pages and

rows, pages and columns, rows and columns with the two–dimensional sums as well as overall independence with the three–dimensional sum.

Exercise 6: Show that:

a) $\Sigma_i \Sigma_j (n_{ij} - n_{i.} n_{.j}/n)^2/(n_{i.} n_{.j}/n)$

$= n \cdot \Sigma_i \Sigma_j n_{ij}^2/n_{i.} n_{.j} - n$;

b) $\Sigma_k \Sigma_i \Sigma_j (n_{kij} - n_{k..} n_{.i.} n_{..j}/n)^2/(n_{k..} n_{.i.} n_{..j}/n)$

$= n^2 \cdot \Sigma_k \Sigma_i \Sigma_j n_{kij}^2/n_{k..} n_{.i.} n_{..j} - n$.

Example: We will apply this to the data on graduate assistants; here all the dof are 1 . Typical critical values are

$\alpha =$	.10	.05	.025	.01	.005
$\chi^2(1) =$	2.706	3.841	5.024	6.635	7.879

a) For independence of algebra, statistics, analysis:

$$65^2[16^2/31(37)45 + 3^2/31(37)20 + 8^2/31(28)45 + 4^2/31(28)20 + 12^2/34(37)45 + 6^6/34(37)20 + 9^2/34(28)45 + 7^2/34(28)20] - 65 \approx 4.08 .$$

The p–value is a bit smaller than .05 so we reject three–way independence.

b) For independence of algebra and analysis:

$$65[24^2/31(45) + 7^2/31(20) + 21^2/34(45) + 13^2/34(20)] - 65 \approx 1.866 .$$

We do not reject this null hypothesis.

c) For independence of algebra and statistics:

$$65[19^2/31(37) + 18^2/34(37) + 12^2/31(28) + 16^2/34(28)] - 65 \approx .461 .$$

We do not reject this independence either.

d) For independence of statistics and analysis:

$$65[28^2/37(45) + 9^2/37(20) + 17^2/28(45) + 11^2/28(20)] - 65 \approx 1.675 .$$

We do not reject this independence.

Note: this is an empirical example of the fact that independence in pairs does not imply mutual independence.

Exercise 7: Test two– and three–way independence for the Aickin, 1983, gun data:

		F0	F1			F0	F1
Y0	G0	126	141	Y1	G0	152	182
	G1	319	290		G1	463	403

Y0 and Y1 are the years 1975–1976; G0 and G1 are responses Opposing–Not opposing gun registration; F0 and F1 are Neutral–Slanted forms of the question on registration in a public survey.

TABLES FOR SOME DISTRIBUTIONS

The tables at the end of this Volume II were generated in OS SAS 5.16 at New Mexico State University by using modifications of the example pages 750–751 in SAS User's Guide: Basics, Version 5 Edition, Copyright 1985 by SAS Institute Inc., Cary, NC, USA.

Here the restriction to three decmal places allows us to skip printing some lines when convenient; for example, if $P(X \leq 7) = 1.000$, so does $P(X \leq a)$ for $a \geq 7$; or, if $P(X \geq 7) = 0.000$, so does $P(X \leq a)$ for $a \leq 7$.

POISSON CUMULATIVE DISTRIBUTION FUNCTION
For example, P(X ≤ 4 | LAMBDA = MEAN = 1) = .996 .

X	LAMBDA .25	.35	.45	.50	.75	1
0	0.779	0.705	0.638	0.607	0.472	0.368
1	0.974	0.951	0.925	0.910	0.827	0.736
2	0.998	0.994	0.989	0.986	0.959	0.920
3	1.000	1.000	0.999	0.998	0.993	0.981
4	1.000	1.000	1.000	1.000	0.999	0.996
5	1.000	1.000	1.000	1.000	1.000	0.999
6	1.000	1.000	1.000	1.000	1.000	1.000
7	1.000	1.000	1.000	1.000	1.000	1.000
8	1.000	1.000	1.000	1.000	1.000	1.000
9	1.000	1.000	1.000	1.000	1.000	1.000

X	LAMBDA 1.5	2	2.5	3	7	10
0	0.223	0.135	0.082	0.050	0.001	0.000
1	0.558	0.406	0.287	0.199	0.007	0.000
2	0.809	0.677	0.544	0.423	0.030	0.003
3	0.934	0.857	0.758	0.647	0.082	0.010
4	0.981	0.947	0.891	0.815	0.173	0.029
5	0.996	0.983	0.958	0.916	0.301	0.067
6	0.999	0.995	0.986	0.966	0.450	0.130
7	1.000	0.999	0.996	0.988	0.599	0.220
8	1.000	1.000	0.999	0.996	0.729	0.333
9	1.000	1.000	1.000	0.999	0.830	0.458
10	1.000	1.000	1.000	1.000	0.901	0.583
11	1.000	1.000	1.000	1.000	0.947	0.697
12	1.000	1.000	1.000	1.000	0.973	0.792
13	1.000	1.000	1.000	1.000	0.987	0.864
14	1.000	1.000	1.000	1.000	0.994	0.917
15	1.000	1.000	1.000	1.000	0.998	0.951
16	1.000	1.000	1.000	1.000	0.999	0.973
17	1.000	1.000	1.000	1.000	1.000	0.986
18	1.000	1.000	1.000	1.000	1.000	0.993
19	1.000	1.000	1.000	1.000	1.000	0.997
20	1.000	1.000	1.000	1.000	1.000	0.998
21	1.000	1.000	1.000	1.000	1.000	0.999
22	1.000	1.000	1.000	1.000	1.000	1.000
23	1.000	1.000	1.000	1.000	1.000	1.000

BINOMIAL CUMULATIVE DISTRIBUTION FUNCTION

For example, P(X < 10 | N = 20, θ = .30) = .983 .

θ	*.01*	*.05*	*.10*	*.15*	*.20*	*.25*	*.30*
X				N = 20			
0	0.818	0.358	0.122	0.039	0.012	0.003	0.001
1	0.983	0.736	0.392	0.176	0.069	0.024	0.008
2	0.999	0.925	0.677	0.405	0.206	0.091	0.035
3	1.000	0.984	0.867	0.648	0.411	0.225	0.107
4	1.000	0.997	0.957	0.830	0.630	0.415	0.238
5	1.000	1.000	0.989	0.933	0.804	0.617	0.416
6	1.000	1.000	0.998	0.978	0.913	0.786	0.608
7	1.000	1.000	1.000	0.994	0.968	0.898	0.772
8	1.000	1.000	1.000	0.999	0.990	0.959	0.887
9	1.000	1.000	1.000	1.000	0.997	0.986	0.952
10	1.000	1.000	1.000	1.000	0.999	0.996	0.983
11	1.000	1.000	1.000	1.000	1.000	0.999	0.995
12	1.000	1.000	1.000	1.000	1.000	1.000	0.999
13	1.000	1.000	1.000	1.000	1.000	1.000	1.000
14	1.000	1.000	1.000	1.000	1.000	1.000	1.000
15	1.000	1.000	1.000	1.000	1.000	1.000	1.000
X				N = 25			
0	0.778	0.277	0.072	0.017	0.004	0.001	0.000
1	0.974	0.642	0.271	0.093	0.027	0.007	0.002
2	0.998	0.873	0.537	0.254	0.098	0.032	0.009
3	1.000	0.966	0.764	0.471	0.234	0.096	0.033
4	1.000	0.993	0.902	0.682	0.421	0.214	0.090
5	1.000	0.999	0.967	0.838	0.617	0.378	0.193
6	1.000	1.000	0.991	0.930	0.780	0.561	0.341
7	1.000	1.000	0.998	0.975	0.891	0.727	0.512
8	1.000	1.000	1.000	0.992	0.953	0.851	0.677
9	1.000	1.000	1.000	0.998	0.983	0.929	0.811
10	1.000	1.000	1.000	1.000	0.994	0.970	0.902
11	1.000	1.000	1.000	1.000	0.998	0.989	0.956
12	1.000	1.000	1.000	1.000	1.000	0.997	0.983
13	1.000	1.000	1.000	1.000	1.000	0.999	0.994
14	1.000	1.000	1.000	1.000	1.000	1.000	0.998
15	1.000	1.000	1.000	1.000	1.000	1.000	1.000
16	1.000	1.000	1.000	1.000	1.000	1.000	1.000
17	1.000	1.000	1.000	1.000	1.000	1.000	1.000
18	1.000	1.000	1.000	1.000	1.000	1.000	1.000

θ	.01	.05	.10	.15	.20	.25	.30
X				N = 30			
0	0.740	0.215	0.042	0.008	0.001	0.000	0.000
1	0.964	0.554	0.184	0.048	0.011	0.002	0.000
2	0.997	0.812	0.411	0.151	0.044	0.011	0.002
3	1.000	0.939	0.647	0.322	0.123	0.037	0.009
4	1.000	0.984	0.825	0.524	0.255	0.098	0.030
5	1.000	0.997	0.927	0.711	0.428	0.203	0.077
6	1.000	0.999	0.974	0.847	0.607	0.348	0.160
7	1.000	1.000	0.992	0.930	0.761	0.514	0.281
8	1.000	1.000	0.998	0.972	0.871	0.674	0.432
9	1.000	1.000	1.000	0.990	0.939	0.803	0.589
10	1.000	1.000	1.000	0.997	0.974	0.894	0.730
11	1.000	1.000	1.000	0.999	0.991	0.949	0.841
12	1.000	1.000	1.000	1.000	0.997	0.978	0.916
13	1.000	1.000	1.000	1.000	0.999	0.992	0.960
14	1.000	1.000	1.000	1.000	1.000	0.997	0.983
15	1.000	1.000	1.000	1.000	1.000	0.999	0.994
16	1.000	1.000	1.000	1.000	1.000	1.000	0.998
17	1.000	1.000	1.000	1.000	1.000	1.000	0.999

θ	.35	.40	.45	.50	.55	.60	.65
X				N= 20			
0	0.000	0.000	0.000	0.000	0.000	0.000	0.000
1	0.002	0.001	0.000	0.000	0.000	0.000	0.000
2	0.012	0.004	0.001	0.000	0.000	0.000	0.000
3	0.044	0.016	0.005	0.001	0.000	0.000	0.000
4	0.118	0.051	0.019	0.006	0.002	0.000	0.000
5	0.245	0.126	0.055	0.021	0.006	0.002	0.000
6	0.417	0.250	0.130	0.058	0.021	0.006	0.002
7	0.601	0.416	0.252	0.132	0.058	0.021	0.006
8	0.762	0.596	0.414	0.252	0.131	0.057	0.020
9	0.878	0.755	0.591	0.412	0.249	0.128	0.053
10	0.947	0.872	0.751	0.588	0.409	0.245	0.122
11	0.980	0.943	0.869	0.748	0.586	0.404	0.238
12	0.994	0.979	0.942	0.868	0.748	0.584	0.399
13	0.998	0.994	0.979	0.942	0.870	0.750	0.583
14	1.000	0.998	0.994	0.979	0.945	0.874	0.755
15	1.000	1.000	0.998	0.994	0.981	0.949	0.882
16	1.000	1.000	1.000	0.999	0.995	0.984	0.956
17	1.000	1.000	1.000	1.000	0.999	0.996	0.988
18	1.000	1.000	1.000	1.000	1.000	0.999	0.998
19	1.000	1.000	1.000	1.000	1.000	1.000	1.000

θ	*.35*	*.40*	*.45*	*.50*	*.55*	*.60*	*.65*
X				N = 25			
0	0.000	0.000	0.000	0.000	0.000	0.000	0.000
1	0.000	0.000	0.000	0.000	0.000	0.000	0.000
2	0.002	0.000	0.000	0.000	0.000	0.000	0.000
3	0.010	0.002	0.000	0.000	0.000	0.000	0.000
4	0.032	0.009	0.002	0.000	0.000	0.000	0.000
5	0.083	0.029	0.009	0.002	0.000	0.000	0.000
6	0.173	0.074	0.026	0.007	0.002	0.000	0.000
7	0.306	0.154	0.064	0.022	0.006	0.001	0.000
8	0.467	0.274	0.134	0.054	0.017	0.004	0.001
9	0.630	0.425	0.242	0.115	0.044	0.013	0.003
10	0.771	0.586	0.384	0.212	0.096	0.034	0.009
11	0.875	0.732	0.543	0.345	0.183	0.078	0.025
12	0.940	0.846	0.694	0.500	0.306	0.154	0.060
13	0.975	0.922	0.817	0.655	0.457	0.268	0.125
14	0.991	0.966	0.904	0.788	0.616	0.414	0.229
15	0.997	0.987	0.956	0.885	0.758	0.575	0.370
16	0.999	0.996	0.983	0.946	0.866	0.726	0.533
17	1.000	0.999	0.994	0.978	0.936	0.846	0.694
18	1.000	1.000	0.998	0.993	0.974	0.926	0.827
19	1.000	1.000	1.000	0.998	0.991	0.971	0.917
20	1.000	1.000	1.000	1.000	0.998	0.991	0.968
X				N = 30			
2	0.000	0.000	0.000	0.000	0.000	0.000	0.000
3	0.002	0.000	0.000	0.000	0.000	0.000	0.000
4	0.008	0.002	0.000	0.000	0.000	0.000	0.000
5	0.023	0.006	0.001	0.000	0.000	0.000	0.000
6	0.059	0.017	0.004	0.001	0.000	0.000	0.000
7	0.124	0.044	0.012	0.003	0.000	0.000	0.000
8	0.225	0.094	0.031	0.008	0.002	0.000	0.000
9	0.358	0.176	0.069	0.021	0.005	0.001	0.000
10	0.508	0.291	0.135	0.049	0.014	0.003	0.000
11	0.655	0.431	0.233	0.100	0.033	0.008	0.001
12	0.780	0.578	0.359	0.181	0.071	0.021	0.005
13	0.874	0.715	0.502	0.292	0.136	0.048	0.012
14	0.935	0.825	0.645	0.428	0.231	0.097	0.030
15	0.970	0.903	0.769	0.572	0.355	0.175	0.065
16	0.988	0.952	0.864	0.708	0.498	0.285	0.126
17	0.995	0.979	0.929	0.819	0.641	0.422	0.220
18	0.999	0.992	0.967	0.900	0.767	0.569	0.345
19	1.000	0.997	0.986	0.951	0.865	0.709	0.492
20	1.000	0.999	0.995	0.979	0.931	0.824	0.642

θ	.70	.75	.80	.85	.90	.95	.99
X				N = 20			
4	0.000	0.000	0.000	0.000	0.000	0.000	0.000
5	0.000	0.000	0.000	0.000	0.000	0.000	0.000
6	0.000	0.000	0.000	0.000	0.000	0.000	0.000
7	0.001	0.000	0.000	0.000	0.000	0.000	0.000
8	0.005	0.001	0.000	0.000	0.000	0.000	0.000
9	0.017	0.004	0.001	0.000	0.000	0.000	0.000
10	0.048	0.014	0.003	0.000	0.000	0.000	0.000
11	0.113	0.041	0.010	0.001	0.000	0.000	0.000
12	0.228	0.102	0.032	0.006	0.000	0.000	0.000
13	0.392	0.214	0.087	0.022	0.002	0.000	0.000
14	0.584	0.383	0.196	0.067	0.011	0.000	0.000
15	0.762	0.585	0.370	0.170	0.043	0.003	0.000
16	0.893	0.775	0.589	0.352	0.133	0.016	0.000
17	0.965	0.909	0.794	0.595	0.323	0.075	0.001
18	0.992	0.976	0.931	0.824	0.608	0.264	0.017
19	0.999	0.997	0.988	0.961	0.878	0.642	0.182
20	1.000	1.000	1.000	1.000	1.000	1.000	1.000
X				N = 25			
0	0.000	0.000	0.000	0.000	0.000	0.000	0.000
1	0.000	0.000	0.000	0.000	0.000	0.000	0.000
2	0.000	0.000	0.000	0.000	0.000	0.000	0.000
3	0.000	0.000	0.000	0.000	0.000	0.000	0.000
4	0.000	0.000	0.000	0.000	0.000	0.000	0.000
5	0.000	0.000	0.000	0.000	0.000	0.000	0.000
6	0.000	0.000	0.000	0.000	0.000	0.000	0.000
7	0.000	0.000	0.000	0.000	0.000	0.000	0.000
8	0.000	0.000	0.000	0.000	0.000	0.000	0.000
9	0.000	0.000	0.000	0.000	0.000	0.000	0.000
10	0.002	0.000	0.000	0.000	0.000	0.000	0.000
11	0.006	0.001	0.000	0.000	0.000	0.000	0.000
12	0.017	0.003	0.000	0.000	0.000	0.000	0.000
13	0.044	0.011	0.002	0.000	0.000	0.000	0.000
14	0.098	0.030	0.006	0.000	0.000	0.000	0.000
15	0.189	0.071	0.017	0.002	0.000	0.000	0.000
16	0.323	0.149	0.047	0.008	0.000	0.000	0.000
17	0.488	0.273	0.109	0.025	0.002	0.000	0.000
18	0.659	0.439	0.220	0.070	0.009	0.000	0.000
19	0.807	0.622	0.383	0.162	0.033	0.001	0.000
20	0.910	0.786	0.579	0.318	0.098	0.007	0.000

θ	.70	.75	.80	.85	.90	.95	.99
X				N = 30			
0	0.000	0.000	0.000	0.000	0.000	0.000	0.000
1	0.000	0.000	0.000	0.000	0.000	0.000	0.000
2	0.000	0.000	0.000	0.000	0.000	0.000	0.000
3	0.000	0.000	0.000	0.000	0.000	0.000	0.000
4	0.000	0.000	0.000	0.000	0.000	0.000	0.000
5	0.000	0.000	0.000	0.000	0.000	0.000	0.000
6	0.000	0.000	0.000	0.000	0.000	0.000	0.000
7	0.000	0.000	0.000	0.000	0.000	0.000	0.000
8	0.000	0.000	0.000	0.000	0.000	0.000	0.000
9	0.000	0.000	0.000	0.000	0.000	0.000	0.000
10	0.000	0.000	0.000	0.000	0.000	0.000	0.000
11	0.000	0.000	0.000	0.000	0.000	0.000	0.000
12	0.001	0.000	0.000	0.000	0.000	0.000	0.000
13	0.002	0.000	0.000	0.000	0.000	0.000	0.000
14	0.006	0.001	0.000	0.000	0.000	0.000	0.000
15	0.017	0.003	0.000	0.000	0.000	0.000	0.000
16	0.040	0.008	0.001	0.000	0.000	0.000	0.000
17	0.084	0.022	0.003	0.000	0.000	0.000	0.000
18	0.159	0.051	0.009	0.001	0.000	0.000	0.000
19	0.270	0.106	0.026	0.003	0.000	0.000	0.000
20	0.411	0.197	0.061	0.010	0.000	0.000	0.000

CHISQUARE CUMULATIVE DISTRIBUTION FUNCTION

For example, P(C < 9.591 | DF = 20) = .025

DF	P					
	.001	*.005*	*.01*	*.025*	*.05*	*.10*
1	0.000	0.000	0.000	0.001	0.004	0.016
2	0.002	0.010	0.020	0.051	0.103	0.211
3	0.024	0.072	0.115	0.216	0.352	0.584
4	0.091	0.207	0.297	0.484	0.711	1.064
5	0.210	0.412	0.554	0.831	1.145	1.610
6	0.381	0.676	0.872	1.237	1.635	2.204
7	0.598	0.989	1.239	1.690	2.167	2.833
8	0.857	1.344	1.646	2.180	2.733	3.490
9	1.152	1.735	2.088	2.700	3.325	4.168
10	1.479	2.156	2.558	3.247	3.940	4.865
11	1.834	2.603	3.053	3.816	4.575	5.578
12	2.214	3.074	3.571	4.404	5.226	6.304
13	2.617	3.565	4.107	5.009	5.892	7.042
14	3.041	4.075	4.660	5.629	6.571	7.790
15	3.483	4.601	5.229	6.262	7.261	8.547
16	3.942	5.142	5.812	6.908	7.962	9.312
17	4.416	5.697	6.408	7.564	8.672	10.085
18	4.905	6.265	7.015	8.231	9.390	10.865
19	5.407	6.844	7.633	8.907	10.117	11.651
20	5.921	7.434	8.260	9.591	10.851	12.443
21	6.447	8.034	8.897	10.283	11.591	13.240
22	6.983	8.643	9.542	10.982	12.338	14.041
23	7.529	9.260	10.196	11.689	13.091	14.848
24	8.085	9.886	10.856	12.401	13.848	15.659
25	8.649	10.520	11.524	13.120	14.611	16.473
26	9.222	11.160	12.198	13.844	15.379	17.292
27	9.803	11.808	12.879	14.573	16.151	18.114
28	10.391	12.461	13.565	15.308	16.928	18.939
29	10.986	13.121	14.256	16.047	17.708	19.768
30	11.588	13.787	14.953	16.791	18.493	20.599
35	14.688	17.192	18.509	20.569	22.465	24.797
40	17.916	20.707	22.164	24.433	26.509	29.051
45	21.251	24.311	25.901	28.366	30.612	33.350
50	24.674	27.991	29.707	32.357	34.764	37.689

DF	P .001	.005	.01	.025	.05	.10
60	31.738	35.534	37.485	40.482	43.188	46.459
70	39.036	43.275	45.442	48.758	51.739	55.329
80	46.520	51.172	53.540	57.153	60.391	64.278
90	54.155	59.196	61.754	65.647	69.126	73.291
100	61.918	67.328	70.065	74.222	77.929	82.358
200	143.843	152.241	156.432	162.728	168.279	174.835
300	229.963	240.663	245.972	253.912	260.878	269.068
400	318.260	330.903	337.155	346.482	354.641	364.207
500	407.947	422.303	429.388	439.936	449.147	459.926
600	498.623	514.529	522.365	534.019	544.180	556.056
700	590.048	607.380	615.907	628.577	639.613	652.497
800	682.066	700.725	709.897	723.513	735.362	749.185

DF	P .20	.30	.40	.60	.70	.80
1	0.064	0.148	0.275	0.708	1.074	1.642
2	0.446	0.713	1.022	1.833	2.408	3.219
3	1.005	1.424	1.869	2.946	3.665	4.642
4	1.649	2.195	2.753	4.045	4.878	5.989
5	2.343	3.000	3.655	5.132	6.064	7.289
6	3.070	3.828	4.570	6.211	7.231	8.558
7	3.822	4.671	5.493	7.283	8.383	9.803
8	4.594	5.527	6.423	8.351	9.524	11.030
9	5.380	6.393	7.357	9.414	10.656	12.242
10	6.179	7.267	8.295	10.473	11.781	13.442
11	6.989	8.148	9.237	11.530	12.899	14.631
12	7.807	9.034	10.182	12.584	14.011	15.812
13	8.634	9.926	11.129	13.636	15.119	16.985
14	9.467	10.821	12.078	14.685	16.222	18.151
15	10.307	11.721	13.030	15.733	17.322	19.311
16	11.152	12.624	13.983	16.780	18.418	20.465
17	12.002	13.531	14.937	17.824	19.511	21.615
18	12.857	14.440	15.893	18.868	20.601	22.760
19	13.716	15.352	16.850	19.910	21.689	23.900
20	14.578	16.266	17.809	20.951	22.775	25.038
21	15.445	17.182	18.768	21.991	23.858	26.171
22	16.314	18.101	19.729	23.031	24.939	27.301
23	17.187	19.021	20.690	24.069	26.018	28.429
24	18.062	19.943	21.652	25.106	27.096	29.553
25	18.940	20.867	22.616	26.143	28.172	30.675

DF	P					
	.20	.30	.40	.60	.70	.80
26	19.820	21.792	23.579	27.179	29.246	31.795
27	20.703	22.719	24.544	28.214	30.319	32.912
28	21.588	23.647	25.509	29.249	31.391	34.027
29	22.475	24.577	26.475	30.283	32.461	35.139
30	23.364	25.508	27.442	31.316	33.530	36.250
35	27.836	30.178	32.282	36.475	38.859	41.778
40	32.345	34.872	37.134	41.622	44.165	47.269
45	36.884	39.585	41.995	46.761	49.452	52.729
50	41.449	44.313	46.864	51.892	54.723	58.164
60	50.641	53.809	56.620	62.135	65.227	68.972
70	59.898	63.346	66.396	72.358	75.689	79.715
80	69.207	72.915	76.188	82.566	86.120	90.405
90	78.558	82.511	85.993	92.761	96.524	101.054
100	87.945	92.129	95.808	102.946	106.906	111.667
200	183.003	189.049	194.319	204.434	209.985	216.609
300	279.214	286.688	293.179	305.574	312.346	320.397
400	376.022	384.698	392.217	406.535	414.335	423.590
500	473.210	482.946	491.371	507.382	516.087	526.401
600	570.668	581.362	590.606	608.147	617.671	628.943
700	668.331	679.906	689.902	708.850	719.128	731.280
800	766.155	778.551	789.247	809.505	820.483	833.456

DF	P					
	.90	.95	.975	.99	.995	.999
1	2.706	3.841	5.024	6.635	7.879	10.828
2	4.605	5.991	7.378	9.210	10.597	13.816
3	6.251	7.815	9.348	11.345	12.838	16.266
4	7.779	9.488	11.143	13.277	14.860	18.467
5	9.236	11.070	12.833	15.086	16.750	20.515
6	10.645	12.592	14.449	16.812	18.548	22.458
7	12.017	14.067	16.013	18.475	20.278	24.322
8	13.362	15.507	17.535	20.090	21.955	26.124
9	14.684	16.919	19.023	21.666	23.589	27.877
10	15.987	18.307	20.483	23.209	25.188	29.588
11	17.275	19.675	21.920	24.725	26.757	31.264
12	18.549	21.026	23.337	26.217	28.300	32.909
13	19.812	22.362	24.736	27.688	29.819	34.528
14	21.064	23.685	26.119	29.141	31.319	36.123
15	22.307	24.996	27.488	30.578	32.801	37.697

DF	P .90	.95	.975	.99	.995	.999
16	23.542	26.296	28.845	32.000	34.267	39.252
17	24.769	27.587	30.191	33.409	35.718	40.790
18	25.989	28.869	31.526	34.805	37.156	42.312
19	27.204	30.144	32.852	36.191	38.582	43.820
20	28.412	31.410	34.170	37.566	39.997	45.315
21	29.615	32.671	35.479	38.932	41.401	46.797
22	30.813	33.924	36.781	40.289	42.796	48.268
23	32.007	35.172	38.076	41.638	44.181	49.728
24	33.196	36.415	39.364	42.980	45.559	51.179
25	34.382	37.652	40.646	44.314	46.928	52.620
26	35.563	38.885	41.923	45.642	48.290	54.052
27	36.741	40.113	43.195	46.963	49.645	55.476
28	37.916	41.337	44.461	48.278	50.993	56.892
29	39.087	42.557	45.722	49.588	52.336	58.301
30	40.256	43.773	46.979	50.892	53.672	59.703
35	46.059	49.802	53.203	57.342	60.275	66.619
40	51.805	55.758	59.342	63.691	66.766	73.402
45	57.505	61.656	65.410	69.957	73.166	80.077
50	63.167	67.505	71.420	76.154	79.490	86.661
60	74.397	79.082	83.298	88.379	91.952	99.607
70	85.527	90.531	95.023	100.425	104.215	112.317
80	96.578	101.879	106.629	112.329	116.321	124.839
90	107.565	113.145	118.136	124.116	128.299	137.208
100	118.498	124.342	129.561	135.807	140.169	149.449
200	226.021	233.994	241.058	249.445	255.264	267.541
300	331.789	341.395	349.874	359.906	366.844	381.425
400	436.649	447.632	457.305	468.724	476.606	493.132
500	540.930	553.127	563.852	576.493	585.207	603.446
600	644.800	658.094	669.769	683.516	692.982	712.771
700	748.359	762.661	775.211	789.974	800.131	821.347
800	851.671	866.911	880.275	895.984	906.786	929.329

RIGHT–TAILED CRITICAL VALUES FOR SNEDECOR F

DFN [DFD] denotes degrees of freedom for the numerator [denominator]. For example,

$P(F \geq 3.438 \mid DFN = 5, DFD = 21) = .02$.

DFN	DFD	P .10	.05	.025	.02	.01	.001
5	5	3.453	5.050	7.146	7.953	10.967	29.752
5	9	2.611	3.482	4.484	4.840	6.057	11.714
5	13	2.347	3.025	3.767	4.020	4.862	8.354
5	17	2.218	2.810	3.438	3.649	4.336	7.022
5	21	2.142	2.685	3.250	3.438	4.042	6.318
5	25	2.092	2.603	3.129	3.302	3.855	5.885
5	29	2.057	2.545	3.044	3.207	3.725	5.593
5	33	2.030	2.503	2.981	3.137	3.630	5.382
5	37	2.009	2.470	2.933	3.084	3.558	5.224
5	41	1.993	2.443	2.895	3.041	3.501	5.100
5	45	1.980	2.422	2.864	3.007	3.454	5.001
5	49	1.968	2.404	2.838	2.978	3.416	4.919
9	5	3.316	4.772	6.681	7.415	10.158	27.244
9	9	2.440	3.179	4.026	4.325	5.351	10.107
9	13	2.164	2.714	3.312	3.516	4.191	6.982
9	17	2.028	2.494	2.985	3.149	3.682	5.754
9	21	1.948	2.366	2.798	2.940	3.398	5.109
9	25	1.895	2.282	2.677	2.806	3.217	4.713
9	29	1.857	2.223	2.592	2.712	3.092	4.447
9	33	1.828	2.179	2.529	2.643	3.000	4.255
9	37	1.806	2.145	2.481	2.590	2.930	4.111
9	41	1.789	2.118	2.443	2.548	2.875	3.999
9	45	1.774	2.096	2.412	2.514	2.830	3.909
9	49	1.763	2.077	2.387	2.486	2.793	3.835
13	5	3.257	4.655	6.488	7.192	9.825	26.224
13	9	2.364	3.048	3.831	4.107	5.055	9.443
13	13	2.080	2.577	3.115	3.299	3.905	6.409
13	17	1.940	2.353	2.786	2.931	3.401	5.221
13	21	1.857	2.222	2.598	2.722	3.119	4.597
13	25	1.802	2.136	2.476	2.587	2.939	4.216
13	29	1.762	2.075	2.390	2.492	2.814	3.958
13	33	1.732	2.030	2.327	2.422	2.723	3.773
13	37	1.709	1.995	2.278	2.369	2.653	3.634

DFN	DFD	P .10	.05	.025	.02	.01	.001
13	41	1.691	1.967	2.326	2.598	3.526	3.537
13	45	1.676	1.945	2.208	2.292	2.553	3.439
13	49	1.663	1.926	2.182	2.263	2.517	3.368
17	5	3.223	4.590	6.381	7.070	9.643	25.669
17	9	2.320	2.974	3.722	3.986	4.890	9.079
17	13	2.032	2.499	3.004	3.176	3.745	6.093
17	17	1.889	2.272	2.673	2.807	3.242	4.924
17	21	1.803	2.139	2.483	2.597	2.960	4.311
17	25	1.746	2.051	2.360	2.461	2.780	3.936
17	29	1.705	1.989	2.273	2.365	2.656	3.683
17	33	1.675	1.943	2.209	2.295	2.564	3.501
17	37	1.651	1.907	2.160	2.241	2.494	3.364
17	41	1.632	1.879	2.120	2.198	2.438	3.258
17	45	1.616	1.855	2.088	2.163	2.393	3.172
17	49	1.603	1.836	2.062	2.134	2.356	3.102
21	5	3.202	4.549	6.314	6.993	9.528	25.320
21	9	2.292	2.926	3.652	3.908	4.786	8.848
21	13	2.000	2.448	2.932	3.097	3.643	5.891
21	17	1.855	2.219	2.600	2.727	3.139	4.734
21	21	1.768	2.084	2.409	2.516	2.857	4.127
21	25	1.710	1.995	2.284	2.378	2.677	3.756
21	29	1.668	1.932	2.196	2.282	2.552	3.505
21	33	1.636	1.885	2.131	2.211	2.460	3.325
21	37	1.612	1.848	2.081	2.156	2.389	3.189
21	41	1.592	1.819	2.041	2.113	2.333	3.083
21	45	1.576	1.795	2.009	2.077	2.288	2.998
21	49	1.562	1.775	1.982	2.048	2.251	2.929
25	5	3.187	4.521	6.268	6.940	9.449	25.080
25	9	2.272	2.893	3.604	3.854	4.713	8.689
25	13	1.978	2.412	2.882	3.042	3.571	5.751
25	17	1.831	2.181	2.548	2.671	3.068	4.602
25	21	1.742	2.045	2.356	2.458	2.785	3.999
25	25	1.683	1.955	2.230	2.320	2.604	3.629
25	29	1.640	1.891	2.142	2.223	2.478	3.380
25	33	1.608	1.844	2.076	2.151	2.386	3.200
25	37	1.583	1.806	2.025	2.096	2.315	3.065
25	41	1.563	1.777	1.985	2.052	2.258	2.960
25	45	1.546	1.752	1.952	2.016	2.213	2.875
25	49	1.533	1.732	1.925	1.986	2.175	2.806

DFN	DFD	.10	.05	.025	.02	.01	.001
29	5	3.176	4.500	6.234	6.901	9.391	24.905
29	9	2.258	2.869	3.568	3.815	4.660	8.572
29	13	1.961	2.386	2.845	3.002	3.518	5.647
29	17	1.813	2.154	2.510	2.629	3.014	4.504
29	21	1.723	2.016	2.317	2.415	2.731	3.904
29	25	1.663	1.926	2.190	2.276	2.550	3.535
29	29	1.620	1.861	2.101	2.179	2.423	3.287
29	33	1.587	1.812	2.035	2.106	2.330	3.108
29	37	1.562	1.775	1.983	2.050	2.259	2.973
29	41	1.541	1.744	1.943	2.006	2.202	2.868
29	45	1.524	1.720	1.909	1.970	2.156	2.783
29	49	1.510	1.699	1.881	1.939	2.118	2.714
33	5	3.168	4.484	6.208	6.871	9.347	24.772
33	9	2.247	2.850	3.541	3.784	4.619	8.483
33	13	1.948	2.366	2.817	2.970	3.478	5.568
33	17	1.799	2.132	2.481	2.597	2.973	4.429
33	21	1.708	1.994	2.286	2.382	2.690	3.830
33	25	1.648	1.902	2.159	2.243	2.508	3.463
33	29	1.604	1.837	2.069	2.144	2.381	3.215
33	33	1.571	1.788	2.002	2.071	2.287	3.036
33	37	1.545	1.750	1.951	2.015	2.215	2.901
33	41	1.524	1.719	1.909	1.970	2.158	2.796
33	45	1.506	1.694	1.876	1.933	2.112	2.711
33	49	1.492	1.673	1.847	1.903	2.074	2.642
37	5	3.161	4.472	6.188	6.848	9.313	24.667
37	9	2.238	2.835	3.519	3.760	4.587	8.412
37	13	1.938	2.349	2.794	2.945	3.445	5.506
37	17	1.788	2.115	2.457	2.571	2.941	4.369
37	21	1.696	1.976	2.262	2.356	2.657	3.772
37	25	1.635	1.884	2.134	2.216	2.474	3.405
37	29	1.591	1.818	2.044	2.117	2.347	3.157
37	33	1.557	1.768	1.976	2.043	2.253	2.979
37	37	1.531	1.730	1.924	1.986	2.181	2.844
37	41	1.510	1.699	1.882	1.941	2.123	2.739
37	45	1.492	1.673	1.848	1.904	2.077	2.654
37	49	1.477	1.652	1.820	1.873	2.038	2.585
41	5	3.156	4.461	6.171	6.829	9.285	24.582
41	9	2.230	2.823	3.501	3.741	4.561	8.355
41	13	1.930	2.336	2.775	2.925	3.419	5.455

DFN	DFD	P .10	.05	.025	.02	.01	.001
41	17	1.778	2.101	2.438	2.550	2.914	4.321
41	21	1.687	1.961	2.242	2.334	2.630	3.725
41	25	1.625	1.868	2.114	2.193	2.447	3.358
41	29	1.580	1.802	2.023	2.094	2.319	3.111
41	33	1.546	1.752	1.955	2.020	2.224	2.932
41	37	1.519	1.713	1.902	1.963	2.152	2.797
41	41	1.498	1.682	1.860	1.917	2.094	2.692
41	45	1.480	1.656	1.826	1.880	2.047	2.607
41	49	1.465	1.634	1.797	1.849	2.008	2.537
45	5	3.152	4.453	6.158	6.813	9.262	4.513
45	9	2.224	2.813	3.487	3.725	4.539	8.308
45	13	1.923	2.325	2.760	2.908	3.398	5.413
45	17	1.771	2.089	2.422	2.533	2.892	4.281
45	21	1.678	1.949	2.225	2.316	2.607	3.685
45	25	1.616	1.855	2.096	2.175	2.424	3.319
45	29	1.571	1.789	2.005	2.075	2.296	3.072
45	33	1.537	1.738	1.937	2.001	2.201	2.893
45	37	1.510	1.699	1.884	1.943	2.128	2.758
45	41	1.488	1.667	1.842	1.898	2.070	2.653
45	45	1.470	1.642	1.807	1.860	2.023	2.568
45	49	1.455	1.620	1.778	1.829	1.984	2.498
49	5	3.148	4.446	6.146	6.800	9.242	24.454
49	9	2.219	2.805	3.475	3.711	4.521	8.269
49	13	1.917	2.316	2.747	2.894	3.379	5.378
49	17	1.764	2.079	2.408	2.518	2.874	4.247
49	21	1.672	1.938	2.211	2.301	2.588	3.652
49	25	1.609	1.845	2.082	2.159	2.404	3.286
49	29	1.563	1.777	1.990	2.059	2.276	3.039
49	33	1.529	1.727	1.922	1.985	2.181	2.860
49	37	1.502	1.687	1.869	1.927	2.108	2.725
49	41	1.480	1.655	1.826	1.881	2.049	2.619
49	45	1.462	1.629	1.791	1.843	2.002	2.534
49	49	1.446	1.607	1.762	1.811	1.963	2.465

LEFT–TAILED CRITICAL VALUES FOR SNEDECOR F

DFN [DFD] denotes degrees of freedom for the numerator [denominator]. For example,

$$P(F \leq .143 \mid DFN = 5, DFD = 21) = .02$$

DFN	DFD	P .001	.01	.02	.025	.05	.10
5	5	0.034	0.091	0.126	0.140	0.198	0.290
5	9	0.037	0.098	0.135	0.150	0.210	0.302
5	13	0.038	0.102	0.139	0.154	0.215	0.307
5	17	0.039	0.104	0.141	0.157	0.218	0.310
5	21	0.039	0.105	0.143	0.158	0.220	0.312
5	25	0.040	0.106	0.144	0.160	0.221	0.314
5	29	0.040	0.106	0.145	0.160	0.222	0.315
5	33	0.040	0.107	0.146	0.161	0.223	0.316
5	37	0.041	0.107	0.146	0.162	0.224	0.316
5	41	0.041	0.108	0.146	0.162	0.224	0.317
5	45	0.041	0.108	0.147	0.162	0.225	0.317
5	49	0.041	0.108	0.147	0.163	0.225	0.318
9	5	0.085	0.165	0.207	0.223	0.287	0.383
9	9	0.099	0.187	0.231	0.248	0.315	0.410
9	13	0.106	0.198	0.243	0.261	0.328	0.423
9	17	0.110	0.204	0.251	0.269	0.336	0.431
9	21	0.113	0.209	0.256	0.274	0.342	0.436
9	25	0.115	0.212	0.259	0.278	0.346	0.440
9	29	0.117	0.215	0.262	0.280	0.349	0.443
9	33	0.118	0.216	0.264	0.282	0.351	0.445
9	37	0.119	0.218	0.266	0.284	0.353	0.447
9	41	0.120	0.219	0.267	0.286	0.354	0.448
9	45	0.120	0.220	0.268	0.287	0.355	0.450
9	49	0.121	0.221	0.269	0.288	0.357	0.451
13	5	0.120	0.206	0.249	0.265	0.331	0.426
13	9	0.143	0.239	0.284	0.302	0.368	0.462
13	13	0.156	0.256	0.303	0.321	0.388	0.481
13	17	0.164	0.267	0.315	0.333	0.400	0.492
13	21	0.170	0.275	0.323	0.341	0.409	0.500
13	25	0.174	0.280	0.329	0.347	0.415	0.506
13	29	0.177	0.284	0.333	0.351	0.419	0.510
13	33	0.180	0.288	0.337	0.355	0.423	0.513
13	37	0.182	0.290	0.340	0.358	0.426	0.516

DFN	DFD	P .001	.01	.02	.025	.05	.10
13	41	0.183	0.292	0.342	0.360	0.428	0.518
13	45	0.185	0.294	0.344	0.362	0.430	0.520
13	49	0.186	0.296	0.346	0.364	0.432	0.522
17	5	0.142	0.231	0.274	0.291	0.356	0.451
17	9	0.174	0.272	0.318	0.335	0.401	0.493
17	13	0.192	0.294	0.341	0.359	0.425	0.515
17	17	0.203	0.308	0.356	0.374	0.440	0.529
17	21	0.211	0.319	0.367	0.385	0.451	0.539
17	25	0.217	0.326	0.374	0.392	0.458	0.546
17	29	0.222	0.332	0.380	0.398	0.464	0.552
17	33	0.226	0.336	0.385	0.403	0.469	0.556
17	37	0.229	0.340	0.389	0.407	0.473	0.559
17	41	0.231	0.343	0.392	0.410	0.476	0.562
17	45	0.234	0.346	0.395	0.413	0.479	0.565
17	49	0.235	0.348	0.397	0.415	0.481	0.567
21	5	0.158	0.247	0.291	0.308	0.372	0.467
21	9	0.196	0.294	0.340	0.357	0.423	0.513
21	13	0.218	0.321	0.367	0.385	0.450	0.539
21	17	0.232	0.338	0.385	0.403	0.468	0.555
21	21	0.242	0.350	0.398	0.415	0.480	0.566
21	25	0.250	0.359	0.407	0.424	0.489	0.574
21	29	0.256	0.366	0.414	0.432	0.496	0.580
21	33	0.261	0.372	0.420	0.437	0.502	0.585
21	37	0.265	0.376	0.424	0.442	0.506	0.589
21	41	0.268	0.380	0.428	0.446	0.510	0.593
21	45	0.271	0.384	0.432	0.449	0.513	0.596
21	49	0.274	0.386	0.435	0.452	0.516	0.598
25	5	0.170	0.259	0.303	0.320	0.384	0.478
25	9	0.212	0.311	0.356	0.374	0.438	0.528
25	13	0.237	0.340	0.387	0.4045	0.468	0.555
25	17	0.254	0.360	0.406	0.424	0.487	0.573
25	21	0.266	0.374	0.420	0.438	0.501	0.585
25	25	0.276	0.384	0.431	0.448	0.511	0.594
25	29	0.283	0.392	0.439	0.457	0.519	0.601
25	33	0.289	0.399	0.446	0.463	0.526	0.607
25	37	0.294	0.404	0.451	0.4699	0.531	0.612
25	41	0.298	0.409	0.456	0.473	0.535	0.615
25	45	0.301	0.413	0.460	0.477	0.539	0.619
25	49	0.304	0.416	0.463	0.480	0.542	0.622

DFN	DFD	P .001	.01	.02	.025	.05	.10
29	5	0.179	0.268	0.312	0.329	0.393	0.486
29	9	0.225	0.323	0.369	0.386	0.450	0.539
29	13	0.253	0.355	0.401	0.418	0.482	0.568
29	17	0.272	0.377	0.423	0.440	0.503	0.586
29	21	0.285	0.392	0.438	0.455	0.518	0.600
29	25	0.296	0.404	0.450	0.467	0.529	0.610
29	29	0.304	0.413	0.459	0.476	0.537	0.617
29	33	0.311	0.420	0.466	0.483	0.544	0.624
29	37	0.317	0.426	0.472	0.489	0.550	0.629
29	41	0.321	0.431	0.478	0.494	0.555	0.633
29	45	0.326	0.436	0.482	0.499	0.559	0.637
29	49	0.329	0.439	0.486	0.502	0.563	0.640
33	5	0.186	0.275	0.319	0.335	0.400	0.493
33	9	0.235	0.333	0.378	0.395	0.459	0.547
33	13	0.265	0.367	0.413	0.430	0.493	0.577
33	17	0.286	0.390	0.436	0.453	0.515	0.597
33	21	0.301	0.407	0.452	0.469	0.530	0.611
33	25	0.312	0.419	0.465	0.482	0.542	0.622
33	29	0.322	0.429	0.475	0.491	0.552	0.630
33	33	0.329	0.437	0.483	0.499	0.559	0.637
33	37	0.336	0.444	0.489	0.506	0.566	0.642
33	41	0.341	0.450	0.495	0.512	0.571	0.647
33	45	0.346	0.454	0.500	0.516	0.575	0.651
33	49	0.350	0.459	0.504	0.520	0.579	0.654
37	5	0.191	0.281	0.324	0.341	0.405	0.498
37	9	0.243	0.341	0.386	0.403	0.466	0.554
37	13	0.275	0.377	0.422	0.439	0.501	0.585
37	17	0.297	0.401	0.446	0.463	0.524	0.606
37	21	0.314	0.419	0.464	0.480	0.541	0.620
37	25	0.326	0.432	0.477	0.494	0.554	0.632
37	29	0.336	0.443	0.488	0.504	0.564	0.640
37	33	0.345	0.451	0.496	0.513	0.572	0.647
37	37	0.352	0.459	0.503	0.520	0.578	0.653
37	41	0.358	0.465	0.509	0.526	0.584	0.658
37	45	0.363	0.470	0.515	0.531	0.589	0.662
37	49	0.367	0.474	0.519	0.535	0.593	0.666
41	5	0.196	0.286	0.329	0.345	0.409	0.502
41	9	0.250	0.348	0.392	0.409	0.472	0.559
41	13	0.284	0.385	0.430	0.447	0.508	0.591

DFN	DFD	P .001	.01	.02	.025	.05	.10
41	17	0.307	0.410	0.455	0.472	0.532	0.613
41	21	0.324	0.429	0.473	0.490	0.550	0.628
41	25	0.338	0.443	0.487	0.504	0.563	0.640
41	29	0.349	0.454	0.499	0.515	0.573	0.649
41	33	0.358	0.463	0.508	0.524	0.582	0.656
41	37	0.365	0.471	0.515	0.531	0.589	0.662
41	41	0.372	0.478	0.522	0.538	0.595	0.668
41	45	0.377	0.483	0.527	0.543	0.600	0.672
41	49	0.382	0.488	0.532	0.548	0.604	0.676
45	5	0.200	0.289	0.333	0.349	0.413	0.505
45	9	0.256	0.353	0.398	0.415	0.477	0.564
45	13	0.291	0.392	0.436	0.453	0.514	0.597
45	17	0.315	0.418	0.462	0.479	0.539	0.619
45	21	0.334	0.437	0.481	0.498	0.557	0.635
45	25	0.348	0.452	0.496	0.512	0.571	0.647
45	29	0.359	0.464	0.508	0.524	0.582	0.656
45	33	0.369	0.473	0.517	0.533	0.590	0.664
45	37	0.377	0.482	0.525	0.541	0.598	0.670
45	41	0.384	0.488	0.532	0.548	0.604	0.676
45	45	0.389	0.494	0.538	0.553	0.609	0.680
45	49	0.395	0.500	0.543	0.558	0.614	0.684
49	5	0.203	0.293	0.336	0.352	0.416	0.508
49	9	0.261	0.358	0.402	0.419	0.481	0.567
49	13	0.297	0.397	0.442	0.458	0.519	0.601
49	17	0.322	0.424	0.469	0.485	0.545	0.624
49	21	0.341	0.444	0.488	0.505	0.563	0.640
49	25	0.356	0.460	0.504	0.520	0.577	0.653
49	29	0.369	0.472	0.516	0.532	0.589	0.662
49	33	0.378	0.482	0.526	0.541	0.598	0.670
49	37	0.387	0.491	0.534	0.549	0.605	0.677
49	41	0.394	0.498	0.541	0.556	0.612	0.682
49	45	0.400	0.504	0.547	0.562	0.617	0.687
49	49	0.406	0.510	0.552	0.567	0.622	0.691

NORMAL CUMULATIVE DISTRIBUTION FUNCTION

For example, $P(X \le 1.40) = F(1.40) = .9192$

x	F	x	F	x	F	x	F
0.01	.5040	0.36	.6406	.71	.7611	1.06	.8554
0.02	.5080	0.37	.6443	.72	.7642	1.07	.8577
0.03	.5120	0.38	.6480	.73	.7673	1.08	.8599
0.04	.5160	0.39	.6517	.74	.7704	1.09	.8621
0.05	.5199	0.40	.6554	.75	.7734	1.10	.8643
0.06	.5239	0.41	.6591	.76	.7764	1.11	.8665
0.07	.5279	0.42	.6628	.77	.7794	1.12	.8686
0.08	.5319	0.43	.6664	.78	.7823	1.13	.8708
0.09	.5359	0.44	.6700	.79	.7852	1.14	.8729
0.10	.5398	0.45	.6736	.80	.7881	1.15	.8749
0.11	.5438	0.46	.6772	.81	.7910	1.16	.8770
0.12	.5478	0.47	.6808	.82	.7939	1.17	.8790
0.13	.5517	0.48	.6844	.83	.7967	1.18	.8810
0.14	.5557	0.49	.6879	.84	.7995	1.19	.8830
0.15	.5596	0.50	.6915	.85	.8023	1.20	.8849
0.16	.5636	0.51	.6950	.86	.8051	1.21	.8869
0.17	.5675	0.52	.6985	.87	.8078	1.22	.8888
0.18	.5714	0.53	.7019	.88	.8106	1.23	.8907
0.19	.5753	0.54	.7054	.89	.8133	1.24	.8925
0.20	.5793	0.55	.7088	.90	.8159	1.25	.8944
0.21	.5832	0.56	.7123	.91	.8186	1.26	.8962
0.22	.5871	0.57	.7157	.92	.8212	1.27	.8980
0.23	.5910	0.58	.7190	.93	.8238	1.28	.8997
0.24	.5948	0.59	.7224	.94	.8264	1.29	.9015
0.25	.5987	0.60	.7257	.95	.8289	1.30	.9032
0.26	.6026	0.61	.7291	.96	.8315	1.31	.9049
0.27	.6064	0.62	.7324	.97	.8340	1.32	.9066
0.28	.6103	0.63	.7357	.98	.8365	1.33	.9082
0.29	.6141	0.64	.7389	.99	.8389	1.34	.9099
0.30	.6179	0.65	.7422	1.00	.9413	1.35	.9115
0.31	.6217	0.66	.7454	1.01	.8438	1.36	.9131
0.32	.6255	0.67	.7486	1.02	.8461	1.37	.9147
0.33	.6293	0.68	.7517	1.03	.8485	1.38	.9162
0.34	.6331	0.69	.7549	1.04	.8508	1.39	.9177
0.35	.6368	0.70	.7580	1.05	.8531	1.40	.9192

x	F	x	F	x	F	x	F
1.41	.9207	1.76	.9608	2.11	.9825	2.46	.9931
1.42	.9222	1.77	.9616	2.12	.9830	2.47	.9932
1.43	.9236	1.78	.9625	2.13	.9834	2.48	.9934
1.44	.9251	1.79	.9633	2.14	.9838	2.49	.9936
1.45	.9265	1.80	.9641	2.15	.9842	2.50	.9938
1.46	.9279	1.81	.9649	2.16	.9846	2.51	.9940
1.47	.9292	1.82	.9656	2.17	.9850	2.52	.9941
1.48	.9306	1.83	.9664	2.18	.9854	2.53	.9943
1.49	.9319	1.84	.9671	2.19	.9875	2.54	.9945
1.50	.9332	1.85	.9678	2.20	.9861	2.55	.9946
1.51	.9345	1.86	.9686	2.21	.9864	2.56	.9948
1.52	.9357	1.87	.9693	2.22	.9868	2.57	.9949
1.53	.9370	1.88	.9699	2.23	.9871	2.58	.9951
1.54	.9382	1.89	.9706	2.24	.9875	2.59	.9952
1.55	.9394	1.90	.9713	2.25	.9878	2.60	.9953
1.56	.9406	1.91	.9719	2.26	.9881	2.61	.9955
1.57	.9418	1.92	.9726	2.27	.9884	2.62	.9956
1.58	.9429	1.93	.9732	2.28	.9887	2.63	.9957
1.59	.9441	1.94	.9738	2.29	.9890	2.64	.9959
1.60	.9452	1.95	.9744	2.30	.9893	2.65	.9960
1.61	.9463	1.96	.9750	2.31	.9896	2.66	.9961
1.62	.9474	1.97	.9756	2.32	.9898	2.67	.9962
1.63	.9484	1.98	.9761	2.33	.9901	2.68	.9963
1.64	.9495	1.99	.9767	2.34	.9904	2.69	.9964
1.65	.9505	2.00	.9772	2.35	.9906	2.70	.9965
1.66	.9515	2.01	.9778	2.36	.9909	2.71	.9966
1.67	.9525	2.02	.9783	2.37	.9911	2.72	.9967
1.68	.9535	2.03	.9788	2.38	.9913	2.73	.9968
1.69	.9545	2.04	.9793	2.39	.9916	2.74	.9969
1.70	.9554	2.05	.9798	2.40	.9918	2.75	.9970
1.71	.9564	2.06	.9803	2.41	.9920	2.76	.9971
1.72	.9573	2.07	.9808	2.42	.9922	2.77	.9972
1.73	.9582	2.08	.9812	2.43	.9925	2.78	.9973
1.74	.9591	2.09	.9817	2.44	.9927	2.79	.9974
1.75	.9599	2.10	.9821	2.45	.9929	2.80	.9974

x	F	x	F	x	F
2.81	.9975	3.11	.9991	3.41 – 3.48	.9997
2.82	.9976	3.12	.9991	3.49 – 3.61	.9998
2.83	.9977	3.13	.9991	3.62 – 3.89	.9999
2.84	.9977	3.14	.9992	3.90 and above	1.0000
2.85	.9978	3.15	.9992		
2.86	.9979	3.16	.9992		
2.87	.9979	3.17	.9992		
2.88	.9980	3.18	.9993		
2.89	.9981	3.19	.9993		
2.90	.9981	3.20	.9993		
2.91	.9982	3.21	.9993		
2.92	.9982	3.22	.9994		
2.93	.9983	3.23	.9994		
2.94	.9984	3.24	.9994		
2.95	.9984	3.25	.9994		
2.96	.9985	3.26	.9994		
2.97	.9985	3.27	.9995		
2.98	.9986	3.28	.9995		
2.99	.9986	3.29	.9995		
3.00	.9987	3.30	.9995		
3.01	.9987	3.31	.9995		
3.02	.9988	3.32	.9995		
3.03	.0988	3.33	.9996		
3.04	.9988	3.34	.9996		
3.05	.9989	3.35	.9996		
3.06	.9989	3.36	.9996		
3.07	.9989	3.37	.9996		
3.08	.9990	3.38	.9996		
3.09	.9990	3.39	.9997		
3.10	.9990	3.40	.9997		

TWO–TAILED CRITICAL VALUES FOR STUDENT T

For example, $P(|T| \geq 2.650 \mid DF = 13) = .02$;

$P(T \leq -2.650 \mid DF = 13) = .01$.

DF	ALPHA .10	.05	.04	.025	.02	.01	.001
1	6.314	12.706	15.895	25.452	31.821	63.65	636.6
2	2.920	4.303	4.849	6.205	6.965	9.925	31.60
3	2.353	3.182	3.482	4.177	4.541	5.841	12.92
4	2.132	2.776	2.999	3.495	3.747	4.604	8.610
5	2.015	2.571	2.757	3.163	3.365	4.032	6.869
6	1.943	2.447	2.612	2.969	3.143	3.707	5.959
7	1.895	2.365	2.517	2.841	2.998	3.499	5.408
8	1.860	2.306	2.449	2.752	2.896	3.355	5.041
9	1.833	2.262	2.398	2.685	2.821	3.250	4.781
10	1.812	2.228	2.359	2.634	2.764	3.169	4.587
11	1.796	2.201	2.328	2.593	2.718	3.106	4.437
12	1.782	2.179	2.303	2.560	2.681	3.055	4.318
13	1.771	2.160	2.282	2.533	2.650	3.012	4.221
14	1.761	2.145	2.264	2.510	2.624	2.977	4.140
15	1.753	2.131	2.249	2.490	2.602	2.947	4.073
16	1.746	2.120	2.235	2.473	2.583	2.921	4.015
17	1.740	2.110	2.224	2.458	2.567	2.898	3.965
18	1.734	2.101	2.214	2.445	2.552	2.878	3.922
19	1.729	2.093	2.205	2.433	2.539	2.861	3.883
20	1.725	2.086	2.197	2.423	2.528	2.845	3.850
21	1.721	2.080	2.189	2.414	2.518	2.831	3.819
22	1.717	2.074	2.183	2.405	2.508	2.819	3.792
23	1.714	2.069	2.177	2.398	2.500	2.807	3.768
24	1.711	2.064	2.172	2.391	2.492	2.797	3.745
25	1.708	2.060	2.167	2.385	2.485	2.787	3.725
26	1.706	2.056	2.162	2.379	2.479	2.779	3.707
27	1.703	2.052	2.158	2.373	2.473	2.771	3.690
28	1.701	2.048	2.154	2.368	2.467	2.763	3.674
29	1.699	2.045	2.150	2.364	2.462	2.756	3.659
30	1.697	2.042	2.147	2.360	2.457	2.750	3.646
35	1.690	2.030	2.133	2.342	2.438	2.724	3.591
40	1.684	2.021	2.123	2.329	2.423	2.704	3.551
45	1.679	2.014	2.115	2.319	2.412	2.690	3.520
50	1.676	2.009	2.109	2.311	2.403	2.678	3.496

	.10	.05	.04	.025	.02	.01	.001
60	1.671	2.000	2.099	2.299	2.390	2.660	3.460
70	1.667	1.994	2.093	2.291	2.381	2.648	3.435
80	1.664	1.990	2.088	2.284	2.374	2.639	3.416
90	1.662	1.987	2.084	2.280	2.368	2.632	3.402
100	1.660	1.984	2.081	2.276	2.364	2.626	3.390
200	1.653	1.972	2.067	2.258	2.345	2.601	3.340
300	1.650	1.968	2.063	2.253	2.339	2.592	3.323
400	1.649	1.966	2.060	2.250	2.336	2.588	3.315
500	1.648	1.965	2.059	2.248	2.334	2.586	3.310
600	1.647	1.964	2.058	2.247	2.333	2.584	3.307
700	1.647	1.963	2.058	2.246	2.332	2.583	3.304
800	1.647	1.963	2.057	2.246	2.331	2.582	3.303
900	1.647	1.963	2.057	2.245	2.330	2.581	3.301
∞	1.645	1.960	2.055	2.242	2.326	2.576	3.291

REFERENCES

Aickin, M. (1983). *Linear Statistical Analysis of Discrete Data.* J. Wiley.

Akahira, M. and Takenchi, K. (1981). *Asymptotic Efficiency of Statistical Estimators.* Lecture Notes in Statist., #7, Springer–Verlag.

Apostol, T.M. (1957). *Mathematical Analysis.* Addison–Wesley.

Barndorff–Nielsen, O. and Pedersen, K. (1968). Sufficient data reduction and exponential families. *Math. Scand.*, 22, 197–202.

Bauer, H. (1972). *Probability Theory and Elements of Measure Theory.* Holt, Rinehart and Winston.

Burill, C.W. (1972). *Measure, Integration and Probability.* McGraw–Hill.

Conover, W.J. (1972). A Kolmogorov goodness–of–fit test for discontinuous distributions. *J. Amer. Statist. Assoc.*, 67, 591–596.

Cramer, H. (1946). *Mathematical Methods of Statistics.* Princeton Univ. Press.

Dallal, G.E. and Wilkinson, L. (1986). An analytic approximation to the distribution of Lilliefors' test statistic for normality. *The Amer. Statist.*, 40, 294–296.

David, H.A. (1970). *Order Statistics.* J. Wiley.

Davis, L.J. (1986). Exact tests for 2 × 2 contingency tables. *The. Amer. Statist.*, 40, 139–141.

Fisher, R.A. (1935). *The Design of Experiments.* Oliver and Boyd, Edinburgh.

Fleming, W. (1977). *Functions of Several Variables.* Springer–Verlag.

Guenther, W.G. (1981). Sample size formulas for normal theory t–tests. *The Amer. Statist.*, 35, 243–244.

Hall, P. (1982). *Rate of Convergence in the Central Limit Theorem.* Research Notes in Math., #62, Pitman Advanced Publ. Program. London.

Hocking, R.R. (1985). *The Analysis of Linear Models.* Brooks/Cole.

Hooper, P.M. (1982). Invariant confidence sets with smallest expected measure. *Ann. Statist.*, 10, 1283–1294.

Ibragimov, I.A. and Hasminskii, R.Z. (1981). *Statistical Estimation.* Springer–Verlag.

Karlin, S. (1957). Polya type distributions II, III. *Ann. Math. Statist.*, 28, 281–308, 839–860.

Kendall, M.G. and Stuart, A. (1974). *The Advanced Theory of Statistics*, 2, McMillan.

Kolmogorov, A.N. (1933) Sulla determinazione empirica di una legge di distribuzione. *Giorn. dell'Inst. Ital. Attuari,* 4, 83–91.

Kolmogorov, A.N. (1950). *Foundations of the Theory of Probability.* Chelsea Press.

Kullback, S. (1959). *Information Theory and Statistics.* J. Wiley.

Kulldorff, G. (1957). On the conditions for consistency and asymptotic efficiency of MLE. *Skand. Akt.*, 40, 129–144.

Lecam, L. (1953). On some asymptotic properties of MLE and related Bayes' estimates. *Univ. Calif. Publ. in Statist.*, 1, 277–330.

Lehmann, E.L. (1959). *Testing Statistical Hypotheses.* J. Wiley.

Lehmann, E.L. (1983). *Theory of Point Estimation.* J. Wiley.

Lehmann, E.L. (1975). *Nonparametric Statistical Methods Based on Ranks.* Holden–Day.

Makelainen, T., Schmidt, K. and Styan, G.P.H. (1981). On the existence and uniqueness of the MLE of a vector–valued parameter in fixed–size samples. *Ann. Statist.*, 9, 758–767.

Marsden, J.E. and Tromba, A.J. (1981). *Vector Calculus*. Freedman.

Martin, L.S. (1985). A note on non–unique MLE's and sufficient statistics. *The Amer. Statist.*, 39, p. 66.

Melnick, E.L. and Tenebein, A. (1982). Misspecifications of the normal distribution. *The Amer. Statist.*, 36, 372–373.

Miller, L.H. (1956). Table of percentage points of Kolmorgorov statistics. *J. Amer. Statist. Assoc.*, 51, 111–121.

Moore, D.S. (1978). Chisquare tests. *MAA Studies in Statist.*, R.V. Hogg, Ed.

Neymann, J. and Pearson, E.S. (1936). Sufficient statistics and uniformly most powerful tests of statistical hypotheses. *Statist. Res. Mem.*, Vol. 1, 113–137.

Neymann, J. and Pearson, E.S. (1933). On the problem of the most efficient tests of statistical hypotheses. *Proc. Phil. Trans. Royal Soc. A.*, 231, 289–337.

Nguyen, H.T., Rogers, G.S. and Walker, E.A. (1984). Estimation in change–point hazard rate models. *Biometrika*, 71, 299–304.

Pearson, K. (1900). On a criterion that a system of deviations from the probable in the case of a correlated system of variables is such that it can be reasonably supposed to have arisen in random sampling. *Philo. Magazine*, 50, 157–175.

Pierce, G.W. (1949). *The Songs of Insects*. Harvard Univ. Press.

Rao, C.R. (1965). *Linear Statistical Inference and Its Applications*. J. Wiley.

Robbins, H.E. (1944). On the measure of a random set. *Ann. Math. Statist.*, 15, 70–74.

Rogers, G.S. (1964). An application of a generalized gamma distribution. *Ann. Math. Statist.*, 35, 1368–1370.

Sampson, A. and Spencer, B. (1976). Sufficiency, minimal sufficiency and the lack there of. *The Amer. Statist.*, 30, 34–35.

Scheffe, H. (1959). *The Analysis of Variance.* J. Wiley.

Schmetterer, L. (1974). *Introduction to Mathematical Statistics.* Springer–Verlag.

Serfling, R.J. (1980). *Approximation Theorems of Mathematical Statistics.* J. Wiley.

Snedecor, G.W. and Cochran, W. (1980). *Statistica Methods.* Iowa State Univ. Press.

Stein, C.M. (1981). Estimation of the mean of a multivariate normal distribution. *Ann. Statist.*, 9, 1135–1151.

Stern, S. and Sherwood, E.R. (1966). *The Origin of Genetics.* Freedman.

Suissa, S. and Shuster, J.J. (1984). Are uniformly unbiased most powerful tests really best? *The Amer. Statist.*, 38, 204–207.

Titchmarsh, E.C. (1949). *The Theory of Functions.* Oxford Univ. Press.

Wald, A. (1943). Tests of statistical hypotheses concerning several parameters when the number of observations is large. *Trans. Amer. Math. Soc.*, 54, 426–483.

Wald, A. (1947). *Sequential Analysis.* J. Wiley.

Wackerly, D.D. (1976). On deriving a complete sufficient statistic. *The Amer. Statist.*, 30, 37–38.

Zacks, S. (1971). *The theory of Statistical Inference.* J. Wiley.

INDEX

The numbers to the right refer to the part and the lesson where this concept is first cited.

A

B

C

E

F

G

H

I

J

K

L

M

N

O

P

Q

R

S

Printed in the United States
By Bookmasters

Printed in the United States
By Bookmasters